HOTTER THAN THE SUN

Time To Abolish Nuclear Weapons

Scott Horton interviews Daniel Ellsberg, Seymour Hersh, Gar Alperovitz, Hans Kristensen, Joe Cirincione and more.

HOTTER THAN THE SUN

Time To Abolish Nuclear Weapons

The LIBERTARIAN INSTITUTE

Hotter than the Sun:
Time to Abolish Nuclear Weapons

Edited by Scott Horton

LibertarianInstitute.org/Nukes

Cover Photo:
Redwing Apache thermonuclear test, July 8, 1956
National Nuclear Security Administration
Nevada Site Office Photo Library

Author Photo Credit: Cornbread

Transcriptions by Jared Wall

Cover Design: TheBumperSticker.com.

Published in the United States of America by

The Libertarian Institute
612 W. 34th St.
Austin, TX 78705
LibertarianInstitute.org

ISBN-13: 978-1-7336473-6-6
ISBN-10: 1-7336473-6-8

For Sophie

Table of Contents

Introduction

Who would have thought that 50 years after the end of the Cold War with China and 30 years after the end of the Cold War with the USSR and final dissolution of the Soviet Union, tensions between the United States and Russia and China would again get to such a point where there is open discussion on the nightly news about the possibility of atomic warfare?

Men and women of my generation, who last felt they had to fear nuclear war back when we were kids, are now parents wondering how to explain it all to our children and how to protect them.

But the danger of nuclear war has never left us. Whether through accident, miscalculation or malevolence, the presence of nearly 15,000 of these devices remaining in the world, while serving as deterrents for the sake of our security, also represents a direct and present threat to our civilization that far outweighs any protection they provide.

Republican President Ronald Reagan came within a hair of making a deal with Soviet Premier Mikhail Gorbachev to dismantle all the world's nuclear arsenals in 1986, when the USSR still stood. Retired government officials from both parties, including former secretaries of state and defense and all the military services, have endorsed the goal of ridding the world of nuclear weapons entirely.

How long can we expect for a policy of Mutually Assured Destruction to keep the peace between the major powers? Another 70 years? Forever? There must be a better way.

This book contains interviews conducted over more than a decade with experts of all descriptions — including Daniel Ellsberg, Seymour Hersh, Gar Alperovitz, Hans Kristensen, Gordon Prather, Joe Cirincione and more — about the threat of nuclear war between major and minor powers, the nuclear arms-industrial complex, the nuclear programs and weapons of the so-called "rogue states" of Iraq, Iran, Syria, Israel and North Korea, the bitter truths and eternal lessons of America's nuclear bombing of Japan in World War II and the dedicated activists working to abolish the bomb for all time.

Nuclear bombs burn hotter than the Sun. A war between larger nuclear weapons states could kill tens or hundreds of millions in just days. Worse, the drastic drop in global temperatures that would be caused by the smoke and soot from even a limited nuclear war could kill *billions* of people through crop failure and famine.

It is irresponsible to allow people to possess such devices.

Humanity has been very lucky so far. We must do everything we can to rid the world of this burden before our luck runs out.

Scott Horton, June 2022

Introduction

Part 1

The Threat of Nuclear War

"I am become Death,
the destroyer of worlds."
— J. Robert Oppenheimer, 1945

Hans Kristensen: The Bleak Outlook for Nuclear Arms Control June 26, 2020

Scott Horton: Introducing Hans Kristensen from the Stockholm International Peace Research Institute. He is at the Federation of American Scientists as well. SIPRI has just put out their latest study, the *SIPRI Yearbook 2020*. Part of that, of course, focuses on nuclear weapons. They have a story here at sipri.org: "Nuclear Weapon Modernization Continues, But the Outlook for Arms Control Is Bleak." Welcome to the show, Hans. How are you, sir?

Hans Kristensen: Thanks for having me.

Horton: Very happy to have you here. There are so many important points brought up here. If we could just start with reminding the audience of which all countries are armed with nuclear weapons and approximately how many?

Kristensen: There are now nine countries today that have nuclear weapons: the United States, Russia, France, Britain, China, India, Pakistan, Israel and North Korea. All together, they possess something on the order of 13,400 nuclear warheads. Most of those are in military stockpiles that are ready to use on relatively short notice, but there's also a chunk of them, something in the order of 1,800 to 2,000, that are on high alert. They're ready to fire within just minutes.

Horton: Those are mostly America and Russia's?

Kristensen: The alert weapons are American, Russian, French and British, yes.

Horton: Is there a ratio handy about how many of these are fission bombs versus thermonuclear H-bombs?

Kristensen: Almost all of them are two-stage thermonuclear weapons. Those are the more advanced weapons that countries like the United States, Russia, France, Britain and China have developed over the years. Newer countries that have only conducted relatively a few nuclear tests don't yet have that capability. India, Pakistan, Israel and also North Korea. Although North Korea has demonstrated in their last test that they can produce something thermonuclear, something that can produce a very high yield, it's a little unclear whether that is a two-stage device or some other technology.

Horton: Are most of those still measured in the kilotons, or are they up into the megaton range?

Kristensen: Megaton weapons are sort of becoming more and more rare. Those were things that people built early on. Now, it's in the hundreds of kilotons or even tens of kilotons. It also depends on the mission, of course. If countries have warheads that are intended for blowing up deeply buried underground facilities or knocking out ICBMs or command structures and that type of stuff, then they tend to be higher-yield because they have to do more damage. But if you're looking at weapons that are needed for more war fighting scenarios, against shallower targets or troop formations or bases or something like that, then you can do the job with just a few tens of kilotons.

Horton: Did I hear you right, that you said India does not have H-bombs?

Kristensen: That's correct, although there have been claims in India that they do. Also, one of the devices that was tested back in 1998 apparently was an attempt to make a thermonuclear design, but it fizzled. So we don't anticipate that they have a two-stage thermonuclear device deployed in their arsenal.

Horton: Okay. Because I had read and actually talked to an expert or two about how one of the real problems with the nuclear standoff between India and Pakistan was that Pakistan only had these much smaller-yield, tactical, battlefield-type nuclear weapons that they would use against an armored column or something like that, while the Indians had focused on building higher-yield, strategic nuclear weapons for killing cities with. And because their armor was so desperately outmatched as well, if the Indians launched a conventional attack, the Pakistanis might have to use low-yield nukes to defend themselves. Then the Indians would have no choice but to retaliate with genocidal weapons of destruction because that's essentially all they have. So I wonder about how you conceive of that whole scenario?

Kristensen: Well, the scenario borrows a little from different aspects of reality. It's too simplistic, though. The point is that both countries have developed medium-range ballistic missiles with warheads in several tens of kilotons that can hit each other's cities. But what's unique about Pakistan is that, in addition to that, they have also developed weapons that are more tactical and appear to be intended for use against, like you mentioned, Indian conventional forces massing inside Pakistani territory. So that's a unique feature of the Pakistani arsenal. That, of course, leads to a lot of concern about how they're going to do that. Are they going to delegate launch authority to the local units so that they could use them early if necessary? How is that going to play out? So there are differences between the arsenals, but there are also a lot of similarities.

Horton: The numbers of Chinese nukes are surprisingly low here, in the triple-digits?

Kristensen: Compared to the U.S. and Russia, they're very low, but…

Horton: It's enough to kill us all.

Kristensen: Yeah, the Chinese have had a different approach to their deterrent posture for many decades. They basically didn't buy into the using of nuclear weapons in war-fighting scenarios. They thought that if they had a few hundred in a posture where they could retaliate, they could not be knocked out. They would always be able to retaliate. That should be enough, they thought, for nuclear deterrence. What we're seeing now, of course, is China increasing its arsenal. We have bumped up the number this year to 320 warheads that we estimate are in their stockpile, and it's increasing. But it has been increasing for a long time, just sort of slowly. We'll have to see if they're going to increase faster, but whatever they're doing, they're not sprinting to parity. Whatever you hear about the Chinese nuclear arsenal, it's not like they're trying to catch up to the Russians and the Americans. They still have a fundamentally different perspective on the role of their nuclear weapons.

Horton: It's important to note that, historically speaking, at the height of the Cold War, there were tens of thousands. Was it 40,000, approximately?

Kristensen: Yeah.

Horton: On the American and Soviet Union side each. So we've made a lot of progress since then, right?

Kristensen: That's correct. At the peak, in the mid-1980s, there were 70,000 nuclear weapons on both the Russian and American sides. Ten thousand of those were on high alert, ready to go within a few minutes. Just totally crazy circumstances. So when the Cold War ended, they started slicing a lot of that excess capacity out, and we saw a huge drop there in the early 1990s and also a little later on.

What we're now beginning to see is that the two sides are sort of slowing down significantly, even to some extent reversing that trend and looking to maintain significant arsenals for the indefinite future. All sides are increasing the value that they attribute to nuclear weapons. They're increasing the role of their nuclear weapons in the way they talk about what functions they should serve. So this is a very troublesome development.

Horton: I'm sorry, what did you say the number was at the height? I thought it was much higher.

Kristensen: Seventy thousand.

Horton: Oh, seventy. I thought you said *seventeen*, and I thought, "Wow, I was way off." Yeah, that's more like what I thought it was — 70,000 nukes.

Kristensen: It was crazy. I mean, you can just imagine spending a couple of hours on Google Earth trying to put 70,000 Xs on the map. I mean, what are we going to do with all that stuff?

Horton: There's an anecdote about Dick Cheney back in 1989, when he was first secretary of defense, being shown on the computer screen a simulation of what it would look like. They just nuke Moscow hundreds of times over, and Dick Cheney finally says, "That's enough. Turn this off," and wanted it redone because it was just completely insane. Of course, he's notoriously the greatest American hawk alive, right?

Kristensen: Yeah, it's ironic that you can find some of those realizations in what some characters like Cheney did. He sent one of his officials, Frank Miller, out to Strategic Air Command, as it was called — or STRATCOM as it's now known — they went through the entire targeting list. (It's a very important reading of that episode in the memoirs of the first STRATCOM commander.) They discovered, not surprisingly, that there was enormous overkill because the nuclear planners had essentially been allowed to do this by themselves with very little oversight. Things have changed since then, but even though we've moved beyond some of that stuff, the nuclear planning today is still surprisingly similar to what it was during the Cold War.

Horton: Daniel Ellsberg has talked about how a lot of it is just bureaucratic politics: "'It's not fair that the air force gets to blow up this city. The navy wants a crack at it too.' 'Okay, Navy. You guys can also hit it with missiles.'" So it's just kind of an episode of some sitcom, some bureaucratic politics.

Kristensen: Yeah, there is an element of institutional competition and turf wars and all this stuff. That's part of it, the dynamic. Mind you, early on, the army also had a dream that they wanted exactly 100,000 nuclear weapons. Just the army. I mean, those were crazy days.

Horton: Now, the Israelis. I actually had the opportunity to discuss this briefly with Mordechai Vanunu on Twitter, where he confirmed to me that he stood by his original leak to the British *Sunday Times*, that the Israelis have 200 nukes. I don't think that he would clarify whether that included H-bombs or not, but we know from Grant Smith's FOIA lawsuits that it does include H-bombs as of 1987.* You guys count 90, something like that, but we all know that "Israel doesn't have nukes" — they're completely deniable and not official and so forth — so I wonder where you come up with that number?

Kristensen: That's a long history, this thing about how you come up with the number for the Israelis. Because there's so little factual information about it. The way it happened back in the 1980s, when Vanunu and others came out with the estimates of the Israeli arsenal, was that people looked at how their reactors operated and calculated from that how many units of plutonium they could have produced over the years. Then they translated

---

* This is your host-editor's error. Documents obtained by Smith only show Israeli research into thermonuclear weapons: Grant F. Smith, "US Confirmed Existence of Israeli H-Bomb Program in 1987," Antiwar.com, February 14, 2015, https://original.antiwar.com/smith-grant/2015/02/13/us-confirmed-existence-of-israeli-h-bomb-program-in-1987/.

that into the number of potential weapons they might have. That's how you got to those high numbers.

The U.S. intelligence community looked at it a little differently. They said that yes, even though that production might have happened, they haven't turned all of that plutonium into warheads. So their number has been much lower over the years, and we've gone with that latter number and said those are the weapons that we think they have actually assembled, although they keep them partially unassembled under normal circumstances; but that they have more plutonium in stock that they could produce more bombs if they needed to. That's how these differences emerge concerning the numbers.

Horton: That includes second-strike missiles deployed on submarines, too? Do you know how many?

Kristensen: That's a big uncertainty right now. There is a consistent persistent rumor that Israel has developed warheads for some cruise missiles on its conventional attack submarines. We are cautiously including that in our estimate this year. I'm just saying this because there's so little information about exactly what the Israeli arsenal includes. There's also a lot of room for speculation and rumors and even hype. So one has to be a little careful not to get swept up in that kind of excitement and get into all sorts of capabilities.

You mentioned people saying that there was thermonuclear capability. Some people believe that. We tend not to think that the Israelis have developed a functioning two-stage thermonuclear capability. We do believe they have a boosted, single-stage design, but doing that requires development of more technology. It's more complex to do. You have to look at what Israel's intention is with its arsenal. What function does it have to serve? The reason people develop thermonuclear weapons is that their targeting strategies require them to blow up things with great explosive power. At first, that had to do with accuracy, since you couldn't hit precisely enough. So you develop large-yield thermonuclear weapons to compensate for the inaccuracy.

Israel is not in that situation. They have relatively accurate missiles. So their calculation, it seems to me, would have to be a little different. They probably don't need these super-high-yield warheads.

Horton: Yeah, I'd agree with that. I don't think anybody does, but I understand what you mean.

Now, let's talk about North Korea a little bit. As you said, they tested a nuke that was possibly boosted, but nobody really knows for sure, right?

Kristensen: The U.S. intelligence community seemed to say that the last test was about 150, some say 200, kilotons. There is this uncertainty about the specific range because different agencies that had monitored it put out different estimates. But it was big. It was significant. It was in a yield that you would have to have either a very large, very significantly boosted single-stage weapon or some form of thermonuclear design. We've heard that characterization from U.S. officials saying that there was some thermonuclear event involved in this, but whether that means that this is a two-stage thermonuclear warhead of the kind of warheads that the United States and Russia developed over the years, that's a little more uncertain. But it was a significant nuclear yield that was produced by that weapon.

Horton: Supposedly, their missiles can reach D.C. now. In one or two of those tests, they reached a high enough orbit that they say it could have reached D.C. if that's how it had been targeted. Then they also say that even if they do have H-bombs, they haven't been successful in miniaturizing them to the point where they'd be able to marry them to one of these rockets and deliver one to Washington. So we have a little bit of breathing room there. But everybody, all kinds of politicians from both parties, say that's the red line. We can never allow them to have H-bombs — or I guess atom bombs — and the means to deliver them, especially not to our capital city. The West Coast, maybe, but not Washington.

We know they started making nukes right after they withdrew from the treaty in 2003. So, on the timeline of their progress, are you worried that they might be able to miniaturize their nukes and marry them to their missiles sometime within the next few years? Or what do you think about that?

Kristensen: Our sense of where they are is that they've developed ballistic missiles that can reach the U.S. It's a little less clear whether they have a functioning warhead for those missiles that can reach the United States. This is a distinction that's normally lost in the public discussion about the threat from North Korea. It's more likely that they have warheads that developed initially for their shorter-range ballistic missile, for the medium-range systems that they have. So U.S. bases and allies in the region would most certainly be at risk. But, like you mentioned, they have made a lot of progress very quickly, and they seem intent on continuing that.

We've just heard some very strong statements from the North Koreans about continuing to refine and improve their nuclear arsenal. That's one thing we've learned from the North Koreans, that you can pretty much trust what they say on this issue. If they say they'll do it, that means something real. So they're not done, and we're likely to see more things coming in the future.

Horton: Now, let's talk about this modernization. Part of this, I think, is just a welfare program for the nuclear arms industry. It was part of the negotiations in the Senate to get the New START passed: "Okay, you get a trillion dollars." It's now almost two. It'll probably be four by the time they're done. So I don't know how much of this is just make-work. I know that they've already deployed the new lower-yield cruise missiles, but what else do we need to know about the so-called modernization here, other than just the special interest aspect? What about the actual change in the nuclear forces?

Kristensen: The bulk of the U.S. modernization program is a complete replacement of the entire arsenal. So everything that was developed and built in the 1980s and 1990s is now coming up for renewal. The commitment that has been made is that all elements of what's known as the triad — the sea-based, the land-based ballistic missiles and the long-range bombers — all of that will be replaced. Also, the shorter-range fighter jets will also be upgraded and replaced. In addition to this comes nuclear production facilities, expanded plutonium pit production facilities that are being planned. We see a modernization of the nuclear command and control system that's supposed to support and manage these nuclear forces. It's a very, very broad and comprehensive modernization plan. Like you said, it's going to cost a lot of money.

The question is: Does it change anything fundamentally compared to what we had before? You have to look under the hood and see what kind of capabilities are being built into these new systems. When they build a new ICBM, it's not just a copycat of the old one. They put advanced capabilities on it which improve its effectiveness. Likewise, when they upgrade a nuclear gravity bomb — for example, the B61 that is used both by strategic bombers, but also by tactical fighter wings here in the United States and also in Europe — they don't just repaint and dust off the one that's there. No, they improve it. For instance, they added a guided tail kit so that it can hit its target more accurately. So even though the numbers may not go up and even though you may not have fundamentally new nuclear weapons, you take the chance and the opportunity to improve the capabilities of the weapons they can have in the future.

Horton: What do you make of all the new hypersonics? Ours and the other guys'?

Kristensen: That's the next chapter in the arms race. Everybody is on that bandwagon, and they're trying to get on it. The focus of that is conventional, but there are also nuclear elements of it. The Russians have rushed into the deployment of a few missiles that have a hypersonic glide vehicle with a nuclear warhead. We're seeing them working on other types of hypersonic weapons that have nuclear capability, but they tend to be dual-capable, if nuclear is involved at all. We're seeing the Chinese working on similar systems. They've even deployed what appears to be a glide vehicle of some sort for its rocket force. There's an uncertainty about whether it's nuclear, but they're certainly working on that. And the United States is obviously pointing to them, saying, "Well, they're doing it, so we need to do it."

There's a real crash program on the way to try to develop these capabilities. We are probably going to see some kind of a hyperglide or hypersonic capability for long-range bombers, as well as for submarines and some ships. So this is really happening. Now, how much does that change? Does it make the world more dangerous? It certainly does in the regional scenarios, where the timelines and the reaction time to these weapon systems will be much shorter. That will put all sides on their toes and be more nervous about what's going on, etc. At the strategic level, I think it has less impact compared to the type of forces that are out there already. So I think it's more within a specific region, that you would see this dynamic.

Horton: It seems like reaction time is everything. If we have half an hour to decide if we're really being nuked to death, then that's already not very much time. But if we have five minutes, then essentially, they're almost guaranteed to choose believing the threat and reacting to it. To err on the safe side would be to kill us all.

Kristensen: Right, exactly. It's the worst of scenarios because all sides inevitably fall into this. They corner themselves. They paint themselves into this corner of worst-case scenario. They always have to assume and plan for the worst, etc. Stability becomes much more brittle in that scenario, and it bothers me. I'm really confused as to why military powers want to go down that route, because it decreases their security and that of their allies.

Horton: I'm a bit of an extremist on this topic, but I wonder how far you'll go with me on the idea that after Ronald Reagan and George H. W. Bush — as much as they did to negotiate away our nuclear weapons stocks back then — Bill Clinton could have picked up right where they left off and negotiated an end to the entire global nuclear arms race and complete disarmament at that point. With the threat of the Soviet Union and World Communism over, they could have just called the whole thing off. It didn't have to be this way at all. We could have, I don't know, 10 nukes each, just to make sure nobody fights, and then that's it.

Instead, it was just sort of what we're talking about with this New START deal: "Hey! We get to build a whole new nuclear weapons factory. That'll be expensive," and the whole thing becomes a self-licking ice cream cone, even though we're not talking about M-16s — we're talking about H-bombs.

Kristensen: Yeah, this business has a sort of self-serving dynamic in it. An element of that, absolutely. There was a huge opportunity missed after the Cold War ended where, like you say, we could have fundamentally changed the role nuclear weapons play and reduced arsenals around the world. It didn't happen for a variety of reasons, mistakes were made on all sides about this, and here we are. Now we are seeing an invigoration of the role of nuclear weapons and an increase of them. Countries are rattling the nuclear sword at each other again in a very overt way. Things are definitely going back.

But I can't help but remark also on one curious fact about the way that nuclear reductions happened. I looked at this closely, and it's really interesting to see how the periods where the most cuts or the biggest cuts happened all were during Republican administrations. There's a dynamic between the White House and the Congress about why that is so.

Horton: Only Nixon can go to China. That kind of thing.

Kristensen: Right, and Democrats have to be tough and can't be seen to be weak and so forth.

But I just want to mention one other thing. On the issue here of nuclear weapons, I think it's important to think about the problem or the issue of nuclear weapons, not just as nuclear weapons in isolation, because the role they play and the reasons for why countries have them also have a lot to do with how they perceive the threat from conventional capabilities. So countries will use nuclear weapons, to some extent, to compensate against what they think are inferior, conventional forces. There's a much more complex dynamic going on in terms of what shapes the direction of the nuclear forces will take and what countries think they can do to reduce their rule.

One of the things we're seeing right now is in the context of the Nuclear Non-Proliferation Treaty. The nuclear powers in that, the P5, as they're called — that's the United States, Russia, Britain, France and China — have sort of found a common theme where they're trying to say to the other non-nuclear weapon states, which is the predominant numbers of countries in that treaty, "Wait a minute, guys, it's not just about us. It's not just about nuclear. You also have to work to create the security conditions in the world so that it is possible to reduce the stockpile and eliminate nuclear weapons." So they're trying to pass some of the responsibility on to other countries as well.

Horton: Alright. Thank you so much for your time, Hans. It's really been great.

Kristensen: Great. Thanks for having me.

Chas W. Freeman: The Threat of Nuclear War with China February 22, 2019

Scott Horton: Introducing Chas W. Freeman, Jr. He's a Senior Fellow at Brown University's Watson Institute for International and Public Affairs. He's also a former U.S. Assistant Secretary of Defense, former Ambassador to Saudi Arabia and other important jobs in the federal government. He was the principal American interpreter for Richard Nixon when he went to Beijing in 1972. Welcome to the show. How are you doing, sir?

Chas W. Freeman: I'm glad to be here, Scott.

Horton: Very happy to have you on the show. Let's start with that last part, your trip to Beijing with Nixon. Because, if I have the history right, it seems like this is one of the most important positive events of the 20th century, when Nixon and Kissinger went over there and shook hands with Mao Zedong. Forget strategically breaking them off from the Soviet Union, but more importantly, paving the way for Deng Xiaoping and the reforms that turned China from a communist dictatorship to a fascist one — which is still really bad in a lot of ways, but it has finally stopped starving people to death by the tens of millions, as it was under Mao. So it seems like you played a really important role in possibly saving millions of lives there, by creating the space for the right wing of the Communist Party to come to power after Mao died. Is that pretty much your take?

Freeman: Nixon said at the time, rather tritely, that it was the week that changed the world. But it did. Essentially, the United States had been using Taiwan to contain China. Now, we turn to using China to contain the Soviet Union. And the cooperation we had with the Chinese in the 1980s, not only in Afghanistan — where, for example, in 1987 we bought $630 million worth of Chinese weaponry for the mujahideen — but also in other ways. There were many listening posts in China that enabled us to gain insights into Soviet weapons development and so forth. The Chinese sold us their version of the MiG-21 to use in training our pilots how to combat it. We played together a key role in bringing down the Soviet Union, and the world is a better place for that.

I would say, however, that we really had no inkling at all that what we were doing was going to result in China abandoning the Soviet system internally and moving to join the American-led world order. Which is what happened six or seven years later, when Deng Xiaoping decided to use the United States to de-Mao-ify China.

Horton: So that wasn't part of the plan? Or was it just sort of a happy result that developed later on?

Freeman: Yeah, our engagement with China was very much self-interested. The objective was almost entirely foreign policy-oriented. We weren't trying to change China internally, but we ended up doing so inadvertently. And of course, between December 1978 — when Deng kicked off his reform and opening process — and today, China's poverty rate has fallen from about 90 percent of the population to less than two percent. China has emerged as a huge market for the world's commodities and manufactured goods, as well as a major producer. Living standards have risen incredibly in China. It has very much become part of the world that we helped to create after World War II.

Horton: This is the topic of your recent speech that we republished at Antiwar.com: "After the Trade War, a Real War with China?" which was given to the St. Petersburg, Florida Conference on World Affairs on February 12, 2019. It's about the rise of China and America's reaction to it. I guess there's some who say — and I can't remember who I'm quoting here, but I've read some things that said — "You know what? It turns out it was a big mistake opening up China. We should have let them dwell in their Maoism because now we've created a monster, and they're more powerful than America!" So we have, I guess, everything to lose by the Chinese gaining from the current global system.

Freeman: You used to hear the same argument about the Marshall Plan in Europe. "Why should we help Europeans get back on their feet and run trade surpluses with us?" The world is a better place for that, and it's a better place because the Chinese lifted themselves out of poverty and became a responsible member of the international community. You have to remember, when Nixon went to China in 1972, the Chinese were calling for the overthrow of the World Bank, the predecessor to the World Trade Organization and the American order. They were calling for revolution everywhere. That all stopped, and China instead became part of the order it had originally decided to overthrow. So I would say, yeah sure, we have a powerful rival now, that is true, but there's so many better things that have happened, that you can't reach the conclusion that it was all a mistake.

Horton: Well, of course, those same people would be anti-communist hawks who would point to Maoism as proof that this is the worst regime in the whole world and that anything would have to be better than that, and they'd be right, right? Mao's government probably was quantifiably the worst government that had ever existed in the world.

Freeman: Oh, I think probably North Korea takes that prize. I can remember a Chinese friend who was assigned temporarily to North Korea in the early 1980s coming back and he wouldn't tell me much about it. Finally, I pressed him, and he said, "Oh, you know, there are quite a number of North Koreans, including senior officials, who have sought political asylum here in China." And he said, "Just think about it. What kind of a place would you have to be from to want political asylum here?" This was when China was just beginning to change, but, in many respects, China has continued to open up internally as well as externally. At the moment, unfortunately, under Xi Jinping, I think the trend is in the opposite direction, which is disappointing to many Chinese as well as to those of us who had hoped that China would continue to open up.

Horton: The Trump administration has a policy much different from the centrist status quo on this. It has all these tariffs and is waging this trade war, and — I guess I'll go ahead and mention this because it's the context of your whole article here and this speech that you gave about this trade war — there's a real panic going on about the rise of China in the halls of power in America. And they kind of have determined that they're going to do something to halt the rise of China. Maybe they're not even certain why they need to, but they know they need to and don't know how they're going to. It sounds like they're really kind of throwing a tantrum here. Could you characterize the state of the relationship there and then also talk about the tariffs and what change that really represents?

Freeman: Well, one of the nice things about going to St. Petersburg, Florida to talk about China is that people are not caught up in the hysteria that seems to prevail within the Beltway. I found a lot of very thoughtful people there who are open-minded in a way that people in Washington these days don't seem to be. So I think it's true that on Russia and on China and maybe on Islam we have become a bit deranged as a country.

In the case of China, the Trump administration's not only got a trade war going, but there are China hawks in the inner circles of the administration who are basically trying to contain, smash, suppress China, reverse its rise, somehow prevent it from technological advance, cut us off from the Chinese, deny Chinese students entry to our universities, shut down cultural exchange with China, and so forth and so on. It's been a pretty broad onslaught, and I think it's really quite remarkable that the Chinese response so far has been so moderate. The trade issues are a very good illustration of a problem. We have failed to ask the question: And then what? Before you do something, you ought to ask, "And then what? What's the other side going to do? Where is this going to lead?" We didn't do that.

Our soybean farmers, for example, I think they've lost the market for decades, not just for a short period. Why on Earth would the Chinese mortgage their future to food supplies from the United States that we've shown we can cut off on a whim? Why should they base their technology on chip exports from the United States when we use their dependence on us to shut down their factories and prevent them from operating the supply chains that they do? I think there's going to be a long-term effect from this, regardless of how the current negotiations turn out, and nobody's really thought about that. We really ought to think more before we leap into adventures abroad.

Horton: So I guess soybeans is one thing, but how severe are the sanctions? How wide-ranging is this, and how many other examples are there like the soybeans?

Freeman: There are major examples across the board. For example, pork exports from the United States. Ironically, the Chinese bought a pork producer here, Smithfield Ham, in order to increase U.S. exports to China. But those exports now are subject to retaliatory tariffs, and the result is that the Chinese are investing in places like Spain and Chile to produce pork. They're not investing here.

I mentioned microchips. You know, most of our laptops are assembled in China. There are some Chinese components in them, but they represent a true international production chain, much of which comes from the United States. So we cut off Chinese solar panel exports to the United States, and the main result of that is that the materials from which those were made, which had been exported from the United States, no longer are bought by the Chinese. They're looking elsewhere.

So, we're doing a lot of things that are going to have long-term effects that we haven't really weighed, and we have a constitutional issue in this country. Supposedly, tariffs are under control of Congress, not the president. The reason the Founding Fathers did that is that they wanted a deliberative process before we did this sort of thing we're doing now, announcing policy with a tweet in the middle of the night, which is exactly how some of this has been announced. So I think the Congress — which has defaulted on many, many issues, from the ability to declare war to the ability to set tariffs and regulate trade — needs to do what we pay them to do, which is to debate, decide, have hearings, educate, shape policy and monitor it, and not just to default to the president.

Horton: That's the thing, right? They transferred all that most-favored-nation status authority, among other things, to the president in the era of Bush, Sr., Bill Clinton, Bush, Jr. and Obama, who could be reliably counted on to pursue that centrist agenda. But now they have left that power in the hands of somebody who sees things very differently. Is there any kind of real strategy on the American side? Is that the point, to cut off all this trade in a permanent way, or is that a means to an end that they think they're really accomplishing with this?

Freeman: Well, the president is a real estate mogul, and if you're in the real estate business, you're dealing with fixed assets. There are buildings, there are people who live or work in them, and you look at each building separately and say, "Am I putting more money into this building that I'm getting out?" He looks at trade in those terms, as a bilateral exercise: "Am I spending more money than I'm taking in?" He doesn't look at the overall global picture, and therefore, he started all this with the demand that we import only as much as we export to China. When you think about that, that's crazy! We benefit a lot from the imports that we get from China and other places, and if we don't import from China, we'll import them from somewhere else. We're not going back to the kind of sweatshop labor that we used to have here, in the textile industry, for example.

So, I think that there is no strategy. There is a very medieval, mercantilist economic philosophy at play, and we have not thought about the long term. We don't have a strategy. If we did have a strategy, it would have to start with improving our own competitiveness, not trying to tear down China's. Everything we're doing is designed to tear down our competitors, whether they're Chinese or German or Japanese or whoever. It's not designed to increase our competitiveness, and we need to do that. We need to return to where we once were, which is looking at best practices abroad and seeing how we might apply them to our own benefit at home.

Horton: Well, I certainly agree with you, of course, about the economics not really being a zero-sum game the way Trump seems to imagine it, and all the mutually beneficial policies going on there. The richer they get, the richer the whole world is, and the better off we all are. And yet, next to that is the question of the Communist Party's revenues and their ability to build up the People's Liberation Army, their navy and everything else. So I'd like to get to the nature of the Chinese regime as it is now regarding the panic on the American side about China's rise. Just how massive is their naval buildup? For example, you mentioned their policy of "area denial," but are they trying to build more of a blue-water navy, create a world empire of bases like we have, and challenge our hegemony on the planet?

Freeman: I don't think that's the Chinese aim. We're in their face; they're not in ours. They're not patrolling off Puget Sound or Norfolk, Virginia; we are patrolling right off their coasts and sometimes doing mock attack runs to get them to turn on their radar so that we can see what their defenses are like and be able to break through them. We're very much in an offensive posture there, and I don't think it's at all surprising that they feel the need to be able to control their periphery, their borders. What's different is that, after World War II, we controlled those borders and that periphery. We seem to believe that that is our God-given right forever and forget that the region got along for several thousand years in quite a different mode when we weren't there. So I think that the military dimension of this is far less than people imagine.

The one real question is the Taiwan question because the Chinese civil war is not over. We suspended it with the Seventh Fleet going into the Taiwan Strait back in 1950, but it's still going on in the Chinese mind. And we don't really have a strategy for dealing with it, other than sticking our chin out and maybe inviting the Chinese to take a whack at it — which goes back to the question you asked: Do we have a strategy? The answer clearly is that we don't have a strategy; we have an attitude.

Horton: Would you put a percentage on the chance that the Chinese would try to invade Taiwan in the next 10 to 15 years? And would America necessarily really go to war with China to protect Taiwan?

Freeman: Well, if you talk to people inside the Beltway, the answer is that we do plan to go to war with them if they try to reincorporate Taiwan into China. The Congress passed a law called the Taiwan Relations Act which implied that, and our military is planning for that, so I don't think there's much question about that. I don't think that the Chinese want to do anything militarily to Taiwan, but I think that they do want Taiwan to agree to terms that reduce and maybe end the division of China in some sense. They put forward a number of proposals, none of which were very attractive to Taiwan apparently, but they made it clear that they want a peaceful settlement of the issue and they'll do their best to achieve one. If they can't get a peaceful settlement, they might resort to the use of force, in which case we have the makings of a nuclear war.

Horton: Let me ask you about that because it seems like a lot of talk about Russia and China from the Blob — the foreign policy wonks and all the writers and think tanks and whatever — they seem to leave nukes out of the discussion, sort of like, "Well, of course, it goes without saying that there's nukes," but they leave it out so much that they really act like nukes aren't in question and that somehow you could have an "air-sea battle" against China that wouldn't end up in the loss of Los Angeles.

Freeman: Well, I think that you're right. I think it's actually in some ways worse than that because our withdrawal from the Intermediate Nuclear Forces, or INF Treaty, is premised in part on the desire of the nuclear war managers to build a new class of shorter-range theater nuclear weapons, which they think could persuade some country near China to put on its territory. Of course, nobody in the region wants to do that because then they'd become a target for Chinese retaliation. But we're actually apparently thinking about the use of nuclear weapons in the tactical sense. We're developing what are called "dial-down" weapons, which means that you can adjust the level of the explosiveness of the warhead. If you listen to the rationale for this, it's pretty clear that the nuclear allergy — that is, the aversion to the use of nuclear weapons on the grounds that once you use them, everybody will use them — is in trouble. It's going away. So I think this is a bigger issue than just China. It involves Russia, North Korea and others as well, and it's pretty depressing.

Horton: I read that part of the motivation behind America's withdrawal from the INF Treaty is really that they want to be able to station theater nuclear weapons closer to China. It doesn't have anything to do with Russia's so-called violation at all.

Freeman: Well, I think we're in a posture with the Russians where they have accused us of violating the treaty. We've accused them of it. There's some merit on both sides, I suspect, but it's really kind of childish to be engaging in that kind of argument and using it as an excuse to get ourselves back into a nuclear arms race, which nobody could possibly win. That being said, I think you're right about China being a large part of the argument.

Horton: That our government is getting into this major dispute, this mutual withdrawal from this treaty with Russia over some strategic policies in regard to the containment of China?

Freeman: Right, and so as I said, the idea would be that we would put nuclear missiles somewhere in the region like, let's say, the Philippines — who are quite concerned that we might plan that — and then use them against China in a war. This is a ridiculous theory. The theory that you can use a nuclear weapon against the Chinese and that they won't take out targets in the homeland, it's crazy. This is not sensible. This is worthy of Stanley Kubrick and *Dr. Strangelove*, but that's where we are.

Horton: Are you referring to this so-called "escalate to de-escalate" thing, where you set off a couple of small nukes and that'll teach them that they'd better back down, and then the bet is that they won't retaliate with a bigger nuke, that they'll go ahead and say, "Whoa, the Americans really mean business!" and stop?

Freeman: Really crazy people do have that kind of theory. That is not how human nature works. You take out Shanghai and then say, "Okay, well, we've done enough," and you just sit there and don't do anything further? No politician in China — any more than a politician in the United States after Los Angeles was taken out — could just sit there and do nothing. So this is just crazy stuff that is theoretically impractical and terribly dangerous.

Horton: Let me ask you a little bit more about their economy. Sorry for jumping around so much, but David Stockman, the former Reagan-era Budget Director, refers to the country every time as the "China Ponzi" because he says the whole thing is on a gigantic paper money-inflated bubble and that they have a 2008 coming due to them, the likes of which nobody's seen in a long, long time. Then, part of his argument is that therefore all the scaremongering about them can really just be put on hold for a little while because the current system over there is not going to last.

Freeman: So far, the system over there is performing pretty damn well in terms of employment, rising living standards, general growth in the economy and so forth, but yes, there are problems in China. There's debt on the local level that has to be dealt with. De-leveraging has to take place.

But the fact is that the Chinese have a pretty good record of facing up to their problems and dealing with them. Many would argue that they have done a better job of that than our system is currently doing, and I would point out also that, however it came about physically, China's manufacturing sector is one and a half times as large as ours. We have a lot more insurance salesmen and bureaucrats doing health care, and they have a lot more people producing real things. If we got into a contest with them, I suspect that the number of insurance bureaucrats we have is not a great strength on our part.

Horton: Well, you got that right. That's a good profession to pick on, too. Thanks for singling them out. They deserve it. Anyway, back to the navies and things, what about all of this hype about the South China Sea? You say here that Vietnam, Korea and Japan each have their own interests, and they each have their own independent power. I don't know how well it really compares to China's military, but you say there's no power vacuum to fill by America or China. I think that maybe you're saying that if America pulled out of there, not much would change. Is that right?

Freeman: We started out talking about President Nixon. Back in 1969, he gave a speech in Guam, which was called the Guam Doctrine. The basic thesis was that the United States should let our Asian partners be in the lead; let them be on the front line, and we should be prepared to back them, but not out front. We have gotten ourselves out in front of our allies and partners. They're really not allies, in that they don't have any obligation to us at all, but we have assumed the unilateral obligation to protect them against all enemies, at least foreign enemies.

So, I think there is a case to be made that we need to reach an accommodation of some sort with the Chinese. They're entitled to have some role in managing the affairs of their own region rather than being excluded as we have heretofore excluded them. We need to look to the countries in the region — Japan, Vietnam, South Korea, Indonesia, India and others — to do more for their own defense. We should be the last resort, not the first resort, in the event of a conflict between them and China.

Horton: Maybe we should just do a regime change there, and then everything will be fine.

Freeman: [Laughing] Some people would like to have a regime change here, too.

Horton: Yeah, well, put me at the front of that list. But, so the doctrine overall — they call it "Full Spectrum Dominance," right? No near-peer competitors. American hegemony over everyone, and obviously Chinese and Russian independence stands in the way of that policy. Is there a long-term strategy that says that eventually we will have total hegemony even in China and Russia? Or is it accepted that now they'll remain independent from us, but we'll rule the rest? Is that it?

Freeman: We're definitely in a slow retreat. That objective of eternal dominance is totally unrealistic and unachievable. It's the sort of absurd goal that caused the Soviet Union to spend itself into self-destruction.

Horton: Well, that's a fact, but does that mean that the rest of the establishment now finally agrees with you about that?

Freeman: Of course not. We have a military-industrial-congressional complex that lives off inflated threats and inflated responses to those inflated threats. That's just a fact. We use the defense budget like a jobs program; it doesn't derive from specific analyses of issues that we have to deal with. It's a "more is better," basic philosophy. We never could spend enough, and the fact that we have been spending so much money — $6 to $7 trillion on wars in the Middle East that we can't win and won't win, and that produce nothing — is exactly why we have a $4 trillion infrastructure deficit in the United States. It's the reason why we have been disinvesting in our educational system, science and technology, and research and development outside the military sphere. This is not a formula for national success. Our allies, the countries that we protect, like Japan and the Europeans, have been very sensible. We offer them a free ride, and they've taken it. They put their money into stuff other than military equipment and preparations for war, and they're doing pretty well.

Horton: But on the American side, the policy at worst is containment, not rollback, of Russia and China. Not at this point.

Freeman: Totally unrealistic.

Horton: That's certainly the case, but I've never known that to stop them. Listen, I can't tell you how much I appreciate your time on the show today, Chas. It's been great.

Freeman: Well, I'm glad to be of help, and it was a pleasure talking to you.

Gilbert Doctorow: Avoiding Nuclear War with Russia February 22, 2019

Scott Horton: Introducing Gilbert Doctorow. He is a political analyst based in Brussels, and his latest book is *Does Russia Have a Future?* We run him from time to time at Antiwar.com. The latest article is "Putin on National Defense: Threats or A Bid to Negotiate on Arms Control?" Welcome back to the show. How are you doing, Gilbert?

Gilbert Doctorow: Well, it's a pleasure to be with you again.

Horton: Very happy to have you here, and I should mention the second piece, part two of this, about guns and butter. It's Putin's big speech that he gave, essentially his State of the Union speech over there, that he does every year. Is that right?

Doctorow: That's right. This was his 15th edition.

Horton: That's quite a long presidency. So, national defense — you say that most of the speech was about domestic policy, but then he got to the tough part. What did he have to say?

Doctorow: Well, his foreign policy section was the last 10 percent of the speech. It was a very brief mention of Russia's priority partnerships, discussions and negotiations across the world. Then he stopped and went into the issue that was foremost on his mind, which was the United States' withdrawal from the INF Treaty, which goes back to 1987 and is one of the great achievements in the Gorbachev-Reagan summits. It was a part of the Architecture of Arms Control, and even if it's a bit outdated, it was viewed by the Russians as significant because it was a proof of mutual confidence and of an ongoing dialogue with the United States, which is now absent.

So, Putin's remarks were to follow up on what happened a week earlier when the United States withdrew. As Pompeo said, "Yes, we give notice that in six months we will be out of this treaty," and then two days later, the Russians said they also will exercise their option to leave the treaty. That was a perfunctory statement.

What Putin did now was to explain what this really means, and he was unusually tough. His language in the whole speech was conversational. I translated part of the speech as it pertains to the United States and to the INF Treaty. If you look at my translation and not the formal, cleaned-up translation that was issued by the Kremlin, it was definitely conversational, sometimes folksy and sometimes very idiomatic in a way that really tickled the fancy of his Russian audience.

Horton: In other words, he was going off the script and saying some things that weren't in the text as published.

Doctorow: Absolutely. And what was off the script and not really represented in the formal language of the presidential administration's translation of his speech was his great disparagement of Europe and his concentration on relationships with one country in the world for matters of security: the United States. What he said that was disparaging and which tickled the fancy of his audience — they were beaming and applauding — was that the United States has not done the honest thing like George W. Bush did in 2002 when it withdrew from the ABM Treaty saying, "We don't feel this serves our purposes and we're leaving." No, they didn't do that this time. Instead, the United States started accusing Russia of violations to justify the United States' leaving this treaty. That, from the perspective of Putin, was totally dishonest because the United States, from the Russian perspective, has been in violation of the Treaty for almost 10 years.

Horton: In what way?

Doctorow: Well, the negotiations on creating what are called missile defense bases in Romania and in Poland from the very first announcement going back to 2003 or 2004 were seen by the Russians as laying the basis for a violation of the INF Treaty. They understood that the launchers that were being installed in these two bases were dual-purpose and could, within half an hour of reprogramming, turn from the quite acceptable and legal missile defense functionality into a launcher of cruise missiles within a range in direct violation of the treaty. These missiles are not hypothetical or planned development by the States; they were taking the Aegis system, which is capable of launching Tomahawks from naval vessels, and they were putting them on land. So the Russians felt that they were being abused. They've been complaining off and on for the last 10 years about this to no effect. The United States went right ahead and brought the Romanian base up to operational level, and they are close to completing a Polish base.

Horton: By the way, there's an article about this in the *New York Times* today by Theodore Postal. People might remember him as that MIT rocket scientist who debunked two of the three chemical weapons attacks in Syria and was cited by Seymour Hersh and so forth. We talked to him about Khan Sheikhoun on this show. His piece today is about the dual-use nature of those missiles. In fact, he says that Obama was duped about this, that Obama thought they were putting in interceptors and that he didn't really realize what the real plans were. Somebody in the Pentagon must have known, though, that this is about the ability to base Tomahawk missiles there that they could arm with nukes and strike Russia at very short notice. Which is actually kind of superfluous. I guess they want to be redundant in their powers — they have submarines in the Baltics that can launch Polaris missiles and hit Moscow in no time anyway.

So the INF in a sense, as you said, was outdated. I don't know if this is what you're referring to, but in a way, it was ceremonial because we can hit them with submarines at medium range rather than long range anyway, right?

Doctorow: One of the big criticisms of Gorbachev that has come to the surface in the last few years in Russian political discussions is that he had been had. They understood very quickly that he was duped, because the ban was on land-based missiles, and the United States proceeded immediately to put equivalent-functionality missiles on submarines in the Baltic Sea, within easy range of Russia. So, in a way, the threat of very short notification has existed for some time, but the availability of a land-based system means that it could multiply many times the number of missiles being fired.

Horton: This is an important part of the narrative that's really been left out, that America really started this. I just spoke with Chas Freeman a few hours ago, and he was explaining that a big part of this is that the Americans were looking for a pretext to leave this treaty. As you're saying, they've already been violating it without consequence up until now, so why do they want to leave it? They want to leave it so that they can put nukes in the Philippines and point them at China. So they're coming up with this excuse, provoking a fight with Russia, and seeking mutual withdrawal from this important medium-range nuclear missile treaty just so they could get away with escalating against the Chinese.

Doctorow: One of the real problems with the shutdown in public discussion of these security issues in the United States is that an issue like the one we have on the table right now has many possible explanations, but you don't hear them because no one is talking about them. So yes, the explanation you gave is one valid potential explanation.

Another explanation is that Washington's intent is to drive a wedge between China and Russia on this, because the Russians are very worried about the United States' cruise missiles being placed in very large numbers. There will be several hundred of these missiles. The idea is that it would be a massive attack which could not possibly be countered. You can do that from land in a way that you can't really do from a submarine here and a frigate there.

The possibility is that the Russians would get very nervous about these systems that are being put in by the States and that they would urge the Chinese to go along with a new treaty, so then Trump would get what he wants. He has said that he wants for this treaty to be renewed but be extended to include the Chinese, when about 90 percent of all of the nuclear capability of the Chinese is in precisely this range of missiles that this treaty bans. So the Russians and the Chinese would have different interests on that question, and I think it is a reasonable alternative explanation to what you just mentioned a minute ago, to see this as a wedge used by Washington hoping to drive the Russians and the Chinese apart.

Horton: I read that even this new missile that the Russians are fielding — that the Americans say is a violation, to which the Russians say, "No, the limit is 400. We're below that!" — even assuming the worst about these missiles in question for the sake of argument, they're deployed along their southern flank with China rather than being aimed at the U.S. That could change, I guess hypothetically, but there doesn't seem to have been any effort by the Russians to install these things facing west. They're all about China. After all, they're right near each other at medium range. What do they need with three-stage ICBMs?

Doctorow: Well, just looking at what Putin said, which some American media understood properly as threats, this is a change in tone. He was much stronger than he was a year ago when he first rolled out these systems, in making it explicit that Russia is ready for anything the United States might be cooking up, and Russia has already run its arms race. Russia believes that it has completed its side of the arms race and is ahead of the United States.

Some of our media understood that these were serious threats. Other parts of the media don't know what to do with it, and this was something that I brought up, a central feature of my follow-up article that you mentioned a few minutes ago, "Guns and Butter?" There wasn't so much the issue of guns or butter that I had in mind; it was the interpretation of Putin's speech by the lead article in the *New York Times*. The *New York Times* does not know what to do about the INF Treaty. They say they would like to see it continued, but they don't give a reasonable argument as to why it should be. The reasonable argument would be that we're going to be blasted off the face of the Earth if we let the Russians go ahead with their state-of-the-art weaponry and have no arms control discussion with them at all. That is the strongest argument for why we shouldn't leave the INF Treaty. It opens the Pandora's Box, which is the title of Postal's article. That Theodore Postal was given space in the *New York Times* is itself an indication that they are stuck and don't know how to handle this question.

Horton: At the State of the Union address, when Trump brought up withdrawing from the INF Treaty, I thought it was remarkable that all the Democrats sat on their hands. Here he is doing everything he can to hawk it up against Russia, and they're not even grateful. What is it, exactly, that they want from him? They won't be happy until he nukes Moscow and gets D.C. nuked in turn, is that it?

Doctorow: I'm afraid you're right. They really are caught in a dilemma. They would like to say that the administration is correct. The *New York Times* ceded the point that Trump is right; the Russians have been violating it. It's tough to say why you're going to continue with the country that you're otherwise bashing every day and speaking about as so much trash. Russia obviously is not trash, but they don't want to come out and say that.

Still, I do have the feeling that the American public may have gotten a little whiff that there's something dangerous in the air, which a year ago they didn't get, because when Putin first rolled out these new weapons systems, there was at first hesitation in the West as to what to make of it. Then, finally, there was a decision that the Russians really are bluffing; they don't have this stuff. Or, as the *New York Times* article said, the Russians don't have the money to do both, and since Putin needs to give to social programs to stay in power — because we know that his popularity ratings are falling terribly from 80 percent to 60 percent — he's going to have to put money into social programs. He can't possibly afford an arms race with the United States. That's one way of giving yourself the reason not to take the issue seriously, but somehow, I think that the broader parts of the American public are confused and just have a whiff that there's something dangerous in this. Is that how you see it?

Horton: Yeah. I mean, I don't know how awakened people are to the threat of it. You're right, though, it really was remarkable the way that they went from "Wow, that sounds advanced enough to be scary, but not believable," — some of these new weapons that he was announcing last year — very quickly to "He's bluffing, we don't have to worry about that." I saw some of these wonks on Twitter back then talking about that.

This is something I talked about with Chas Freeman, too. There really is an unreality where they talk about the possibility of war with Russia or China in a way where it goes without saying that yes, of course, everyone has H-bombs. But because it goes without saying, it goes without being mentioned. Ever. No one ever says, "Yeah, but this is all just hypothetical because in reality we can never, ever, ever, ever have a war with China, because then we'd lose L.A. or worse." That part never gets brought up, and I am afraid that in D.C. there really is this bubble of unreality where they really entertain the discussions and the plans and strategies for conventional war against these countries, as though nukes weren't an issue. I know how bureaucrats are too, where once these things are on paper and it's just kind of a thing, it's hard to undo it. It's hard to get around the fact that "this is our new concept that we're operating under," or whatever kind of jargon they have for what they're doing. I think that's really the biggest danger, that they're in H-bomb denial.

Doctorow: People are speaking about the "Sanctions from Hell" — and it's nice to catch the news with terms like that — and it's not just Americans; Western Europe also is persuaded that economic warfare is just an alternative to kinetic warfare. What they are ignoring is that in our own past, economic warfare became kinetic warfare. We wouldn't have had Pearl Harbor if we hadn't been waging a fierce economic war against Japan to deny it access to raw materials. That's one thing. A trade war can become a real war, and a conventional war can become a nuclear war. Putin has said very straight that if the survival of Russia is in question, he will go nuclear.

Two days ago, one of the senior Pentagon officials was speaking at the Brookings Institute and describing a new technical possibility, which the Pentagon hopes will get financed to the tune of $135 billion, to move by stealth into either China or Russia and take them over by working through weaknesses in their defenses — a kind of Trojan Horse, which will cost $135 billion if Congress goes along. The problem with that magnificent technical objective is that it ignores the fact that it just takes one finger on the button to respond to that type of knife to the jugular. Conventional warfare can easily become nuclear warfare.

Horton: That's the whole thing. I was just having this conversation about India and Pakistan where so much of this fight over Kashmir is just national pride. It's completely stupid, unquantifiable and essentially worthless emotionalism by a bunch of grown-up babies. They can't just compromise about this kind of thing. It's also a flashpoint where a conventional battle could very quickly turn nuclear because either side, if they have nukes in reserve, what are they going to do? How many divisions are they going to lose before they bust one out? The whole thing is crazy.

Doctorow: I'd like to come back to the very start of our conversation. I never could get to say what was so amusing to the Russian audience in Putin's speech. He used a term which can be translated as a kind of "pointing of little piglets." It's a very farm type of language. Then there were two instances in his speech when Putin was a kind of throwback to Khrushchev, which amused me tremendously, because in writing about 10 days ago, I had argued that the thing missing in Putin's approach to us was threats. He doesn't bang his shoe on the table like Khrushchev did, and he doesn't get our attention, the attention of the broad public.

Horton: Yeah, I really enjoyed that article. You're saying that if he would only raise his voice a little bit, maybe people would pay attention. Instead, he has this kind of deadpan thing, which works really well on a guy like you or a guy like me who's reading closely in between the lines; but to the rest of the people, they just don't even notice what he says at all.

Doctorow: Exactly. There were two points in the speech this past week which were straight Khrushchev. One was the "little piglets." Who were the piglets? They were Western Europe, and he was referencing their agreement with Washington about there being Russian violations of the INF to justify canceling. For Russian patriots to hear the Germans, the French, the Italians and the rest of them all described as little piglets — that was, for them, a real moment of satisfaction.

The other moment that was Khrushchev-like in his speech was in the middle of the speech, when he was describing the scientific achievements that Russia is now experiencing. He likened the development of this Vanguard — I think it's Mach 5 or Mach 6 — Missile as a major achievement because it will operate within its own envelope, and that this scientific breakthrough has some semblance of the nature of Russia's launch of Sputnik.

Well, Sputnik was launched under Khrushchev, and it was awe-inspiring. It changed American psychology about Russia and about their place in the world, and here Putin was saying that they have just achieved something of similar significance. Well, we'll see in the year ahead, as they proceed to demonstrate how this avant-garde operates, but these two throwbacks to Khrushchev days were useful in raising the temperature to a point where people notice that the water is boiling.

Horton: This is the whole thing, right? Where the ideology of American empire comes up against the hard edge of reality. This is something that Pat Buchanan, who's a big critic of the new Cold War, has said all along: that when it really comes to push or shove, do the Americans really intend to go to war over Estonia? No. The NATO treaty can say whatever it wants, but we're not going to trade Denver and Houston and Miami and Charlotte and D.C. and New York for Tallinn and Warsaw. The line, as Pat Buchanan says, used to be drawn halfway across Germany, but now we've moved the line all the way to Russia's borders. On one hand, that could be a real tripwire. On the other hand, we probably don't mean it. It's probably the kind of thing where we can encourage our satellites to act tough and get in fights where they expect that we'll back them up, but at the end of the day, we won't. Kind of like the uprising in Hungary in 1956.

Doctorow: It's not just that we don't have the enthusiasm for Estonians; it's that it is impossible to defend them. You can have all the tripwires you want, but that's all they are — tripwires. They are not self-standing units that could match the Russian forces. All the Russian forces would have to do is just cross the border. You just have to look at the map. Logistics tell you this is impossible for there to be a substantial resistance to any Russian invasion of the Baltics. Yet the Russians haven't the slightest interest in invading the Baltics. But if, hypothetically, Putin were to have an issue, there's no way we could defend those countries. It has nothing to do with our will. We were just idiots ever to sign them into NATO with the collective security requirement of "one for all, and all for one" that is impossible to realize, given the logistics.

Horton: Let me ask you about your opponent's side of the story a little bit more. I am, like you, used to arguing about all of the things that the Americans and the European allies have done to provoke this crisis with the Russians, including the crises in Ukraine and in Syria, and especially in the NATO expansion going back to Bill Clinton. But let's say you asked Stephen Hadley, who worked for George W. Bush, or somebody like him, "Hey, so what's your problem with Russia, anyway?" They would have a litany of accusations against the Russians, along the lines of how the Russians "let them down" and "failed to be partners." But do the hawks have a case at all? I mean, it's not that Vladimir Putin is an angel just because George Bush and Barack Obama are really, really lousy presidents on our side.

Doctorow: Well, the Russians aren't bunny rabbits, and anybody who thinks that they're nice and cuddly is making a serious mistake. But there's no reason why they should be cuddly. They are realists. They're defending their own interests. We are not considering other people's interests, and that is the key error, the key fault, in our foreign policy — that the rest of the world does not exist, that there are only American interests. That is not sustainable. That is not realizable. And it can get us into serious trouble.

As far as where we are today, there have been many comparisons made. What are we living through? What do we look to in the past that has some resemblance to where we are today? People speak about the new Cold War, but there are various phases of a Cold War. Since we're talking about the INF Treaty and medium-range missiles, and we're speaking about their being placed in Europe by the United States potentially, it's very nice to consider where we were in the 1980s, when a very similar problem came up; only it was not the Americans, but the Russians, who were in first. They put in SS-20s (our nomenclature for their missiles) aimed at Western Europe. Western Europe admitted, very grudgingly and with enormous protests, that the United States, to counter this threat, could put in Pershing missiles.

So that sounds a bit like where we are today, but I disagree with the validity of that comparison. It's only the missiles themselves having a similar range which makes it sound like this problem is the same. It's not. Putin's answer made it clear that the situation today much more closely resembles 1962 during the Cuban Missile Crisis. What we're talking about is the period of time for warning that the other side is making an attack, usually meaning a decapitation attack, to knock out the leadership, to knock out the principal structures of the other country and then to take over. That period of time has been an hour or a half hour with ICBMs. With these missiles that the United States is potentially putting into Romania and Poland, the warning time, as Putin said in his speech, comes down to 10 or 12 minutes, which is nonsensical. That really is not a warning time. There's nothing you can do about it. There's no time even to ascertain whether it's a false alarm. The only logical thing you do would be to implement a "push-on-warning" policy. However, what the Russian response has been is that they're putting into the field their Mach 5 Mk-9 missiles, which would shorten the travel distance to Washington to the same 10 or 12 minutes that we would be giving Moscow with our slower-moving missiles based in Romania.

It's a situation very much like 1962, when Khrushchev revolted against the secret basing of American missiles against Russia in Turkey and decided: "Two can play this game. You're shortening the distance and the travel time to Moscow? We'll do the same and we'll put missiles into Cuba."

Horton: But here's the sad part. The Cold War ended 30 years ago. The Soviet Union ceased to exist 27 years ago. So all of this is completely insane and unnecessary. To think that we're even having this conversation at all is, well, shocking but not surprising.

Doctorow: It is, and it isn't. They point to Russia as a hard nut to crack, but we had an opportunity to crack that nut, and we didn't want it. The opportunity was in 1994–1995, when Boris Yeltsin agreed to the Polish request to join NATO, and he let the Americans have their way with Poland in the anticipation that Russia would then be the next country to be admitted to NATO. We forget about all these things. When you go back, you see why the Russians have been so disappointed that NATO was built up not as an integrating force for Europe with their inclusion, but at their expense and as a new wall used to exclude them from security agreements governing Europe.

Horton: Alright. Well listen, thank you very much for your time again on this show, Gilbert. It's great to talk to you again.

Doctorow: Thanks so much for having me.

Lawrence Wittner: Trying a Little Nuclear Sanity March 22, 2012

Scott Horton: Alright, y'all, welcome back to the show. This is *Antiwar Radio*. Our next guest is Lawrence Wittner. He's an award-winning American historian, writer and activist. He is Professor of History Emeritus at SUNY Albany, and his latest book is called *Working for Peace and Justice: Memoirs of an Activist Intellectual.* His website is LawrenceWittner.com, and he's got this great piece at CommonDreams.org called "Try a Little Nuclear Sanity." Welcome to the show, Lawrence. How are you?

Lawrence Wittner: Fine, Scott. How are you doing today?

Horton: I'm doing great. I appreciate you joining us here. I would like some nuclear sanity, thanks for offering.

Wittner: So would I!

Horton: There's a thing called the Sane Act. What a great name for an act, assuming that it's not like the "Patriot" Act and pushing something *in*sane. But I'll take it on face value for now. What do we have here?

Wittner: Well, it's legislation introduced by Congressman Ed Markey of Massachusetts on February 8th of this year. It's the Smarter Approach to Nuclear Expenditures Act (HR 3974), and it would cut $100 billion from the U.S. nuclear weapons budget over the next 10 years by reducing the current fleet of U.S. nuclear submarines, delaying the purchase of new nuclear submarines, reducing the number of ICBMs, delaying a new bomber program and ending the nuclear mission of air bombers. So basically, it calls for a major cutback in the U.S. nuclear weapons budget, which is a very large budget.

Horton: $100 billion over 10 years. So I guess if we extrapolate it out and assume that it's evenly spread to $10 billion per year, what is the total nuclear weapons budget every year? Do you know?

Wittner: Well, it's something like $60 billion. That money is spent not only on maintaining current nuclear weapons and replacing them with more modern weapons, but also with maintaining and indeed building new nuclear weapons labs at tremendous cost. One of the compromises that the Obama administration made with Republican opponents of the New START Treaty a year or so back was that he would spend a lot — over $100 billion — of federal government money on building up nuclear labs, building new ones and getting ready for a new surge of nuclear weapons. So the government, despite talk of the goal of a nuclear-free world, is looking into the distant future in terms of maintaining and upgrading its nuclear weapons program.

Horton: I don't remember if I read this in your article or somewhere else, but somebody was saying that Russia is now being brought into the missile shield. I guess the excuse was that this is to protect Poland from Iran. So the Russians are calling our bluff and saying, "Okay, well, we want to help protect Poland from Iran, too. Since it's not about Russia at all, you won't mind." And I wonder, did they get away with calling our bluff on that? And are they, in fact, being brought in? Do you know?

Wittner: As far as I know, they're not. The Russians certainly have made the point that if the U.S. government needs this missile defense shield to defend Europe against Iranian missiles, then the Russians would be glad to cooperate. But the U.S. government won't go for that and insists on having a missile defense shield that it controls and that the Russians have no part of. So there's a real conflict between Russia and the United States as to this missile defense program, and it's not going away.

Horton: Is it even fair for us to call it a missile defense program? Maybe it's *kind of* one, or maybe you have an opinion about this. It seems to me like it's not a missile defense program; it's just one part of the first-strike posture, where we're trying to build up, in the minds of our war commanders, such an ability to overwhelm the Russians that the Mutually Assured Destruction balance is thrown off. And then it would be assured that we will destroy them and we'll be able to shoot down enough of their nukes on the way up with our airborne lasers and our satellite weapons, where we would be able to succeed in a first strike. I don't know how many destroyed cities they think is acceptable, but that's really what they're going for: a first-strike capability and to cancel out Russia's ability to credibly deter us.

Wittner: Well, that may be what the U.S. government is going for, but it's hard to fathom motives. Certainly, it would give the United States — if it works, which is quite questionable — an advantage, since it could neutralize a Russian nuclear deterrence of the United States and allow the United States to have a first strike, or a more effective strike, against Russia. For many years, people have debated this whole idea of missile defense — called during the Reagan years "Star Wars" — and the defenders of the system say, "Well, it's merely defensive; it's not aggressive in any way." But the critics of the missile defense system say that it's part of warfare, that just as soldiers of old would carry not only a sword but a shield, so this shield enables the swordsmen to fight more effectively in the future conflict. Therefore, when discussing defense and offense, you have to take into account the fact that the shield will help with warlike behavior.

Horton: Well, I'm kind of amazed by this whole thing. When I was elementary school age in the 1980s, it was still the Cold War, *détente* was over, and it was the renewed brinksmanship of the Reagan years. It seemed like there was at least some kind of probability that we could actually get into a full-scale humankind, extinction-level nuclear war with the USSR. Yet the USSR has been gone for more than 20 years now, and it just seems like we must be stuck in the wrong dimension or something if we still have thousands of these things pointed at each other, when Soviet communism is over. The Russian military has pulled back, and they set all their satellites free — well, most of their satellites free.

Wittner: It's certainly clear that the current nuclear weapons budget and nuclear weapons arsenal that the United States maintains is part of Cold War thinking. That is, there's been no upgrading in thought since that time, no understanding that it's time to reassess what's going on in the world. So Congressman Markey, at the time of introducing the Sane Act, said, "America's nuclear weapons budget is locked in a Cold War time machine."

Horton: Tell us a little bit more about Edward Markey and how much support he has, how big of a deal this is perceived to be inside the Congress, whether you think it's really going anywhere, those kinds of things.

Wittner: It's still somewhat marginal, but it's growing in support. There are 45 co-sponsors to the Sane Act. It's been endorsed by a variety of peace groups. It has the support of the National Council of Churches, the Project on Government Oversight and the Congressional Progressive Caucus. So it has some possibilities of getting traction in this Congress. Congressman Markey, I should note, has long been a critic of nuclear weapons. He's no newcomer to this cause, and during the 1980s he was converted to it by the late Randy Forsberg, who founded a nuclear weapons freeze campaign trying to stop the nuclear arms race during the early 1980s. And SANE was the Committee for a Sane Nuclear Policy running from the late 1950s through 1987, when it merged with the nuclear weapons freeze campaign, so the name "Sane Act" is designed to ring a bell with people who remember that history and understand that there's a long-term citizens' campaign against nuclear weapons and for saving the world from nuclear war.

Horton: Isn't it strange how the most important issue ever, the future survival of all of mankind, is seen most of the time like the province of some hippies and some elbow-patch professors? Nobody else seems to care, really, or think that this matters as an issue very much. Of course, there's the Friends Committee, but other than that, if you ask people what the most important issues are in America? Getting rid of the nukes is probably 100th on the list.

Wittner: Well, you know, if you ask people whether they would like to get rid of nuclear weapons, they say yes by overwhelming majorities. There's tremendous support for a nuclear-free world. But on the other hand, I think you're right that it's not the issue that's on the front burner or in the front of their minds. They're more concerned about a variety of other immediate matters. So it's only when a nuclear crisis looms, such as the Reagan administration threats of nuclear war against the Soviet Union, that people get out in the streets. They demonstrate, and they join mass campaigns against nuclear weapons. At this point, I think that the nuclear question has receded from the forefront of public debate, but nonetheless, it should be there, since nuclear weapons are the means to destroy the world and all living things.

Horton: You also make a good point in your article when you say that it's really been the Non-Proliferation Treaty that has done the most, which was an American-created treaty and pushed by the United States on the rest of the world: "You guys promise not to get nukes, and we'll promise to get rid of ours one day." For the most part, it's really worked, even if they're trying to use it as a tripwire for war against Iran right now.

Wittner: Right. The United States didn't want that kind of treaty initially. It got together with the Soviet Union during the early 1960s and began to work at a treaty to prevent other nations from developing nuclear weapons and to get the other nations to sign on and say they would not build the bomb. But the other nations said, "Well, we're not going to sign that sort of treaty. How about you'll get rid of your nuclear weapons?" So eventually, the treaty was a compromise and provided for the non-nuclear powers to pledge not to develop nuclear weapons and for the nuclear powers to swear that they would get rid of theirs. Naturally, the nuclear powers have been dragging their heels ever since, and they don't really want to get rid of them. But when pressed, they're willing to make some progress on that score. In fact, this agreement has worked somewhat. There are far fewer nuclear weapons in the world today. There are about 20,000 of them still remaining, as compared to the height of the Cold War, when there were some 70,000 in existence. Nonetheless, those last 20,000 are going to be abandoned only with great reluctance by their possessors. So the struggle goes on, and I would suggest that your listeners, if they're really concerned about the future survival of the world against this nuclear menace, should contact their members of Congress and urge them to get behind the Sane Act and other measures that will get rid of nuclear weapons for good and for all.

Horton: I can imagine a bunch of congressmen being shocked to find out that, "Wow, did you guys get a bunch of phone calls about this Sane Act today? Me too! Isn't it shocking all of a sudden how many people are getting behind this issue that we usually don't hear about?" Wouldn't it be nice if we could shock them a little bit and get some of those phones to ring, get some of those Congresspeople at least to take notice?

Wittner: I think so. It really has been popular pressure that led to arms control treaties such as the Non-Proliferation Treaty. Without that popular pressure, governments are not going to move on this score. Governments drift very easily into war, especially when they have very powerful weapons, and they become very dangerous at that point. Therefore, the only way we can ensure that they don't fight nuclear wars is to force them to get rid of their nuclear weapons.

Horton: I saw a thing on the Science Channel or something years ago. Some mathematician said, "Well, I'm a mathematician, and as long as there are thousands of nuclear weapons in the world, then there's a probability greater than zero that they will be used at some point." That seems like a convincing enough argument right there to get rid of all of them.

Wittner: Absolutely. You know, terrorism is a danger, and Iran developing nuclear weapons is a danger, but those are our dangers that would not exist in a nuclear-free world. Only by getting rid of nuclear weapons and the materials for making them will we ever be safe from nuclear war.

Horton: You know, even though I was bashing him earlier for his brinksmanship, Ronald Reagan — even though he failed to pull it off, because apparently he believed in the myth of Star Wars under 1980s technology — believed that he was on some kind of mission from God to abolish nuclear weapons, to make a deal with the Soviet Union to get rid of all of them; then to use our moral superiority to insist on the abolition of the rest of the nukes on Earth too, in the hands of the Europeans, the Chinese, the Israelis, etc. But he blew it, and he didn't pull it off. But maybe the anti-nuclear activists could improve their spin a little bit if they adopted and used Ronald Reagan as their mascot, had Ronald Reagan's face on all of their propaganda and said, "We must create a nuclear weapons-free world like Ronald Reagan wanted." Then that way you're attacking the right from the right and you're showing the broad-based, bipartisan consensus on this issue. Think of all the space between you and Ronald Reagan — we all agree on getting rid of the nukes! See what I mean?

Wittner: Well, I think you're right. I think nuclear weapons are not an issue for the right or the left. They're an issue for all people who value their families, friends and loved ones and don't want to see them consumed in nuclear war. Arms control and disarmament treaties have been sponsored by both Republicans and Democrats, and I think they should be.

Horton: Okay, well, right on. Thank you very much for your time on the show today. We're sure with you.

Tom Collina: America's Dangerous "Nuclear Sponge" December 6, 2019

Scott Horton: Alright, you guys. Introducing Tom Collina. He is a Policy Director at the Ploughshares Fund, and here he is writing in *The National Interest*: "Why Are We Rebuilding the 'Nuclear Sponge'?" Welcome back to the show. How are you doing?

Tom Collina: I'm great. Great to be here.

Horton: Very happy to have you here. This is the same subject we talked about last time, three or four years ago. I sure never have forgotten it. And people ask me about that all the time: "What's that nuclear sponge thing?" And I send them that piece by you at Defense News, "Welcome to America's 'Nuclear Sponge,'" So okay, first of all, the question — what is a "nuclear sponge," Tom?

Collina: Yeah, great question. The people in the U.S. Upper Midwest know what it is. It's where the United States deploys its land-based ballistic missiles. And just to back up, the United States has a nuclear arsenal made up of land-based ballistic missiles, submarine-launched missiles, as well as long-range bombers. But the land-based ballistic missiles, which we call ICBMs, are based in the Upper Midwest of the United States, and their mission is a strange one. Their mission is not really to be launched; their mission is to be targets. And if there were ever an attack from Russia, which is highly unlikely, then Russia would have to take out those targets, those missiles in the Upper Midwest, in order to avoid those missiles coming back at Russia. So that's why it's called a "nuclear sponge." Those missiles are there to soak up a Russian attack and to prevent, presumably, the Russian missiles from being targeted at, say, New York, San Francisco or Austin, or at submarines that are deployed out at sea. The problem with this is that it puts a target on the good people of the Upper Midwest, and completely unnecessarily. There's no reason why we have to draw that attack to the United States. In fact, if you think about it, wouldn't we rather draw an attack outside? Why are we drawing an attack towards the United States? The whole mission simply makes no sense.

Horton: In other words, as Ron Paul put it back when he was running for president in 2008, "We could protect this country with a couple of good submarines." And as you're saying, if anybody's got to be a nuclear sponge, let it be the fish out in the Pacific Ocean somewhere, right?

Collina: Exactly. If you didn't have the ICBMs out in the American heartland, then if the Russians were going to attack us, they would choose some other sites. But they'd also have to worry about the submarines and so they'd have to go after those. The thing is that the submarines are invulnerable. When they're out at sea, no one can find them. So point number one here is that Russia has no incentive to attack the United States. It would be a suicidal mission, regardless of whether the United States has these ICBMs out in the Upper Midwest. So before we rebuild this force, to the cost of about $100 billion — I mean, that's quite a lot of money — we should be asking the question: Do we need these things? Remember, the Cold War ended 30 years ago. Why are we rebuilding these weapons that are tremendously expensive, that we don't need, and that put a target on the back of the good people of the Upper Midwest?

Horton: Well, as you just said, it's because they're incredibly expensive, as you write in the article about the economics of the H-bombs. But just on the strategy thing, help me understand this because I think America and Russia both have around 2,000 deployed and around 6–7,000 total H-bombs, right? And so, in the event of an emergency, never even mind the ones in storage, just the deployed H-bombs, it sounds like the Russians have enough to hit all the fields of silos in the Midwest, and New York, San Francisco, Denver, Austin, Miami and wherever else — Washington, D.C. if we're lucky — and also to try to hit whatever military bases overseas where nuclear bombers are stationed, Japan or wherever. It still seems like 1,600 is a lot of H-bombs. You could hit all the important military bases and all the important cities, and American forces around the world, too, probably, with that.

Collina: Right, and there's also a question here. At some point, it wouldn't matter where you are or whether you've been directly hit, because that'd be enough nuclear force to create something that is known as nuclear winter, where once cities start burning, you create so much fire and soot and smoke and ash that all goes up into the upper atmosphere, and then there's basically a huge cloud that covers most of the Earth. And then guess what? The sunlight can't get to the ground, and we actually destroy our food supply. So, in some ways, if you're in one of the cities that are attacked, you're in some ways the lucky ones. Because it's going to be a year after that when things really start to go south for the rest of the world.

Horton: Yeah, that's what my grandpa always said: "If there is a nuclear war, I hope the first one lands right on my head."

Collina: [Laughing] Exactly, but these are dark conversations.

Horton: Yeah, well, hey, it's a funny thing about the nuclear weapons because everybody knows we have them. It's not like it's some crazy, fringe topic necessarily, but it sounds kind of fringy because if you're concerned about it, then you must be some kind of alarmist claiming that there's going to be a nuclear war that is going to break out any day now. That's obviously not what you're saying or what really any of the anti-nuclear weapons activists are really talking about. But otherwise, it just goes without saying that this is life in the 21st century, same as the last, that we've got nukes and we're always going to have nukes. That's how we keep the peace, they say, and to bring it up at all is really kind of the fringe position. For everyone else, it's like the Sun rising in the east, that's all.

Collina: Right, and it's too bad, because you're right, we're not saying that there's going to be a war tomorrow by intent. And in fact, what I'm trying to say is that there is no reason why the United States would attack Russia or why Russia would attack the United States. It would be a suicidal thing to do, so in that sense, deterrence works because neither side has any rationale or reason to attack the other. But that doesn't mean accidents can't happen, right? You can't deter accidents, you can't deter mistakes, and you can't deter madness. We have a president in the White House who is right now undergoing impeachment, which is a tremendously stressful process, and who is quite impulsive, and some people would question his judgment. But the president has sole authority to launch these ICBMs and all the rest of the U.S. nuclear arsenal on his sole prerogative with no checks or balances from anybody else in the administration or in Congress. So mistakes can happen, misjudgments can happen. To me, that's the most likely way we get to a nuclear catastrophe, is by a blunder or by a mistake, so we need to reorient U.S. nuclear policy to prevent the mistakes and the blunders, not to deter an intentional attack, which in my mind has a vanishingly small chance of ever happening.

Horton: So there's two things that you mentioned there. You have accidents, like, "Oops, we accidentally dropped an H-bomb on North Carolina!" where eight out of nine fail-safes failed. One of them kept it from going off, but if it had gone off, it might have been blamed on Russia. They wouldn't be able to admit it was them. They'd just go ahead and start launching bombs at Moscow. And there's been, what, 20-something of those? Able Archer and the Norway missile in 1993, you know, a few of those kinds of mistakes.

Then also, you say misjudgments. I think that's the thing really to clue in on, is bad assumptions by politicians. As you say, the worst hawk in the world doesn't want to start a nuclear war with Russia that extinguishes our species. Yet there's nobody as uniquely wrong about everything as a politician in the national government of the United States. I mean, these people are absolutely nuts, and if the current nut-ball wasn't in power right now, we might have had an air war with Russia to protect al Qaeda in Syria under Hillary Clinton's no-fly zone. Because the consensus is that the guy in the three-piece suit and the cleanshaven chin is a genocidal monster, and the al Qaeda suicide-bomber terrorists backed by Turkey and Saudi Arabia and the CIA are the moderate rebel heroes trying to defend themselves. And they really believe their own lies about that! I mean, you could read their tweets. They're still certain of that crazy narrative, and they were really talking about confronting the Russian air force for bombing the "favored" terrorists in al-Nusra instead of the disfavored terrorists in the Islamic State.

Or take Ukraine, where America overthrew the government, and then when the people in the far east of the country said, "We don't recognize the authority of the new government," the new government attacked them. And yet, every single day since March 2014, the people in Washington, D.C. have said that Russia invaded Ukraine and attacked Ukraine, when all they did is send a couple special forces guys to help the people of the east defend themselves from attack. It's just a matter of chronology. It doesn't matter whose side you're on. This is a matter of fact. Yet everyone in D.C. is wrong about that, and I think they really believe that Putin wants to invade and conquer Eastern Europe. They say it all the time, then they nod, and at the impeachment hearings, they say Ukraine is fighting them there so we don't have to fight him in America — two different impeachment witnesses claimed that the Russians are coming to attack the United States. That is how stupid and horrible and wrong the people with the power in this country are. So now tell me we couldn't blunder into a nuclear war? I don't know, man, I am concerned.

Collina: The part that worries me the most is when you bring nuclear weapons together with cyber-attacks. Because what we're learning is that the U.S. system of command and control over nuclear weapons is a computer-based system, and you might think, "Well, that system is not going to be vulnerable. They would take all the precautions to prevent that." But no, it's vulnerable like any other computer system, and we know that the system has been hacked in the past. It can be hacked again. There are vulnerabilities with this system. So when you have warnings that an attack may be underway, you have to be very careful about responding to that. The United States has a policy called "launch on warning," where the president could launch those ICBMs out in the Upper Midwest when there's warning of an attack based on radar satellites and computer information. But if that information is wrong and you launch your forces, then you just started a nuclear war potentially by mistake. This is another reason why we need to get rid of the ICBMs out in the Upper Midwest, because the only purpose of those weapons is, as we said, to be a nuclear sponge. There is no rationale for launching them other than that, because you're launching them before you know that an attack is really coming. So again, all they are is targets, and from that perspective, they're simply unnecessary.

Horton: Who came up with the strategy of the nuclear sponge? Because I think at first the missiles were put there to be used. Now they're there, I guess you're saying for a retaliatory strike only, but not even then. It seems like their job really shifted. Also, that raises the question: Why would the Russians target them if the Russians know that we're not going to use them, that they're just there to be targeted?

Collina: Right. Back in the day, we only had the ICBMs, and the policy was to launch them on warning when we thought there was a Russian attack coming. But then we got the submarines, and so we didn't have to do that. Now there are real concerns with launch on warning, launching before you know a real attack is coming. We should not do that. We should wait to see if we're really under attack before we launch our nuclear weapons. In that case, the ICBMs will all be gone because they're completely vulnerable. They're sitting out there in their silos. The Russians know exactly where they are. Again, in the highly unlikely event that Russians wanted to attack us, it would make sense for them to target those missiles because that's a threat to them if they're attacking us. So they'd need to take them out. The problem is that they still wouldn't attack because they can't take out the submarines. The submarines are still going to be there. We have most of the U.S. nuclear forces on submarines, about a thousand warheads, so Russia will be decimated by the U.S. submarine response.

Horton: What about cruise missiles? Is that a fourth part of this triad now? Or that just counts as ICBMs?

Collina: When we talk about the bombers, they have two weapons on them. They have gravity bombs that drop out from the bottom, and then they have cruise missiles that are shot off. Those are both included in the bomber numbers.

Horton: I've got you. So part of the controversy over the anti-missile missiles being stationed in Poland was that the same launcher could shoot some kind of cruise missile that could very well be H-bomb-capable. Is that right?

Collina: Exactly. One of the problems with the situation with Russia right now is that Russia has good reason to be frustrated with the West, as you were saying. The Cold War ended 30 years ago. If your listeners remember back then, there was really hope that the West, NATO, and Russia could create a partnership, put the Cold War behind us and build a much more positive relationship. One thing led to another, and for a number of major reasons — for example, NATO deciding to expand towards Russia — Russia was made to feel tremendously uncomfortable. Another thing the United States did with NATO is to deploy, as you just mentioned, missile defenses in Eastern Europe that Russia saw as aimed at Russia. And more to the point, they feared that the United States could deploy offensive nuclear weapons in those same launchers. So these are all the reasons why, in my opinion, we are now at such a terrible state of relations with Russia when, with the end of the Cold War so long ago, we should have been in a much, much better place.

Horton: On that previous thought, because those launchers could be used, does that not mean the Russians have to assume that there are nuclear weapons there and that's what they would be used for? Why would we use a dual-use launcher?

Collina: My guess is that the Russians have good enough intelligence that they can see what we're doing there, and they probably know we don't have those cruise missiles there. But if we're so confident that we don't have offensive missiles in those launchers, the United States and NATO should let the Russians come in and do an inspection, if we've got nothing to hide.

Horton: Well, now the INF Treaty is gone, so it looks like there's probably more chance they're actually going to deploy nukes there, right?

Collina: Yes. The plan is not necessarily to deploy them there, but the United States is now developing a nuclear Tomahawk missile that could fit in that launcher. I believe Russia did violate the INF Treaty, unfortunately, by deploying prohibited short-range missiles, but certainly their fear of the U.S. putting similar missiles into these launchers in Eastern Europe could have driven them to do that. Both sides had concerns about the INF Treaty, but this treaty could have been saved, in my view, if the two sides had gotten together and explained to each other their concerns — U.S. concerns about Russian violations, and Russia's concerns about what we might use these launchers for in Eastern Europe.

Horton: I had read two things on that. One, that the Russians were really deploying these new treaty-busting missiles in their south to deter China, rather than in their west to deter American or NATO forces in Europe. And then the second thing was, as Chas Freeman explained on the show, that the reason the Americans didn't want to try to save the treaty wasn't about Russia at all; it was about their wanting to deploy treaty-busting missiles around China as well. But as a result, we're going to end up having a buildup and a game of brinksmanship between America and Russia in Europe as a side effect of both Russia and America's China policy here. Is that close?

Collina: Yes. Let me put it this way. The Obama administration knew about the Russian violation, and Obama wisely did not withdraw from the INF Treaty because he saw value in the treaty, and he wanted to work with Russia to try to get them back inside. Then President Trump comes along, and he sees a treaty that the Russians are cheating on, and he just decides, "I'm not a big fan of arms control anyway, let's just get out of the treaty," and he throws it by the wayside, which was a huge mistake.

I really do think they could have found a way for everyone to stay inside and improve the treaty. But the Trump administration decided to get rid of it in a very short-sighted move. So now what you have is that the Russians are unconstrained in what they can do, and the United States may now be seeking to put intermediate-range nuclear and conventional missiles back in Europe. I don't see anybody in Europe wanting these missiles, right? I mean, I don't think any of the NATO allies are clamoring to have a redeployment of these missiles. Back in the day, before the INF Treaty, there were huge protests in Europe because the European citizens did not want these nuclear missiles in Europe, and that pressure, in part, led to the INF Treaty in the first place. Now, as you were mentioning, the Trump administration wanting to deploy these missiles in the context of China, to me, is simply unnecessary. A lot of what happens in Washington — and this happened when the Cold War ended — is people thought that the Russian threat was going away. We needed a new threat to justify the military contracts, the defense spending, all the rest, so China became the new threat, right? China became the new monster under the bed.

Horton: It seems like if we didn't have that whole skewed system, that any reasonable government would have just said, "Hey, Russia, I have a great idea, let's invite Beijing to join the treaty with us!"

Collina: Interestingly, that's what President Trump is trying to do with a different treaty, the New START Treaty, where he's trying to say, "Let's bring China into it." The problem with that, although I understand the spirit of it, is that, as you said at the beginning of the show, the United States and Russia have thousands of nuclear weapons, while China has a few hundred. So China says, "Well, I'm not interested in joining your talks. We're a responsible, minimalist country when it comes to nuclear weapons. When you all get down to our level, we'll talk." Right? There simply isn't a parallel set of interests to allow that to happen. We're really still at the point where there's two mega-powers in the nuclear sphere, the United States and Russia. They need to get their act together to lower these forces that are not giving them security, that are only costing them a lot of money. Bring that down, and then China might be interested in talking. But at this point, China says, "Look, this is not our problem, you guys have got to figure this out."

Horton: I know I sounded crazy earlier when I talked about how we could have had an air war with Russia over Syria, but that was what the Chairman of the Joint Chiefs of Staff, Joseph Dunford, testified to Congress as to why we needed to avoid having a no-fly zone in Syria. So that was crazy-me paraphrasing this very sober guy in charge when he was explaining why he was against it.

I wanted to bring up this old George Carlin bit about how everybody's got the NIMBY syndrome: If you want to open up a halfway house, or a group home for retarded people, or something, Not In My Back Yard. "Not in my back yard! Except military bases! We love military bases, even if they're loaded with nuclear weapons. I'll take a little radiation as long as I can get a job." That's how desperate people are for work, that they're willing to make those kinds of compromises. They don't even know that it's crazy and wrong that they should have to. This is a big part of your article again at *The National Interest*: "Why Are We Rebuilding the 'Nuclear Sponge'?" Money.

Collina: Right, and it comes in on other levels, too. So here in Washington, where I am, you've got defense contractors who want the new ICBM program, which is going to be a $100 billion program. That's a lot of money. To cancel that program would cost a lot of jobs and a lot of careers in the nuclear defense space. But when you bring it down to the local level, you've got states, you've got military bases, missile bases in these states, say, North Dakota, and there are jobs associated with those bases. So any time Congress wants to try to reduce or close the base or change its mission, you have a huge pushback from the congressional constituency of that state that says, "You can't do this to us, we need those jobs." Regardless of whether the mission of that military base is valid, or whether they'd be a lot safer without them. In the case of the ICBMs, these states are literally putting themselves in the crosshairs of nuclear war by hosting these missiles, and in my opinion, they'd be much better off if those weapons weren't there.

At the same time, we have to understand that these things mean jobs, so if states are going to support an effort to remove these weapons, there has to be some other viable economic plan, right? You just can't kick the people out on the street and say, "Go do what you're going to do." So this is where economic redevelopment comes in, and that would be really great. People have been thinking about how to bring different, better, more forward-leaning jobs to these states. I mean, I truly believe there's not much future in the industry of nuclear weapons. I think it's on the wrong side of history, and that's not where you want to be building your expertise.

Horton: Well, mostly when they close down other military bases, nuclear or otherwise, usually it just benefits the local economy anyway, right?

Collina: Right. So once you get this state on board working with the state or the community to find other ways to do things, it works out great. I spent a lot of time in San Francisco, where Ploughshares Fund has its other office. In San Francisco, they had this military base out there called the Presidio, and then they converted it from a naval base to office space and all kinds of stuff. That is creating so much more economic benefit for the community than when it was a military base. So there are all kinds of things you could do. You just have to plan for it and convince the community that they will be better off once that transition is made. It's change, and people are scared of change, I get it, but we need to work with people and convince them that this is part of a better future for them.

Horton: When you're talking about that kind of money in terms of the contracts, I mean, I think you're right, as hard as it seems, probably closing down the base is against the will of the local tax recipients there. As difficult as that is, that seems within the realm of possibility. But how do you break Congress away from Northrop Grumman paying them chump change to pass these appropriations bills and to continue the status quo? I remember when they were passing the New START Treaty, the Senate tacked so many riders on the thing, it was like a pro-weapons treaty. You know, that was where a big part of the new $1 trillion, now $2 trillion, plan to revamp the whole industry and weapons infrastructure came from.

Collina: Right, and we really do have to call this what it is, which is corruption. These defense contractors have way too much influence. They take a lot of the money that they get out of contracts, and then they turn around and spend it on lobbying to make sure they get the next contract or that this current contract stays in place. So they're giving money directly to members of Congress, they're hiring government officials as they come out of government, and then, when they go back into government, they've been on the payroll of these corporations. The whole thing is in desperate need of change, and unfortunately, because of corporate corruption in these ways, the government is stuck. People say, "Shouldn't we replace these ICBMs?" If you come at it from a rational perspective, the answer is: "No, we should not do this." But that's not the way the situation is dealt with in Washington, right? It's all about money, politics, connections, vested interests, parochialism, and so we wind up getting trapped in bad decisions and wasting tons of money. It's a corporate corruption problem that really can only be solved from the very top. We need a president to come in and say, "It's going to be this way, and if you don't like it, deal with it, but this is the way it's going to be."

Horton: In a Carlinian way, it really is a riot, isn't it? That this could possibly be our system? We could be talking about any other corporate welfare scheme for bankers or for arms dealers, for that matter, but we're talking about H-bombs here, Tom. Like it's just another public corruption scam, which it is, we're talking about bombs that could kill a whole city, machines that are made for the purpose of committing genocide.

Collina: Right, and, like anything else, it has become just another jobs-generating industry, regardless of the consequences. So we're really coming to the point where we all know what has to happen. We all know what we need to do, but the way the government is run is such that we can't solve the problems with the answers that are right there. If we were only smart enough to grab them.

Colleen Moore: America's Nuclear Arms Nightmare April 10, 2020

Scott Horton: Introducing Colleen Moore. She is the Digital Engagement Manager at Beyond the Bomb and Global Zero. She has this very important article, co-authored with our friend Ben Freeman, running in *The National Interest*: "Nuclear Arms Nightmare: Don't Let New START Die." Welcome to the show, Colleen. How are you doing?

Colleen Moore: I'm doing well. Thanks so much for having me.

Horton: Thanks so much for doing this show and for writing such an important piece here, that you've done with Ben Freeman, about the START Treaty. This was "New START," which was signed in the Obama years and is about to expire, is that right?

Moore: Yes, New START was signed in April 2010 by President Obama and Medvedev of Russia. It followed the original START Treaty and continued this pattern of U.S.-Russia cooperation on arms control caps for both countries' strategically deployed weapons, and it provides for inspections and data-sharing. It's such an important treaty. It's basically the last guardrail on the nuclear arsenals of U.S. and Russia after the Trump administration withdrew from the INF Treaty, the Intermediate Range Nuclear Forces Treaty, last year. So this is literally the only guardrail against a full-blown nuclear arms race, and there's less than a year to extend this treaty. Vladimir Putin of Russia has said that they're prepared to extend it and they want to negotiate, so it's kind of in Donald Trump's court right now, really, to commit to extending this key treaty.

Horton: Is he under pressure from the Democrats to let the thing expire, because otherwise that proves he's a pro-Putin traitor?

Moore: That's kind of what we covered in the piece on the foreign influence. I think there are some key pieces of legislation that Congress has put forth, like Senator Ed Markey of Massachusetts, his SAVE Act to preserve and extend New START. Congressman Eliot Engel and Senator Van Hollen also have legislation that actually has some bipartisan support to encourage the U.S. to pursue negotiations to extend the treaty.

That being said, I do think, like we covered in our piece in *The National Interest*, that Congress is skeptical of just talking about the need to cooperate with Russia, because, like you said, they don't want to come off as pro-Putin. But the facts are there. The American people want to extend this treaty.

According to a study by the Nuclear Threat Initiative, 80 percent of Americans want to extend this treaty, and so it is up to Congress to follow through on what their constituents want.

Horton: Donald Trump ought to be completely impervious to any of this pressure at this point, and he ought to be able to rally the Republicans against it. Because it was a year ago that Mueller admitted that the whole Russiagate thing was a hoax; there was no truth to it at all. They even admitted in the Mueller report that they had no chain of custody from the Russians to WikiLeaks for the emails. All of the different accusations about Papadopoulos, Page, Mike Flynn, Senator Sessions, and every bit of the Russiagate story fell apart, because it was all a lie that was drummed up by the CIA and the FBI in the first place to try to prevent Trump from taking office and then secondly, as they put it to CNN, to "rein him in" from moving forward with improving American relations with Russia. As he said, he wanted to get along with Russia. Now that that's all been dead and buried for a year, Trump ought to be dancing on Russiagate's grave and celebrating. He should say, "Not only am I going to re-sign New START, but I want three more of them." Why not? The whole thing wasn't true. He ought to be shoving that down the Democrats' throats and forcing them to oppose him publicly on a new arms deal and see how that plays in an election year.

Moore: I think that you make a good point. He has this opportunity, but he likes his name on things. I think that might be one of the reasons he hasn't extended New START, because it was an Obama-era agreement. I think that he just doesn't like Obama's name on it. I mean, I'd be fine with him saying, "This is a Trump agreement that we've done." Go for it. As long as you extend it, put your name on it, that's fine.

We just want to keep this pivotal agreement here because it has led to reductions in nuclear arsenals. It has given us insight into each other's nuclear arsenals, and that's important. If that's what it takes for him to put his name on it, then I'm okay with that.

Horton: There are a couple of Democratic senators of import who are good on this stuff. Dianne Feinstein, for instance, is for continuing this kind of diplomacy, all the hype about Vladimir Putin notwithstanding, correct?

Moore: Absolutely. I think there are some key champions in Congress right now. Another avenue that we're focusing on is passing a "no-first-use" policy here in the United States, and we definitely have a lot of champions in Congress for that, such as Representative Adam Smith — who's chairman of the House Armed Services Committee — Senator Elizabeth Warren, and Senator Bernie Sanders. They're definitely champions of no-first-use and nuclear disarmament efforts. So there are people that we can really rely on to hone our issue.

Horton: For the Republicans: If any Republican congressman is having any trouble with this, just invoke Ronald Reagan and George H. W. Bush. They were the ones who signed the INF Treaty. They were the ones who reduced America's nuclear stockpile from 30,000 to 7,000. That's huge. And they're Republicans. It wasn't Bill Clinton and Barack Obama who did that. They can just invoke the legacy of the great Ronald Reagan. That's what every Republican wants to hear, so go ahead and give them that.

Moore: You're right, Ronald Reagan is the one that said a nuclear war can never be won, so it must never be fought. I'm not really sure why that's not being invoked more often by Republicans, because, like you said, they do love the legacy of Ronald Reagan. It was Reagan and Gorbachev that negotiated the Intermediate Range Nuclear Forces Treaty in 1987, and that's really what paved the way towards further agreements. That was really the first agreement that eliminated a whole class of nuclear weapons and led to the original START Treaty, which was the first one really to home in on deep reductions in our nuclear arsenal.

Horton: Right, that was under Bush I in the early 1990s. I remember reading that after he lost and was still just a lame duck in December 1992 and January 1993, he took two more trips to Russia to sign new and expanded deals, even just with days ticking away on his administration. It's funny to think about: George H. W. Bush might, in a sense, be quantitatively the greatest man who ever lived. He went and signed deals that eliminated tens of thousands of nuclear weapons from our arsenal and from the Russian arsenal as well. That's actually, if you think about it, the greatest accomplishment of any two men in a room that ever happened. They got rid of a grand total of something like 60,000 H-bombs. It's incredible to think that the American and Soviet arsenals were ever built up to those kinds of numbers in the first place, but what a legacy, and what a legacy to invoke!

Moore: Like you said, we've come such a long way. During the Cold War in the 1980s, we had more than 70,000 nuclear weapons, and now, in 2020, we have just under 14,000. So we've come a very long way, and a lot of those leaders have been Republican. I think that that's the most important thing that I want to home in on, that Trump is destroying this legacy. The U.S. might not be able to call themselves the leader in nuclear disarmament anymore. We have this history all the way from Eisenhower, that all presidents have committed to some aspect of nuclear disarmament. Now we're kind of seeing that history fall through before our eyes, starting with the INF Treaty, and now with not a lot of movement towards extending New START.

Horton: On the INF, explain what that treaty did and what it means that Trump has already gotten us out of that one.

Moore: The INF Treaty eliminated all ground-launched missiles with a range of 500 to 5,500 kilometers. It was the first treaty that eliminated an entire class of weapons, and it led to the original START Treaty. Trump withdrew from the INF Treaty in 2019 unfortunately, and I think it really was because he had John Bolton, the former national security adviser, in his ear. John Bolton has built his entire career on destroying any kind of avenue to diplomacy, especially as it relates to arms control, starting with the Anti-Ballistic Missile Treaty in the early 2000s. He was in Trump's ear about the Iran deal right when Trump came into office, and now the INF Treaty. Even though John Bolton's out, I still see his legacy in the Trump administration. Without the INF Treaty, it's really only New START left that is containing a full-blown nuclear arms race.

Horton: I've read that the Russians may have technically violated the deal but that they were only making their medium-range missiles for their frontier with China. They weren't trying to mess around in Europe at all. They needed these medium-range missiles to deter China, but the Americans essentially took advantage of that and said, "Aha! See? They're developing these rockets that we think are technically in violation of the treaty," and then they used that as an excuse to get America out of the treaty in the name of Russia breaking it, not because their motive was to reintroduce a bunch of Pershing medium-range missiles into Europe. Rather, they want medium-range missiles for China, too, and they want to put them in Korea or on Navy ships in the Pacific in order for them to hem China in.

So, in order to threaten China, both Russia and America were willing to break this treaty that was keeping H-bombs out of Europe. Now, of course, we're going to end up with a buildup of medium-range missiles over the ability to threaten China, when we already have half our missiles pointed at them anyway.

Moore: Yeah, when we're talking about the hundreds of nuclear weapons we have on hair-trigger alert, I mean, that's insane. The purpose of the U.S. nuclear arsenal right now is war-fighting, and it should be deterrence. That should be how we're paving the way towards the elimination of nuclear weapons. We have these nuclear weapons on hair-trigger alert that could be launched within minutes. The president, who has sole authority over these weapons, has to decide within mere minutes if it's a credible threat and if they should respond.

Horton: That's the other thing, right? If you're talking about ICBMs coming over the Pole, then you have maybe 15 or 20 minutes to decide whether it's a real attack, what to do in response, and to pick up the red phone and try to do something. But when you're talking about these medium-range missiles, the time is reduced for decision-making, and you probably get much itchier trigger fingers and much more emotional responses.

We've seen plenty of close calls already, most famously in 1993, when the Norwegians launched a rocket. They had told the Russians that they were launching a satellite, but the word didn't get through the chain of command, so Russia assumed that it was an American first strike on Moscow and almost went to nuclear war. That was two years after the Soviet Union was gone. Three to four years after the Cold War was over, we almost had an H-bomb exchange over an accident, and this is exactly the kind of thing that could lead to that sort of crisis, totally unnecessarily.

Moore: That's why Global Zero's view is that as long as these weapons exist, we're not safe. You pointed out just one example of a possible nuclear attack that we came really close to. It's really scary how many accidents there have been. In the United States — this is detailed really well in Eric Schlosser's book *Command and Control*, all about the Damascus Incident, which was just a technical problem where somebody dropped a wrench in the silo and it led to this awful disaster. There was another incident where a nuclear bomb was dropped on North Carolina.

Just with the existence of these weapons, you don't know what can happen, as these weapons continue to exist with the nuclear posture that we have right now and the fact that, like you pointed out, we have minutes to decide if it's a credible threat. That example in the 1990s, it was one person in Russia, I think named Petrov, who saw that it wasn't really a credible threat, and he paused. Thank God for him, but it shouldn't have to come down to that.

Horton: Yeah. If it had been not him, but the guy on the night shift, we'd all be dead. And in fact, he was disobeying orders. His orders were that you're definitely supposed to turn the key and only then call your boss. He just essentially refused to do what he was supposed to do, because he knew what would happen. In fact, I think he said that he knew that his superiors, if he had told them, would have gone ahead and launched. That was why he deliberately kept it from them, because he knew that they would have just gone ahead without thinking it through.

By the way, I forgot what it's called, but there's a documentary on Netflix about that incident where they dropped the wrench in a missile silo in Arkansas. The guy dropped the wrench down in the silo, and it almost led to an H-bomb going off. The one in North Carolina, where the plane accidentally dropped the bomb, when they found it, eight out of nine fail-safes had failed. It was just one little switch that kept that thing from detonating. I mean, it's pretty easy to imagine what they would do if an H-bomb went off in North Carolina. Unless the air force was already on the phone with the president telling him, "It was our mistake, sir, sorry," then they almost certainly would start bombing somebody. They would assume we were under attack and go nuts, and that could be the end of everyone everywhere, over a fluke.

Moore: There's a big argument right now over so-called low-yield, smaller nuclear weapons, which is just a crazy argument. There's no such thing as a "small" nuclear weapon or "limited" nuclear war because, even if it is one of these "low-yield" nuclear weapons detonated, we'd have global consequences. The atomic bombs we dropped on Hiroshima and Nagasaki were considered low-yield by today's standards, and those destroyed cities and caused long-term effects for generations of the Japanese, so it's insane that we're even having this argument about new low-yield nuclear weapons in the United States.

Horton: Back to the New START Treaty, what were the limits? And what do we face if they're lifted here? Just total free-for-all, back to 40,000 nukes each?

Moore: It could be, yeah. This is really the only guardrail. It was specifically on strategically deployed nuclear weapons, and it limited them to a specific number. But the other part of it is just the data-sharing and the inspections. That's such a key part of giving each other insight into our nuclear arsenals, and it's forcing us to cooperate and talk about it. The limitations set by the treaty are 1,550 deployed nuclear warheads: approximately 700 deployed submarine-launched ballistic missiles and deployed nuclear-capable bombers, and then 800 total non-deployed land-based launchers. So, like you said, without New START, there's really nothing to contain the nuclear arms race. We don't have the Intermediate Range Nuclear Forces Treaty anymore, and we don't have limits on strategically deployed nuclear weapons, so we're already in the midst of a nuclear arms race. It's pretty scary to think of what could happen without this.

That's why Senator Markey's legislation is really key, because it urges the U.S. to negotiate to extend the treaty, but if it is not extended, that would permit funding to increase the U.S. strategic nuclear arsenal above the New START limits through 2026. It's trying to use the power of the purse in Congress to prevent that. That doesn't necessarily call on Russia to do anything, and there is a provision in there that if Russia starts to build up their strategic nuclear arsenal, then that's a moot point from the U.S. perspective. That's why it's so important, and there would be no guardrail left on the nuclear arms race without New START.

Horton: When you talk about the number of nuclear weapons that we have, even under the New START Treaty, that's more than enough for the American military to kill every Russian and Chinese military base and city in one day. If that can't deter you, nothing can, even if you believe that we absolutely must retain H-bombs in some numbers to deter Russia and China, which I think is a huge false assumption. But even if you buy that, you don't need more than we already have. There are more than enough to kill every last Russian and Chinese in the world.

Moore: Global Zero released an Alternative Nuclear Posture Review a few years ago, and one of the things that we advocate is eliminating intercontinental ballistic missiles. We don't need ICBMs, and it would save us so much money. We spend billions of dollars on ICBMs every year. Between 2017 and 2046, we're going to be spending $149 billion on just ICBMs, and we don't need them. They're there for a first strike. The idea is that they would absorb that first strike; that's why they're known as the "nuclear sponge." They just make us more susceptible, especially when we're in a time of crisis right now with the COVID pandemic and we don't even have enough money for ventilators, masks, beds, nurses and doctors, and we're spending billions every year not only on ICBMs specifically, but also on our nuclear weapons arsenal as a whole and the modernization of our nuclear forces.

Horton: Tom Collina has written a lot about this nuclear sponge thing. This is something that ought to be probably not surprising, but still shocking, to Americans: that our government's official policy is to keep Colorado, Nebraska, the Dakotas and Wyoming stocked with these Minuteman missiles deep in their silos, not to use them, but to force the Russians to waste all their nukes nuking Colorado, Nebraska, the Dakotas and Wyoming in order to spare the coasts, where the "important" people live — which is a joke, because they'd all be dead in a war with Russia, too. It's not like they're going to use up all their last nukes on Colorado and then not be able to hit New York or Washington. That's preposterous. Yet somehow, it's still the plan that "we've got to make sure that they waste as many nukes as possible killing the good people of Nebraska." It's just completely nuts. Who comes up with this stuff?

Moore: That's a great question. In Global Zero's Alternative Nuclear Posture Review, we advocate eliminating ICBMs, taking those weapons off hair-trigger alert, and no-first-use policy. These are very common-sense policies.

Like you said, what crazy person came up with this? I don't think that advocating disarmament is something that's crazy radical. It's thought of as, "It's just these hippies who are advocating disarmament," but to me, it makes so much sense. Something like no-first-use and restricting sole authority, I don't know why that's such a crazy idea not to have one person be able to launch a nuclear first strike. It's pretty common-sense to me.

Horton: By the way, speaking of hippies and Global Zero, just like we were talking about Reagan and Bush, on this issue you have to attack the right from the right. You've got an entire staff full of Republicans, including even Henry Kissinger, who supports this group, correct?

Moore: Yeah, we have over 300 political leaders, military commanders, and national security experts from all across the political spectrum. We have Republicans. We have Democrats. We have people from all over the world, in every nuclear-armed region of the world, advocating our plan to eliminate nuclear weapons. Like you said, there are people on the right who are advocating this. We have Ambassador Richard Burt, who advocates our mission. It's not a crazy far-left movement; we have people from all across the spectrum.

Horton: And George Shultz, Reagan's Secretary of State, right? I'm not sure if he's still alive, but he was on the list back when, anyway. William Perry, too — who was Bill Clinton's Secretary of Defense but is famously a nonpartisan expert genius type who wrote a book about how we have to get rid of the nukes — I know he supports what you guys are doing. That should be enough right there. If William Perry says that we can do without the H-bomb, then we can do without the H-bomb.

Anyway, there's another major treaty that is going up in smoke here, the Open Skies Treaty. It looks like Pompeo and Esper have convinced Trump to abandon that one, too. Can you tell us about that?

Moore: Admittedly, I'm not an expert on Open Skies, and Global Zero doesn't talk about that too much. From my understanding, that was such a key treaty that has been around since the early 1990s. It permits flights over each other's territories to collect data, and the nuclear risk reduction community has spoken in favor of this treaty because it opens up cooperation and data-sharing. It's another one that, like you said, the Trump administration just announced that they're going to withdraw from. It's definitely within this pattern of not wanting to cooperate with Russia and not valuing diplomacy, clearly not valuing our global security.

Horton: Even the nationalists have said, "Never mind cooperation as an ideal or anything like that; we get more out of this treaty than they do." All the different things that we get to discover about them, whatever their criteria are in their mind, by far outweighs whatever the Russians get to find out by flying over the U.S.A., so it seems like a pretty easy one to stay in if they wanted to.

Moore: Both of these treaties were definitely within the self-interests and national security interests of the U.S. So when we're talking about that audience of nationalists, people more on the right, Republican lawmakers, that's what we really have to home in on, that it's in our self-interest. We're getting so much out of both of these treaties. Without Open Skies, we don't have that data-sharing on their nuclear arsenal.

Horton: Like you say here in the piece, 80 percent of the American people are for this, so if ever there was a time for the right in America to attack the right from the right; to go after your congressmen, go after your senators and make it clear that you support this stuff and that Ronald Reagan would want you to also; and for the liberals to go after members of the Senate and the House from the left as well and say, "You know what? I don't care what you think about Trump and Russia, this is far more important than that. This is the future of humanity at stake. We absolutely, first of all, need to shut up about criticizing Trump for wanting to negotiate with the Russians, and secondly, we need to get behind him and support him on this."

When Nixon came home from China, the Democratic leaders of the House and the Senate and the editorial page of the *New York Times* supported him and said, "Great job, Nixon." It wasn't because they leaned Red; it was just because this was the right thing to do, and they weren't going to let their partisan hatred of Nixon get in the way of him opening up China and essentially eliminating the Cold War tension with them at that time. That's the perfect example to invoke. Go after the left from the left and the right from the right, and just insist that this is the consensus of the American people. We demand that they stay within these treaties. For any political challengers out there in this campaign season who want an issue to beat your opponent over the head with, here you go. The American people agree with you, the swing voters agree with you on this, so go ahead.

Moore: The other organization that is a partner with Global Zero, Beyond the Bomb, is a U.S. grassroots movement to reduce the threat of nuclear violence, specifically no-first-use. We're really trying to insert our issues into the presidential election and other elections happening around the country all down the ballot this year.

Unfortunately, it's not a salient issue yet. I think Elizabeth Warren referenced no-first-use once during a debate last year. New START was brought up briefly, and it seemed like most of the Democrats that were in the race pretty much agreed that New START is pivotal to extend. But still, we're not seeing national security, foreign policy or nuclear weapons policy being brought up on the national stage as much as it should. Global Zero, before Beyond the Bomb existed, we got Hillary Clinton to bring it up in a debate with Trump in 2016, and that was really Global Zero's doing. That got our message of eliminating nuclear weapons and other nuclear risk reduction policies out there. That's what Beyond the Bomb is really trying to do this year. Like you said, the voters support this. Anyone that's running in any election, it will be in their best interest to talk about nuclear disarmament, to talk about nuclear risk reduction.

Horton: Absolutely. What you say about it not exactly being a salient issue right now in terms of media attention or whatever, that still should just be seen as an opportunity for people to lead. Just like with the war in Yemen — you might not even know that we're at war in Yemen, but we are. It's horrible, and we've got to stop it. Go ahead and take a position of leadership on the issue. Make it an issue. And if you can succeed in that, then you must be a successful politician. I don't know how many people running for office are ever going to hear this, but people who have influence over them can make their voices known and suggest, "Here's a great issue that your idiot opponent has probably never heard of. It's one that you could really beat him over the head with and score some points and lead the people toward a better future on this."

After all, as I like to say, this is, quantitatively speaking, opinions aside, the most important issue in the world: America's relationship with Russia, specifically as it relates to our hydrogen bombs and what we're going to do with them. This is the fate of our species hanging in the balance here. Nothing else matters at all compared to this. This is the first hundred most important things in the world, and everything else is after that. So it should be an opportunity for anybody who wants to follow your lead, Colleen, and lead people on this and try to make a positive difference.

Moore: I 100 percent agree. It's the most important issue of our time right now.

Horton: Listen, I can't tell you how much I appreciate the fact that you work for Global Zero and Beyond the Bomb, that you've made this your primary interest, and that you made your career out of this. It is the most important thing in the world, and this article is a hell of an important article, too. It's called "Nuclear Arms Nightmare: Don't Let New START Die," and it's co-written with our friend Ben Freeman over at *The National Interest.* Thank you so much again, Colleen.

Moore: Thanks so much for having me.

Conn Hallinan: The New Nuclear Arms Race May 8, 2017

Scott Horton: Introducing our friend Conn Hallinan from Foreign Policy in Focus, and his own blog is "Dispatches from the Edge" at WordPress.com. He writes great stuff, including this one from last week: "These Nuclear Breakthroughs Are Endangering the World." What a great job you did on this one. Welcome back to the show. How are you doing, Conn?

Conn Hallinan: I'm fine, Scott. But yeah, it's scary stuff.

Horton: Yeah, you're completely freaking me out here. You say Mutually Assured Destruction was never the policy. The policy has always been first-strike; it's just that now they can actually get away with it. At least they think they can.

Hallinan: Exactly. This started in 2009. The Obama administration presented what they were going to do as being that they were going to modernize the warheads to make sure that they would work. It was a bad idea to begin with, but what they didn't tell people they were doing was that they were creating this new super-fuse, which allows for a smaller warhead fired from a submarine to be able to take out a hardened missile silo.

The way that nuclear war has traditionally worked is that land-based missiles are much more accurate than sea-based ICBMs, but they're also more vulnerable because they're stable and everybody knows where they are. So they've always been considered a little iffy, but they always had a backup, which was your submarines. Submarines weren't terribly accurate, but they weren't supposed to be. They were not designed to take out other nuclear weapons; they were designed to retaliate, destroying cities or military formations.

With this new super-fuse, however, they now have the ability for a submarine to take out hardened ICBM silos in a first strike. Keep in mind that an *Ohio*-class submarine carries 24 Trident II missiles. That means 192 warheads, and those warheads each have an explosive force of anywhere from 100 to about 475 kilotons. The bomb that destroyed Hiroshima was about 15 kilotons, and the bomb that destroyed Nagasaki was about 18 kilotons, so these are enormously powerful weapons. The *Bulletin of the Atomic Scientists* put together an article saying that these new changes were "exactly what one would expect to see if a nuclear-armed state were planning to have the capacity to fight and win a nuclear war by disarming enemies with a surprise first strike."

This is at a time of rising tension between the Russians and NATO, and certainly rising tensions in the Far East with China. So they're putting out a warning, nobody's picking it up, and it's really disturbing. These are things that people kind of steer away from because they can't really imagine it, and I think that's one of the real disadvantages of nuclear war. Nuclear weapons are either so destructive that people say, "Oh, no, they'd never use them," or they don't really understand how destructive they are.

And, you know, the Indians and Pakistanis are constantly threatening one another, and a nuclear war between India and Pakistan is a real possibility. What people don't necessarily understand is that if there were such a war, it would have worldwide consequences. It would produce enough smoke that would shut out sunlight so that you couldn't grow wheat in Russia and Canada. It would also deeply affect the Asian monsoon. They figure that approximately 100 million people would probably starve to death as a result of a war between India and Pakistan. A war between the United States and Russia or China, with vastly larger weapons, there's no way to contemplate really what that would mean.

Horton: [Sarcastically] Hey, at least a few 100 million humans would survive probably underground in Australia, in the Outback somewhere.

Hallinan: Well, you know, they're moving to New Zealand. There's this big movement in the last year of billionaires buying up land in New Zealand, and 15 percent of them are Americans, and they're literally thinking in terms of survival.

Horton: Not that they should spend any of that money trying to roll back the empire or anything like that.

Anyway, I have so many follow-up questions. First of all, tell me exactly, what is a super-fuse? I guess you're saying it means that they can set off the nuke closer to the target?

Hallinan: That's what it is, but it doesn't make the missile more accurate. The problem with submarine missiles can be explained like this. When you're on land, you know exactly where you are. If you know exactly where you are, then it's easy to figure out a trajectory to drop a weapon exactly where you want it to hit, but when you're in the ocean, you're never quite certain where you are. Now, if the goal is to knock out a city, it doesn't make any difference if you're four blocks off; the city is gone regardless. But if the goal is to knock out a silo, you really need to be right on target because you have to be able to produce an explosion of at least 10,000 pounds per square inch in order to knock out a reinforced silo. What the new super-fuses do is allow you to be much more precise regarding exactly where you're going to detonate the weapon.

There are two basic warheads in the U.S. arsenal: the W-76 and the W-88. The W-76 is the most common, and it's 100 kilotons. The W-88 is much bigger, and it's about 475 kilotons. Until they developed the super-fuse, knocking out reinforced silos was the job of the W-88, but they don't have that many of them. They have lots of W-76s. With the new super-fuse, the W-76 can knock out a reinforced silo.

Horton: What you're saying is that the number of these nuclear warheads, as determined by these previous treaties, had maintained a certain level of balance between the U.S. and Russia. But now, just because of the increased accuracy, we can devote all the lower-yield nukes to taking out those silos, which means that we have all these higher-yield nukes to hold in reserve. Then, as you write in your article, the theory would be we can hit virtually all of their silos, and then we can say, "Aha! We still have thousands of high-yield, multi-megaton H-bombs to hold over all your cities if you even dare to think about retaliating. Don't even try it, or else we'll completely erase what's left of your civilization."

Hallinan: Exactly. In other words, with the super-fuse, they could knock out almost all of the Chinese and Russian missile silos and still retain 80 percent of their nuclear weapons. Therefore, as you said, they'd now be in a position of: "Okay, you want to retaliate? We still have 80 percent of our nukes."

It's scary because one of the things about putting these very accurate missiles on submarines is that submarines can get a lot closer to your shoreline. In other words, if you're firing a Minuteman III missile from Kansas and it goes over the Pole and strikes the Russians or China, it takes about 30 minutes for a weapon to do that. But if you had a Minuteman II or a Trident missile on a submarine, you can get within 50 or 100 miles of somebody's shoreline, and now they've only got 10 minutes to decide whether or not they're under attack.

The other thing that I think is a little scary is that there really is a technological gap between the United States and the Russians. The U.S. has very sophisticated space-based sensors, and they can pick up a missile launch anywhere in the world, track it and make a decision. They can much more easily answer the questions: Is this the real thing? Where is it going? Etc. The Russians don't have a very sophisticated space sensor system. What they've done instead is build a big sensor system on land. The problem with it being on land is that they don't have over-the-horizon visibility, so they only have 15 minutes to figure out whether or not somebody is starting a nuclear war. That doesn't give them a lot of time to check on things. This came up about a decade ago when the Russians suddenly went to full alert and there was a lot of puzzlement in the West. What happened was that the Norwegians fired a missile that was headed not for Russia, but for the North Pole. It was a weather satellite, but the Russians on the ground couldn't determine exactly where it was headed, and they saw what they thought was a Minuteman II coming in at high altitude.

And one of the strategies for a first-strike nuclear war is that instead of launching a vast number of missiles, you take one missile and detonate it at about 800 miles high over a country. It produces a huge electromagnetic radiation pulse that fries everything. It fries all your cars, all your ambulances, all your hospitals, the electromagnetic radiation just absolutely blows anything which is not protected. Nobody has been able to really develop really good protection against electromagnetic pulses.

Horton: I have to push back a little bit there, though. My friend Gordon Prather is a nuclear weapons scientist from back when, and he says that in order to do that, you need neutron bombs, enhanced radiation H-bombs with real thin shells on them so most of their energy is released as radiation instead of as heat. And he says that it's actually not that effective. That doesn't negate what you're saying about how the Russians would fear such a thing, but it wouldn't be an effective way of launching any kind of attack.

We hear this kind of scaremongering about North Korea and Iran all the time, of course, and about the dangers of the EMP pulse and all that, but really, only America has neutron bombs capable of producing such an effect anyway.

Hallinan: Well, the Americans actually have an EMP bomb that's not a nuclear weapon.

Horton: But they're made for missile defense, for taking out incoming nukes coming over the Pole.

Hallinan: Right. But if you detonated a hydrogen bomb at 800 miles over Russia, you would do a lot of damage. One bomb wouldn't take everything out, but they're quite surprised by how effective those electrical magnetic pulses are.

In any case, one of the problems here is that the other side doesn't know what you're up to, and one of the things that's happened is that the Obama administration has been deploying anti-missile systems in Poland and Romania, and there's also a plan to produce anti-missile systems aboard ships — the Aegis system.

Horton: I'm sorry to interrupt again, but can you please be more detailed about the status of that, because I thought that Obama had backed down? This was the famous hot-mic moment, when he told Medvedev, "Just let me get re-elected and then I will chill that out," you know?

Hallinan: Yeah, no. They actually are deployed now in Poland and Romania. They say it's for the Iranians. Well, the Iranians don't have a rocket that can get there, and they don't have nuclear weapons.

It's the same reason why the Chinese and the Russians are upset about the THAAD system that is being set up in South Korea. They see it not as directed at North Korea, but directed at them, so they're very, very unhappy. The Obama administration could have gone back and re-signed the Anti-Ballistic Missile Treaty that the Bush administration withdrew from in 2002, but they didn't do that.

So the idea behind the anti-missile system is this. Somebody launches a first strike at you, and they don't get everything. Then you launch a counterstrike, but it's obviously an enfeebled counterstrike because most of your missiles have been destroyed. You launch this counterstrike, and you have an anti-missile system which picks up the few missiles that you're able to get in the air.

Now, between us, you know that the most famous thing about anti-missile systems is that they can't hit the side of a barn. The systems that we have deployed in Alaska and San Diego, I wouldn't depend on them to do much of anything. But can a potential enemy take that chance? From the point of view of the Chinese and the point of view of the Russians, they see the United States developing the ability to launch a first strike and increasing their number of anti-missile systems. The most logical response to that, if you're the Chinese or the Russians, is: "To hell with nuclear treaties!" They'll just build a whole lot more rockets and a whole lot more warheads, and you won't be able to get them all. This, I think, is going to ignite a nuclear arms race.

Horton: To hear Trump tell it, he says, "Look, the Russians are embarking on improving all of their nuclear weapons, and we have to keep up with them." He's up there defending and expanding upon the policy he inherited from Barack Obama.

Hallinan: Right, and it's really in place at this point. What's not in place is the infrastructure aspect. That is, basically everything is armed now with a super-fuse. What they're also contemplating is building up the infrastructure, which means more aircraft carriers, more missile-firing cruisers, destroyers, submarines, etc. That's where the trillion dollars comes in. If they do that, it's going to cost a trillion-plus dollars to do it, and it will probably take another 10 to 15 years.

Horton: But what about just the waste of all of these most brilliant people who can figure out how to split atoms and put them back together again, whose talents are being wasted on plotting ultimate speciecide here? Just think of the opportunity costs. As Frédéric Bastiat would say, we *see* this, but what's *unseen* is what these people could be producing in terms of goods and services if they had to get real jobs. They could be saving humanity. There's still like one-tenth of humanity that goes to bed hungry at night, maybe one-fifth. These problems aren't solved yet, and yet these guys are sitting around still figuring out how to fuse hydrogen atoms together over my city.

Hallinan: Yeah, Scott, it's one of those things that you look at, and it's almost as hard to contemplate as nuclear war. It doesn't make any sense. Oxfam just did that study that found that one percent of the world's population controls 50 percent of the world's wealth, and the top 20 percent controls 96.4 percent of the world's wealth, which means that 80 percent of the world gets by on less than five percent of the world's wealth. That's an unstable situation, and the idea of wasting all of this scientific knowledge building these things — which, if you use them, will destroy the world instead of dealing with these realities — is hard to get your head around.

Horton: We should note that when they talk about those very fewest people at the top, most of it is all new wealth that they had created. It wasn't like these were all the great-great-grandsons of John D. Rockefeller. But you're right in that we don't have a natural order of seeing where these capital investments would go if the system wasn't rigged. Bill Hicks used to joke that if everybody woke up and realized how insane this all is, it would destroy the economy. The whole thing would fall apart. It's all built on the arms industry. It would be this major adjustment. Of course, the joke is that all this stuff is a waste and there might be a huge disruption in the economy for a minute, but no more than all the people who lost jobs shoveling horseshit when automobiles were invented. It's still an improvement. And not only that, just think not necessarily about all the brains and the engineers, but about all the brawn that's just wasted being infantry in the military, too. The amount of waste that goes into this is just incalculable.

Hallinan: There are two things which I think people should really press for, actually something you can get your hands around. First, take nuclear weapons off of the hair-trigger status. In other words, remove the warheads from the launchers. You can watch this stuff. You know when people are loading up their nuclear weapons, as we knew when Pakistan did it in 1999. Take all of the nuclear weapons off of hair-trigger status. It wouldn't be difficult to do. The second thing — I think it's harder to do but is the more sensible — is to make a pledge of no-first-use. Now, what that means is that if you have a situation where there's some kind of dust-up between the Russians and the Americans, or the Chinese and the Americans, or NATO and the Russians, people know that you're not going to resort to nuclear weapons. That's a really important thing. I want to get rid of them. That's what Article 6 of the Non-Proliferation Treaty says, that you're supposed to get rid of nuclear weapons. That may take a while, but the idea that you can't get rid of them is nonsense. There are all sorts of things that were very effective in warfare but are now banned: gas, dumdum bullets, etc. It can be done. The only problem is that there is, as you said, an establishment, and this is a vast undertaking. A trillion dollars, that's a lot of money. A lot of people are making a lot of money on this, and it's very hard to get the scientific community to back away from these things.

Bob Sheer wrote a book called *Thinking Tuna Fish, Talking Death*, and it's really a fascinating book in which he interviewed all of the nuclear scientists at Livermore and Los Alamos. The interesting thing about it is that these guys are fascinated by nuclear weapons. I mean, they really love working with nuclear weapons. They don't even think of them as weapons; they think of them as puzzles, and they're solving these puzzles. They're not thinking about what it is that they're actually making.

Horton: I asked my friend Dr. Prather about that. I asked, "Hey, Doc, you're such a great guy. I love you. How could you make H-bombs? This is your thing? These can't be used except indiscriminately, my man. What the hell?" And he told me, "Well, at least back then I believed, and I guess I kind of still believe, that we were keeping the Reds from crossing through the Fulda Gap. But now? Forget it. There's no justification for this now. The Soviet Union's 25 years gone." You and I aren't buying it, but at least then there was something like a rationalization.

Hallinan: There was the Cold War. And Scott, probably you and I are the only people that understand what the Fulda Gap is, but that was a basic of NATO training.

Horton: It was the Soviet path into Western Europe.

Hallinan: Right, there was a certain fear that half the armor of the USSR was going to pour through the Fulda Gap. But at this point, we've increased the accuracy of these things to the point where people could contemplate trying it, and it scares the hell out of me. I've got four grandkids. If there's going to be a nuclear war, I personally think that a first strike is an illusion. I think that the other side would have enough nuclear warheads to use them, and it would destroy civilization. Everybody that we know and love and everything that we like would be gone. It's hard to get your head around.

Horton: Yeah, and again, it's just like you said about how unthinkable and therefore unimaginable it is, but we've just passed the 100th anniversary of entering into World War I, where tens of millions of people were killed. World War II where — according to a thing I read recently, maybe 50 million died just in China — we're pushing 100 million for all of World War II killed, probably 90 percent or more of them were civilians. The unthinkable has been thunk and has been carried out before. Daniel Ellsberg printed something in his nuclear weapons series that he did for TruthDig a few years ago. He published the charts from the DOD of estimates of how many hundreds of millions of people are expected to die on the first day, the first week, the second week, the third week, etc. And these are, as you say, scientific men in their uniforms and their smocks and doing their work, looking at a thing out of context. Ultimately, it's almost a sure bet at this point that they're dooming humanity to extinction, modern civilization anyway. Humanity is going to have to start all over again because of these guys.

Thank you very much, I appreciate it.

Hallinan: Okay, brother. I'll talk to you soon.

Horton: Okay, guys, that's the great Conn Hallinan. Over at Foreign Policy in Focus, FPIF.org, this is a really good one where he goes into all the great details for you and explains all this stuff — if you read this, you'll be able to explain to your friends and your family: "These Nuclear Breakthroughs Are Endangering the World." Yeah, that's no overstatement whatsoever.

Daniel Ellsberg: The Doomsday Machine and Nuclear Winter
August 31, 2018

Scott Horton: Alright, you guys, introducing Daniel Ellsberg. If it wasn't for him, we'd probably still be fighting the Vietnam War. But with some help, he leaked the Pentagon Papers that helped undermine support for that war. And of course, he's the author of the book *Secrets: A Memoir of Vietnam and the Pentagon Papers*, and the latest is *The Doomsday Machine: Confessions of a Nuclear War Planner*. Welcome back to the show. How are you doing, Dan?

Daniel Ellsberg: Good to be here, Scott.

Horton: I'm very happy to have you on the show, and I wanted to say thank you. This is the first time I've had the chance to officially thank you for endorsing my book *Fool's Errand*, so thanks, it makes me look smart.

Ellsberg: It's a very good book, I was glad to have the chance. It's a great book, people should read it.

Horton: I appreciate that. So it turns out that you formerly were a complete madman helping run America's nuclear war policy — one of the saner among the madmen, that comes through clearly in the book here, *The Doomsday Machine*. I almost don't know where to begin, but I guess I'll start with asking you to describe the RAND Corporation and your role there as the background for the story you tell here.

Ellsberg: RAND is an acronym for "research and development," R & D, although some people used to say it's "Research And No Development." It did analysis and research for the Air Force right after World War II, where the Air Force wanted to keep getting advice from people like operations analysts and systems analysts. I was in the economics department, which was doing a lot of analysis of the problems of supposedly deterring a surprise attack from Russia during a period when a lot of people in the intelligence community were predicting that the Russians would have a much greater arsenal of ICBMs than we would. They did test one before we could, and that made that claim plausible.

My colleagues and I were of the belief that the Russians were on a crash program and were moving toward hundreds of missiles which could disarm the United States of any ability to retaliate to a nuclear attack by destroying all of the Strategic Air Command bases that could retaliate. So what I was working on was in hopes of preventing any nuclear war by deterring it, by having an ability to retaliate which the Soviets could not destroy. That turned out to be a delusional framework because when we finally got reconnaissance satellites over all of the Soviet Union, which was not until the fall of 1961, after I had been working on war plans to deter such an attack, it was discovered that they had not built, and had acquired no capability for a surprise attack at all. They had very heavy medium-range and intermediate-range missiles and bombers that could reach Europe and could annihilate Europe in any nuclear war, but as opposed to the estimate that I had heard at the Strategic Air Command in August 1961 that they had 1,000 ICBMs, a month later, the conclusive result of photographs was that they had four. The SAC commander Thomas Power had been off by 250 times. Khrushchev had not built on his initial tests. They had really bypassed first-generation ICBMs waiting for more advancements. This means they hadn't really sought an ability to destroy the United States. That would have been Khrushchev's opportunity, if there ever was one, to destroy the U.S. nuclear forces in a surprise attack without experiencing total devastation in Russia.

They did end up building up their ICBMs in the mid-'60s, mainly as a result of the humiliation in the Cuban Missile Crisis. Having disposed of Khrushchev, who had backed down in that crisis, Brezhnev, who had replaced him, told their military, "You can have what you want," and what they wanted was what the U.S. had. What the U.S. had was the ability to annihilate the Soviet Union with effects that they didn't understand at the time would have killed nearly everyone on Earth. The Russians proceeded to imitate that and acquire their Doomsday Machine that could kill nearly everyone. And that's what we've lived with ever since.

Horton: In the book, you talk about when you were made aware of this new CIA intelligence that said that they really only had four nuclear weapons. You understood immediately that that meant they did not mean to get a first-strike capability, but you had a hard time convincing anybody else of that.

Ellsberg: RAND, as a corporation, as a private NGO, they were a nonprofit organization, but they got their money from the Air Force and later from the Department of Defense. Now, they get more than half of their money from people other than the U.S. national government; it comes from state governments, cities and various other things for civilian work. That's all after I left RAND in 1970. At that time, I was a consultant to the Defense Department, to the White House, and to the State Department, to some extent, for war planning. I came back, having learned this estimate in Washington, which was not available to RAND because Eisenhower had cut out the private contractors from National Intelligence Estimates. So the RAND people had essentially been preserved in amber in their beliefs of 1958, when they were cut off from the Intelligence Estimates, that we were far behind the Soviets, and they couldn't adjust to this change that I was reporting. They said, "How would they know that there were so few? What's that based on?" It was totally implausible to them. Nobody at RAND believed it, except a handful who had special clearances and who did know about the reconnaissance. So it was very hard for them to realize that what they had been working on night and day for years with a very great sense of urgency — they were saving the world from the effects of a Russian surprise attack on the U.S. — had been a delusion; that we had been working on a false problem and that we were making the problem worse by our recommendations for how to meet this false problem.

Horton: The way you describe it, your knee jerked the wrong way in response to this, and you wrote a speech that was very consequential in basically calling Khrushchev's bluff publicly. You say that that helped to precipitate the Cuban Missile Crisis, that then precipitated the Soviets obtaining their own Doomsday Machine.

Ellsberg: Not to be grandiose about this — a lot of factors went into the Cuban Missile Crisis — but the speech that I proposed, which was picked up by Deputy Secretary of Defense Roswell Gilpatric, I had first proposed that it be given by the president. But the idea was that in the midst of the Berlin Crisis — which was still very hot at that time — we should inform the Soviets that they had been running a hoax, that we knew it, and that they could stop talking about their ability to destroy Britain, Paris, the U.S., and their lines about, as Khrushchev had put it, putting out missiles like sausages on a production line. This was perhaps true, in retrospect, with respect to their missiles against Germany and against Europe, but not against the U.S. There was no production line running against the U.S. So I said we should let them know that we know he has been bluffing all this time. It was approved at all the highest levels, and the speech was given.

As I say, a number of factors went into the Cuban Missile Crisis. In particular, Khrushchev's correct estimate that the U.S. intended to overthrow the regime in Cuba after failing at the Bay of Pigs, and that we were running a huge covert — and we did begin covert operations pretty much at that time — but he believed we would run a huge covert program of assassination, sabotage and spying against Cuba, and use that as an excuse for a chain of events that would lead to a U.S. invasion of Cuba. That was quite realistic on his part, and that was a major — I knew nothing of that covert program at the time — inducement for him to put missiles in Cuba. But an additional inducement, as he said, was our talking about the fact that we were so much bigger than they were in strategic forces, and I had contributed to that. Not only me, but, more importantly, President John F. Kennedy and Secretary of Defense McNamara made speeches to that effect quite a bit. But it was the Gilpatric speech which was the first expression of that, and I have to acknowledge that I definitely played a role in the process of encouraging and provoking Khrushchev to feel that he had to match our missile capabilities.

Horton: And that was the point behind it? You weren't just beating your chest, there was a strategy behind writing the speech in this way?

Ellsberg: It was a question of trying to keep Khrushchev from beating his chest and saying, "Come off it on Berlin. We're going to maintain our presence there in West Berlin, and your threats of keeping us out are ill-based."

Horton: So did that speech help to contribute to him backing down and building the wall rather than invading Western Berlin?

Ellsberg: Oh no, he had already built the wall. That was in August, I believe, and the crisis wasn't over. We came very close, I think it was in October after the Gilpatric speech, where tanks were confronting each other at Checkpoint Charlie in Berlin. Russian tanks and American tanks looking at each other. If someone had fired, if someone had had a slippery trigger finger, we might not be here because the chance of war coming out of that was very great. That's the kind of abyss we have been living next to for the last 70 years.

Horton: Well, I want to get back to the full devastation of nuclear war a little bit later, but on the history stories of the Kennedy years here, I'm interested in hearing your tales of your role during the Cuban Missile Crisis. You describe that you were pushing a pretty hawkish line, particularly in terms of refusing to pull missiles out of Turkey.

Ellsberg: It wasn't a question of pushing it; it was a question of me agreeing. I was a very small cog in there, although I was at a staff level just below the Executive Committee of the National Security Council. I was helping put in reports to that committee. My opinion was that he didn't need to pull missiles out of Turkey in a trade, and that we had him outgunned. Remember, when you say I was a madman, I was mad in the sense that I was a Cold Warrior who believed all the premises of the Cold War, some of which were true and others which were entirely untrue and made the world very dangerous.

So I was a Cold Warrior along with virtually every member of the Executive Committee, and nearly all of them were opposed to Kennedy's willingness to trade at that point. I definitely disagreed with him on that point. I was wrong, as I make it very clear in the book. I was 31 years old. I was four years out of the Marine Corps. Fortunately, I was not making these decisions. On the other hand, these people, twice my age and very distinguished and famous members of the Executive Committee, were foolish and unwise. As Bobby Kennedy later said, "If half of them had been making the decision, the world would have blown up." So I don't want, even in confession here, to be grandiose and say I was the bad one, but I was a participant in a policy that was madly reckless.

Horton: The way you tell it in the book is that you realized your mistake immediately after the crisis was over. That actually the U.S. and the USSR were much closer to war than you thought.

Ellsberg: I know what you're referring to there, but that's a little misleading. What I'd discovered, to my horror, was not that the situation had been so dangerous — I did discover that later, that's in the book in detail — but some of that we didn't learn until 30 and 40 years later. What I discovered was that the EXCOMM people, starting with Paul Nitze, for whom I was working, had believed that it was very dangerous and had gone ahead anyway. That was shocking to me, not that they were right, but that they in fact were more right than they knew. But I had believed that Khrushchev had to back down and therefore it wasn't that dangerous. That's why I had been going along with all this, but if I had thought, as Paul Nitze said he did, that there was a 10 percent chance of nuclear war, I couldn't have possibly participated in that. I couldn't understand how these people could have been depth bombing or mock-depth bombing Russian submarines, putting our nuclear weapons on alert, flying them in the vicinity of the Soviet Union for purposes of intimidation, making threats and preparing for an invasion, when they thought there was a 10 percent chance of nuclear war. And I thought to myself, as I say in the book, "Who are these guys? Who am I working for? Are they all crazy?" The irony is that as the years went by, we realized that Khrushchev had managed to get nuclear weapons into Cuba despite the blockade. He got them in before the blockade. So the nuclear weapons were there, they didn't have locks on them that could prevent them from firing. A Soviet colonel in charge of surface-to-air missiles had shot down the U-2 on Saturday, October 27, 1962, without orders, in fact, against orders from Moscow, and the same could have been done with these missiles.

A puzzle that I don't go into in the book but intrigues me very much is how Khrushchev could have failed to tell us that he had those warheads there. That would have virtually ruled out the possibility of invasion. By the way, it wasn't just the warheads, but also the medium-range ballistic missiles that he had which could reach Miami and Washington. He also had short-range missiles which could be used against an invasion fleet, and since they couldn't reach the U.S., the Politburo had, before Kennedy made his threats, delegated authority to local commanders to use those missiles. Now, we didn't dream that Khrushchev would give that much control to lower commanders and take that risk. We didn't know that short-range missiles were there. We didn't know it for 30 years until the Russian archives began to open. Had we known all that, U.S. threats of invasion essentially would have been off. Why Khrushchev didn't tell us that is inscrutable to me. He could have won the Crisis if he'd told us that. My only explanation is that in a regime that secretive, even more secretive than ours, the habit of secrecy was such that when it was in his interest to tell us something, he didn't do it.

Horton: Well, that's really the story, how close we came to complete nuclear devastation, while at the same time so many bad assumptions were being held as gospel by the men in charge, including the people you were directly working for.

My dad was at UCLA, and his political science professor was gone during that time. When he came back after a couple weeks — he evidently had been advising the government — he said, "We've never been closer to nuclear war. There's no such thing as *closer* to nuclear war. That was the absolute brink."

Ellsberg: That wasn't the only time. People think of that as being the one time we were close to nuclear war. Actually, there were a number of times. One in particular that I wish I had space in the book to go into in detail was in 1983, when the Russians got a false alarm in the midst of what they thought of as a crisis situation. Andropov, the president at that time, believed that a nuclear exercise of ours, Able Archer, was actually cover for a first strike by Reagan. In the midst of this, he was preparing himself and getting ready to anticipate exactly when it would come and take the extremely foolish move of trying to strike first — "Strike second first," they say — by hitting our missiles before ours could arrive. When I say foolish, that's just what we would have done under the same circumstances. In other words, neither side has shown anything like wisdom in these preparations and in the threats, and we have in fact been very close, as in 1983 and other occasions, to blowing the world up.

Horton: I want to rewind a little bit and talk about the beginning of the book, about your investigations into the current war plan as it was under the Ike Eisenhower administration. I want to hear about what you found in the old War Plan, the plan for general nuclear war against the Communist Bloc, and how you tried to change it for the better.

Ellsberg: I had drafted a question, which ended up being given to the Joint Chiefs of Staff in the name of President Kennedy: "If your war plans were carried out as planned, how many people would be killed in the Soviet Union and China?" I had anticipated that they didn't really have a ready answer to that. Strangely enough, my friends in the air staff were under the belief that the Joint Chiefs never actually calculated how many people would be killed as opposed to how many bases would be hit, how many cities, how many industrial areas and so forth. But my friends were wrong. The Joint Chiefs came back with an estimate very quickly, within a week, and it was shown to me because I had drafted the question. I saw the answer in the White House, and the answer was 325 million people.

Since they therefore had a calculation, I sent another question down through the Deputy National Security Assistant: "How many would be killed altogether in the world by our strikes?" The answer to that was another 100 million in East Europe, the so-called captive nations of the Warsaw Pact; another 100 million in West Europe, our allies, not with any warheads landing there, but because of fallout in East Europe and Russia; and another 100 million in contiguous areas like India or Pakistan; and in neutrals, like Austria, Afghanistan, Finland and parts of Japan, which would be destroyed by fallout. Finland, for example, would be annihilated by fallout from our attacks on submarine pens near Leningrad. So the total would be 600 million.

Horton: You mention Korea in the book as well.

Ellsberg: Korea would have been destroyed by fallout from our attacks on China. That would be just as true today. War with Korea would put a lot of fallout into China, being next to each other as they are. So 600 million. That was 100 Holocausts. My reaction in the first week and the second week when I read their answers was, in both cases, just horror. I thought, "These are the most evil plans and preparations that had ever been made in the history of humanity." Not to mention that they were also insane.

I knew that this was not a bluff like the ones Khrushchev had made in the 1950s, when he threatened to use missiles that he didn't have. We had all these missiles, and they were operational. These were plans for hypothetical war five or ten years in the future. This is what would happen with the weapons we had on alert, many of which I'd seen on research trips to the Pacific. I had even touched a nuclear weapon at one point. It was on the ground, ready to be loaded. It was an interesting experience because it was a cold day and the weapon was warm from the radioactivity, like an animal body. It was an eerie feeling, actually to feel this metal skin warm to the touch and know that that was a 1.1 megaton bomb — that's a million plus 100,000 tons' equivalent of TNT. We dropped two million tons in all of World War II, and this was one bomb that could drop half a World War II in one bomb. These bombs were carried by single pilot planes at that time and were also loaded onto B-52s. So I knew that we had the capability. We were threatening it and were prepared. It was on alert, and this could happen. That's been true to this day.

Something that I didn't know then and nobody knew until 1983, the same year as the Able Archer crisis, when scientists, physicists, aerospace people, and a number of other scientists discovered the cities we were planning to hit. By the way, the way you got to 600 million dead, aside from the fallout, was that every city in Russia and China was to be hit with nuclear weapons. That, of course, brought your casualties up into the hundreds of millions of deaths. But that turned out to be a small fraction of the people who would actually die from these attacks aside from the fallout, because smoke from these burning cities — as physicists realized in 1983, 40 years into the nuclear era — would be lofted by firestorms and intense up-drafts from widespread fires, the kind of which has happened only three times. In World War II, we tried to get firestorms and widespread fires that would get to an especially intense temperature. We tried to do this in Germany and Japan, but only three times was the weather right to do that, once in Dresden, once in Hamburg and once in Tokyo. In Tokyo, the effect of that was to kill 100,000 people in one night, March 9th and 10th, 1945. Almost no Americans realize that that was the most murderous terrorist attack, the killing of civilians deliberately and for political effect, more than had ever been done. More people were killed in the firebombing of Tokyo than in either Hiroshima or Nagasaki.

The point is that every nuclear weapon would cause a firestorm, as happened in Hiroshima and, to a lesser extent, Nagasaki. That meant that these firestorms would not only cause intense temperatures, which is what they were aiming at in World War II, but that these firestorms would inadvertently cause up-drafts that would loft smoke, soot and the toxins of various kinds up from the burning cities into the stratosphere. At that altitude, it wouldn't rain out; it would stay there and would go around the globe very quickly, within days. That smoke and soot would absorb or block out most of the sunlight from reaching the Earth's surface. More than 70 percent of the sunlight would be lost in a war with Russia. The effect of that would be ice age conditions on the surface of the Earth, with frozen lakes even in the spring and summer. It would also kill all the harvests. Those conditions would last — as further calculations in the last 10 years have shown — more than a decade. That would kill nearly everyone on Earth. So the answer as to the effect of a war — the very question that I was asking in 1961, to which without offering to resign, apologizing or expressing anguish they gave the answer of 600 million people dead — was wrong. There were then three billion people, and it would have starved to death nearly all of them.

Now there's 7.4 billion people. A nuclear war today probably wouldn't mean full extinction — some people would live on fish and mollusks in the Southern Hemisphere in Australia and New Zealand — but 98 or 99 percent of the people would starve, including the attacker, whether it was the U.S. or Russia, whether they went first or second. So it is properly called a "Doomsday Machine" in the Herman Kahn sense of a system that could not only threaten but actually carry out the death of nearly everyone. Herman Kahn, a colleague of mine at RAND, suggested that as a hypothetical device which he said could not exist and almost surely never would exist. He thought no one would build such a system, but he was wrong. It existed right then in 1960 and '61 in our Strategic Air Command, and within a few years, the Russians had one, too. Now there's two "Doomsday Machines" coupled together by their fear of a surprise attack from the other. There are warning systems and a readiness to get off their ICBMs within minutes, 10 minutes at the most, but actually several minutes on receipt of a warning. These warning systems could be false and have been false a number of times on both sides. One time in 1995 after the Cold War, where Yeltsin actually was posed with his nuclear computer, his so-called football or briefcase, that could send off his missiles. It had been opened for him in the belief that a missile was heading toward Moscow at that time. It wasn't, and fortunately he didn't press the button, or we wouldn't be here. But that's what we've been living with ever since.

Horton: I want to talk about the improvements that you tried to make, because it is important to show just how much or how little you were able to ratchet the program down. So explain how it could have gotten that way in the first place. I understand the fear of a first strike like you're saying, but kill every single city in all of China and all of Russia? How could good ol' Ike Eisenhower have approved a plan like that?

Ellsberg: He did approve it, albeit with misgivings. He called it overkill. His main concern was too many missiles per city. To him, that seemed inefficient and wasteful, that was his major concern. There were two military men, including — I'm glad to say, as a former Marine — the Commandant of the Marine Corps, General Shoup, who said, "To kill all the Chinese when it hasn't even been their fight, when we've started a fight with Russia? This is not the American way. This is not a good plan." But it was the American plan, and it didn't change, despite his sensible objection to it. Others didn't object, because they were afraid of a Sino-Soviet pact, which actually didn't exist at that time in 1961; they'd been moving apart steadily toward an adversarial relationship from about 1959 or so, but we hadn't recognized that yet. In any case, the idea was that in case of a war with the other superpower, we're not going to leave the Chinese aside to collect all the gains and be in charge of the world. No, they're going to go, too. I'm afraid that is very likely the planning to this day, although it's very secret and I don't know the current planning.

What I do know is that you asked the question of whether, in my staff role, I tried to guide drafting of war plans in 1961. Did I try to improve that? The very simple answer was yes, and I tried to introduce the option of withholding attacks on China in a conflict with the Soviet Union. To make it so that you didn't necessarily combine them irrevocably as our then-existing plans did. So you weren't always getting Russians and Chinese in the event of any conflict over Berlin, Yugoslavia, Iran or anywhere else, or in an uprising in the East or an attack on West Europe. To withhold from hitting China in such a scenario seemed a simple proposal, and it was given the okay by the Joint Chiefs. All of my guidance was sent to the Joint Chiefs as I drafted it, including to withhold from China. Another one was the option of withholding attacks on cities. I could hardly conceive of why, in what was supposedly a preemptive attack or even a first strike for an escalation in

Europe, they would want to hit cities? It didn't make any sense to me. It turned out that this was just a carryover which lasted indefinitely from World War II.

General LeMay, who was in charge of the attack against Tokyo that killed 100,000 people in one night, concluded from this that that was the way to fight and win a war against civilians: crack their morale, crack their existence. LeMay said to a colleague of mine, Sam Cohen at RAND, on one occasion, "War is killing people. When you kill enough of them, the other guy quits." That's what he had learned. It was a very misled notion, but that's what he put in when he became the first commander of Strategic Air Command, and the first targets for our nuclear weapons, atomic weapons, fusion weapons, were cities. Eight Russian cities, then twelve, twenty, as we got more bombs. The whole arsenal essentially was against cities.

Then, when the Soviets tested a weapon in August 1949, every airfield in the Soviet bloc became a potential target. Now you had to have weapons for them, and planes for them, and it was very good for the Air Force, which was getting the lion's share of the budget. So those estimates of a threat from Russia were good for the Air Force, and RAND worked for the Air Force. Without thinking of themselves as corrupt in that sense, they — perhaps I should say we — *we* didn't appreciate how the estimates we were getting from the Air Force were very self-serving and very biased. We managed not to notice that or think of that.

Anyway, that was LeMay's way of fighting a war. He was succeeded as commander of SAC by General Thomas Power, who had been the actual man in charge of the flight and of observing the firebombing in Tokyo while LeMay was back at the base. So Power had been in charge of the greatest massacre in human history. He became commander of SAC, and he was commander of SAC during the Cuban Missile Crisis when he and LeMay were among very few people in the world who would have actually liked to see a nuclear war come; who thought better now rather than later, before the Russians build up, and we'll get rid of the Communist menace and so forth. These were dangerous, dangerous people. But we lucked out.

What I'm saying is that these preparations have been very profitable to the American military-industrial complex. I don't think we would have made these choices, taken these risks, taken these gambles and been mistaken in the same way, always in favor of building up forces, if it wasn't very profitable to do so. Not only profitable for Boeing and Lockheed and Northrop Grumman and General Dynamics, but the other members of the aerospace industry and electronics industry. They have subcontractors in virtually every state in the union. That's why we're rebuilding all these things again, for no other reason that I can see other than the fact that the production lines need to be kept going at those corporations.

It's not only profits, but it's votes. It's jobs; the unions have generally supported it. Congress has supported all this, and there's basically no opposition. When congressmen and senators from the states such as Montana, Wyoming and North Dakota that now have ICBMs write letters to the president about keeping those missiles on alert for the benefit of the local jobs, other senators don't attack them. If they need bases in Minot, North Dakota for ICBMs, the others figure, "Okay, I need bridges, I need whatever else, in my state," and they get in line after each other, and that's our political economy. But remember, the Russians now are the same. They used to have their bureaucratic motives for not being humiliated and not being less than the U.S. There was enough to get them to build a Doomsday Machine which no one has ever needed or even planned as such, but they have a lot of missiles aimed at our cities. Now they have profits just like we do. They have their Boeing and their Northrop which need to be kept in business and make a lot of profit. And both sides are rebuilding their Doomsday Machines right now.

Horton: In the book, you talk about some stuff that just seems like mindless bureaucratic inertia. With the advent of the hydrogen bomb, for example, they just replaced the atom bombs with hydrogen bombs in all the same war plans, even though one H-bomb is worth however many atom bombs, but they didn't adjust. They just switched up.

Ellsberg: That's right, and the plans that I found while working on war plans up through the mid-'50s, those plans were horrific in their way, but they talked about a million dead, maybe five million or 10 million. That's of course terrible, but that's less than World War II. We killed almost two million civilians by air power: 600,000 Germans as well as 900,000 Japanese before Hiroshima and Nagasaki, which added another 300,000 or more. So close to two million civilians in World War II. That was over a matter of years. Now they were planning to do it in weeks, but still they were at the level of the number of civilians killed by the U.S. in World War II. Of course, the Germans had done the same, not so much with bombing, although they set the precedent in the Blitz and in other places, and the Japanese even earlier. In the end, though, we ended up sending 10 bombs for every bomb that they had dropped on Britain.

It stayed down at that level, and then all of a sudden, in the mid-to-late 1960s, the estimated casualties came to 100 million, 200 million, 300 million, and as I said, ultimately 600 million, when it comes to the total body count that they had in mind. That was by substituting H-bombs (hydrogen fusion bombs) for the fission bombs that were used on Hiroshima and Nagasaki. Most Americans to this day don't understand the difference between those two. They know that one is larger than the other, probably that the H-bomb is larger, but by how much? What they don't understand is that the first droppable H-bomb in 1954 had an explosive power 1,000 times more than the Nagasaki bomb. The Nagasaki bomb was used as its trigger, and every thermonuclear weapon has a Nagasaki-type plutonium fission bomb as its detonator, its percussion cap. These bombs can be 1,000 times more powerful than the original atom bombs. Nowadays, for efficiency, we carry smaller weapons on top of missiles which are very accurate and can create much more devastation than we could earlier with large, droppable bombs from planes. So the individual weapons are smaller than that, but are still 10, 20, 50 times more powerful than the Nagasaki weapon.

One reason for mentioning this is to say that as dangerous as things are, they really can get worse pretty quickly. India and Pakistan right now have only A-bombs, the kind that we use as triggers for H-bombs. They don't have H-bombs, although India claims to have made a test. If they did, it was kind of a fizzle. The same is true of North Korea. They claim they have made a test. They may or may not have an H-bomb. In any case, it takes a lot of tests to have a working H-bomb. So of the nine nuclear weapons countries, you have these three countries — North Korea, India and Pakistan — that do not have H-bombs. Here in the U.S., a lot of people have been in favor of re-starting nuclear weapons testing for some time. Russian labs are said to want to restart testing. If that happened, India, Pakistan and North Korea would each get bombs very quickly, within a couple of years. The difference that that would make is this. Right now, with just a few hundred A-bombs between them — India and Pakistan each has individually more than 100 A-bombs — and if they used 50 each against cities, they would put enough smoke into the stratosphere that it would cause this nuclear winter effect. It would reduce sunlight by 17 percent. It wouldn't be a full nuclear winter, it wouldn't starve everyone, but it would starve one-third of the Earth's population, or a little less, perhaps one to two billion people. With H-bombs, they would have a "Doomsday Machine," and it would be *three*-thirds. India and Pakistan have been recurrently at war over Kashmir, and over that, everybody in the world has a stake in whether that occurs. That capability should not exist. The other nuclear weapons countries' A-bombs, even though they don't have as many as the U.S. and Russia, could create more than two billion dead. The U.S. and Russia can, with a small fraction of their alert forces, starve everyone on Earth. That's an unconscionable, immoral, evil, dangerous — we don't have words for this possibility. No matter how low the probability is, it isn't zero.

The possibility of killing everybody on Earth has simply never existed, except for the last roughly 70 years of the nuclear era. It's very hard for the human mind even to comprehend something like that. God knows we haven't acted to remove that capability, which could be done even without giving up deterrence altogether. No country has a real justification for having the 100 weapons that India and Pakistan each have, and that Britain and

France have, or the 300 that China has. Let alone the thousands that the U.S. and Russia have. I don't have any justification for it, but they do exist.

Conceivably, we could mobilize an effort to remove those Doomsday Machines. You can't uninvent nuclear weapons altogether, but you can eliminate Doomsday Machines, and you could do it very quickly. Dismantling these weapons could be done in a matter of months and without a lot of cost; in fact, you'd be saving a lot of cost. You'll be saving close to a trillion dollars in the U.S. alone, with similar savings in Russia. We don't buy our weapons from them, and they don't buy their weapons from us. Our Boeing and our Lockheed make a lot of money from these. A lot of people have jobs relating to it. Likewise in Russia and likewise, to a much lesser degree, in the other nuclear weapons states. It's not irrational, from this short-term, very narrow point of view, that people have interests in maintaining the Doomsday Machine, but their interests do not justify the risk.

Horton: That's the thing of it, right? If you completely dismiss the corporate interests and the bureaucratic interests, and you look at it from just what the average civilian might imagine would be the national interest of the United States, do we need to have these nuclear weapons at all? Just how reasonable would it be to go ahead and get rid of all but ten of them? Or better yet, all but two?

Ellsberg: If you imagine our disarming entirely while Russia retained many weapons, or some weapons, leaving Russia with a monopoly of nuclear weapons, most people would not feel that served world peace or world order. In other words, the idea of leaving another adversary or rival with a monopoly of nuclear weapons, most people would not find that served national security, and they wouldn't be wrong in my opinion. I would probably be among them.

How many does it take to deter nuclear attack? Dr. Herbert York, who was the first head of Livermore Nuclear Weapons Design Laboratory along with Los Alamos Laboratories for many years, once asked that question. His speculative answer was one. Or ten. Or, from another point of view, we made a calculation, a hundred. He got to that by asking, "What's the largest number of people we would like as a ceiling for one person to be able to annihilate in a brief period of time, with nuclear weapons?" Let's just suppose that ceiling was World War II-level: 60 million. He said that could be done by 100 thermonuclear weapons of a medium to a moderate size of 100 kilotons. One hundred weapons would give you 60 million dead. But that was without counting smoke. We now know the death toll from that would be closer to one or two billion from starvation.

Putting all that together, he saw that a justifiable level, in a world where others had nuclear weapons and you had to fear attack to some extent, was something between one, 10, and a 100 being the maximum, with the ideal number being closer to one than 100. He was the head of research and engineering in the Defense Department for years, and later was a major negotiator and a nuclear physicist who was very deeply into the design of nuclear weapons, so he was extremely authoritative.

Every country but North Korea has more than that now. North Korea has 10 or 20 but has enough material for 60. It's very hard to justify any country having more than that, and the other eight nuclear weapons countries all have more than that. Israel is said to have 80, but by other estimates they have a couple hundred. How does Israel need as many as 80 nuclear weapons? But anyway, all the others have more than that. India and Pakistan each have more than 100. China has something like 300. Russia and the U.S. each have on the shelves, in terms of weapons that are supposed to be dismantled that haven't been, something like 7,000. In terms of operational weapons, each has about 4,000 weapons with 1,500 on alert, which is far more than enough to cause nuclear winter if their plans were carried out.

Since the Cold War ended 35 years ago, the number of weapons between the U.S. and Russia has gone down enormously, by 85 percent. That sounds very impressive, except that the effect of these has not changed at all. When you've starved everyone, which you could do with a slight fraction of the remaining weapons, it really doesn't affect the final account. The stakes have not gone down since the end of the Cold War at all. Of course, the probability of war went down with the ending of the Cold War, because these fears that led to preemption were diminished. The weapons should have been diminished then, not by 85 percent but by 90 percent of where they are now, down by 99 percent at the least, and to zero eventually. But they haven't. And now, with a new Cold War revving up, these fears are revving up again, so the risk of nuclear war happening is going up, not down.

Horton: In the last couple of minutes here, I really wanted to ask you about this. It sounds like science fiction, but you talk in the book about how the men on the Manhattan Project, when they were setting off the first test bomb in New Mexico, had done some calculations and figured that there was a real risk that they would literally ignite the entire atmosphere, burn it off and kill every single human being and living cell on the planet. Yet they did it anyway. Is that really right?

Ellsberg: Yes, it turns out that one or two very qualified people, in particular Hans Bethe, thought the risk was zero. That was his intuition. Almost none of the others agreed with them. It was not zero, and most of them thought it was very small, except the most authoritative experimental physicists in the world at that time, Enrico Fermi, was not at all sure that it was as small as the others thought because he thought, "There could be something here we haven't calculated. This is an experiment that has never been done or even anything close to it. It could be in the interaction of some kind we haven't figured in." He thought the risk of that was 10 percent, which is what he said on the day of the test, and he offered odds. He made bets, supposedly jokingly, on the end of the world, on whether New Mexico would burn altogether, or the entire world. The reason I went into that so much in the book was that it's kind of a forecast on the gambles we've been making ever since.

Horton: Wait, before you make that point, can you explain how that would supposedly work with the nitrogen and whatnot?

Ellsberg: Sure. What they feared was the heat of the atom bomb. They knew that the bomb would work. They knew they could make an atom bomb work right from the very beginning in 1942 when they started on the Manhattan Project as a big organizational effort. They realized that heat of that degree could be the trigger for a fusion reaction in hydrogen isotopes, which is what ended up being later developed into the H-bomb. They saw that as a possibility, and it turned out to be true. But before the H-bomb was developed, they also saw that the heat from an A-bomb might ignite the nitrogen in the atmosphere and the hydrogen in the oceans, and in a fraction of a second, a millisecond, the entire atmosphere and the oceans would ignite entirely, and the Earth would from then on be a barren rock.

It wouldn't destroy the Earth; it would just destroy all life on the Earth. That was a possibility that they saw right from the beginning. It scared the man in charge of the project, Compton, enough that he first thought, "We've got to stop all research on this if there's even a possibility that we could end life on Earth." Hitler was made aware of this, by the way, at the same time, and as Albert Speer reports, he was surprisingly not enthralled by the idea of being the author of the destruction of all life on Earth. Even as a capability, he didn't like that idea. He held it against scientists that they were doing this sort of thing. He decided not to go ahead with the bomb.

We went ahead with the research, hoping that it would turn out to be impossible that this would happen. But it didn't turn out to be impossible by the time they were ready for the Trinity test of 1945. In fact, they didn't really know it until after new data from that test, the explosions of Hiroshima and Nagasaki, and other tests made it clear that it was not a possibility. The data had been right at the beginning that it was not possible. Still, they went ahead with the test and the explosions not knowing that.

The reason I dwell on that piece of history is that there were scientists then taking a gamble in 1945, at a time when the war against Germany was over and there was no question of victory in the war against Japan. Even at that point, they were still making a gamble which in Fermi's eyes was a 10 percent gamble, a 10 percent chance of losing everything on Earth. Not just a billion people or two billion people, and not just humanity. You know, if humanity goes with nuclear winter, most life will survive because most biomass is microscopic and microbial, and most of that will survive. Perhaps some very small animals survived when the dinosaurs went extinct 65 million years ago. So in a nuclear winter situation, humans would go, life would go, but what they were gambling with was even the microbes would go. If you care, if you identify as a part of life, or if you're concerned that life in some form goes on, they were gambling with that.

What I'm saying is that this shows the human propensity for pursuing short-run benefits. Narrow benefits of various kinds, which can include just keeping Boeing's production lines running, keeping the votes in on that, keeping our status as protector of West Europe — even when in the last 35 years West Europe has not needed protection — and keeping that first-use threat that we've relied on for over 50 years now. Those kinds of incentives had been enough to gamble with the possibility of ending civilization. What's most disturbing is that there have been times when that wasn't just a three-in-a-million chance, or even 10 percent as Fermi said. In the Cuban Missile Crisis, on October 27, the chance of destroying civilization was a lot higher than 10 percent, as I described in the book.

This was also true in 1983, had Colonel Petrov not decided to say that it was a false alarm that his satellite warning system was giving him. That system said that there was a 50 percent chance that they were being attacked, which is what he actually believed, but he chose not to report that. If he had reported a 50 percent chance, they would have preempted at that time, and we wouldn't be here. So we've relied on the prudence and wisdom of people like Petrov, or a sub commander named Arkhipov in the Cuban Missile Crisis, to keep us from launching the Doomsday Machines we have built.

So the story about atmospheric ignition in the book is a parable for the human willingness to gamble on total destruction, not only of ourselves, but of killing everybody. There's a willingness to gamble on that for the short-term benefits of prestige, of staying in office, of being the head of an alliance, or of winning a particular confrontation. India and Pakistan fighting over Kashmir, for example.

Horton: When they made the H-bomb, were they worried about the same effect of atmospheric ignition again?

Ellsberg: Yes, they were worried. They had discussions about it and again decided to go ahead. I've never been able to find out whether at some point they decided that it was not possible with the H-bomb, whether it was before they exploded one, or not until the test, like what had happened with the A-bomb.

Horton: They were tested in the atmosphere over and over again, so they sure were pretty confident, I guess. It still sounds crazy, but possible.

Ellsberg: What this shows is that being crazy is something very accessible to humans. Same goes for people who are very smart and intelligent. They can also do crazy things, they can be crazy, they can have crazy beliefs, and they can certainly have wrongheaded beliefs. It happens all the time. You're wrong about lots of things all the time, but persisting in that craziness — as the Republican Party is about climate right now — persisting against all the scientific evidence that there's no manmade climate crisis. That's what has led Noam Chomsky to call the Republican Party the most dangerous organization in history.

Horton: Well, stick with the H-bombs here for a second. We have a situation with Trump, where he's under such attack for supposedly being a Manchurian Candidate and all this. Do you think it's possible that he could have a nuclear deal with Russia? Seems to me like it'd be good politics for him to just turn the tables and do a giant deal.

Ellsberg: Trump is an execrable character, from almost every point of view. He has one sane policy that I'm aware of, and that is what he's been saying during the campaign elsewhere: "Why can't we be friends with Russia, or at least have normal negotiations, collaborate, cooperate?" That's the more sane and prudent policy than the opposite, which is that we can't negotiate with Putin any more than we thought we could negotiate with Stalin, Khrushchev or any of the others. It was wrong then, and I would say it's wrong now.

So, he's on the right path on that one, and he's opposed by the establishments, both Democrat and Republican, who seem to want a cold war for no reason other than maintaining a trillion-dollar defense buildup of our Doomsday Machines, which is good for the corporations in the U.S., in Russia, and in other countries who have black budgets for their nuclear capabilities which are very prone to corruption. They're all pursuing this.

On this point, I would say Trump is sane, and he talked about the possibility early in his administration of a grand bargain with Putin to reduce the number of nuclear weapons. Two things about that: He hasn't pursued that, and I see no evidence that Putin wants to do that. Putin doesn't want to disappoint his military-industrial complex any more than any of our candidates have wanted to disappoint our military-industrial complex. That includes Trump, but we're talking about a buildup that was scheduled under Obama and that Hillary Clinton was proposing, too.

Why is Trump saying this? Well, maybe it's just because he's sane, but that's not the answer that comes first to mind with Donald J. Trump. Why is he right on this one subject when many other people are wrong? My own guess is that on this point, they do have blackmail on him of various kinds. I think it will turn out to be, although it has definitely not been proven yet, that there was a lot of money laundering done by the Trump Organization for German and Russian oligarchs, criminals and various people. They have something on him. Even so, the policy of negotiating for reducing weapons, which he is no longer talking about, would be a good one. We have to face the fact that politicians, like other humans, are mixed bags who can be very wrong on many subjects while being right on one or two. We have to use our best judgment. Any judgment that leads toward war with the other nuclear superpower, Russia, is a very bad judgment and a very unwise judgment. We see that in people that I otherwise respect very much. Indeed, when Trump went to Helsinki, a lot of Democrats were denouncing that. That was not my opinion. I favored those negotiations, and hope something comes of them.

Horton: Let me ask you about what you think about what's going on at the RAND Corporation now.

Ellsberg: I've had no contact with them for 40 years. I have no idea what they're doing now.

Horton: Do you read their papers or anything?

Ellsberg: Not really, because I don't respect that process that much, and I don't have time.

Horton: Well, I'm kind of curious about the way you describe your work there at the time, and how convinced you were about your beliefs as to what was going on and what needed to be done about it. I wondered how cockamamie their reports come out now compared to your reports from back then, when the USSR is gone, when Soviet communism is defeated, and this is all just a make-work program for these arms dealer companies.

Ellsberg: I don't have the impression that RAND is as big a player anymore as they used to be, when they were one of the first so-called think tanks. That's another reason why I haven't kept in touch with them. There's no question at all that there are many think tanks that support the Cold War. That being said, RAND has done some reasonable reports that I've seen in the paper from time to time. They strongly said years ago that, from all the research they could do, gays in the military would not be harmful to national security, contrary to General Powell, for example. They've had various reports on this and that, and when they come up with something good, I don't find it to be implemented very quickly, if at all.

Horton: [Laughing] Well, they're good on the Mujahideen-e-Khalq. Anyways, thank you, Dan. I really appreciated every word of this book. It's really great, and I hope everyone will read it.

Ellsberg: Okay, good. Thanks.

Andrew Cockburn: How Easy It Is to Start a Nuclear War
July 20, 2018

Scott Horton: Alright, you guys, introducing the great Andrew Cockburn. He's the Washington Editor at *Harper's Magazine*, author of the book *Kill Chain* and, most famously now, the book *Dangerous Liaison* about America's relationship with Israel. He's also written about Iraq. What was before *Kill Chain*? Andrew, welcome back.

Andrew Cockburn: Before *Kill Chain* was *The Threat*, about the height of the Cold War when the Russians were meant to be all over us and 10 feet tall. I pointed out that this wasn't really true.

Horton: And one about Iraq that you wrote with Patrick back in the 1990s, right?

Cockburn: Yeah, it was called *Out of the Ashes*. It was about Iraq. It was a biography of Saddam Hussein. Remember him?

Horton: [Laughing] Yeah, he was going to attack us if we didn't attack him first.

Cockburn: Absolutely, because he had all those nuclear weapons, chemical weapons and weapons of mass destruction, which he was ready to rain down on us. Anyway, that shows that you can always believe American intelligence.

Horton: You sure can, and you'd better not contradict them, or you're guilty of treason.

Cockburn: Absolutely, and they'll haul you off in chains, which is probably what's about to happen.

Horton: Yeah, I guess we'll see. I'm going to be in the concentration camp with all my favorite writers, so I don't really mind that much. Listen, you write great stuff, man. I'm a big fan of your brother Patrick. I really liked your late brother, Alex Cockburn, who was a founder of CounterPunch. But Patrick, of course, is the most important journalist in the world in covering the wars for us, at *The Independent*. I interview him all the time. But you really write some great stuff, too. It tends to be more long-form journalism rather than war dispatches like Patrick's. You guys complement each other very well. This one is in Harpers.org: "How to Start a Nuclear War." Apparently, it's really easy.

Cockburn: Yeah, it's kind of horrifying. It starts off with this guy Bruce Blair, who was in a missile silo in the early 1970s, bored out of his mind like most of them. He figured out how to launch not just his own unit, his own squadron, which was 50 missiles — 60 megatons, which would have killed a few hundred million people — but how to launch the entire U.S. nuclear arsenal. Basically, how to blow up the world in one easy motion.

Horton: You say he was a first lieutenant in the air force?

Cockburn: Right. I should have made it clearer that he was a first lieutenant, but he pointed out to me that there were people who were in the same position who were second lieutenants straight out of college. You graduate, you throw your hat in the air, and a few months later, you're sitting in front of a missile silo with the means to blow up the world.

Horton: So if I know TV, then I know that that means that he would have had to pull a gun on his buddy and force him to turn the key at the same time, and they'd be able to launch maybe one missile. So what are you talking about?

Cockburn: Well, he would have had to knock him out, unless his partner in the silo was in on the scheme. He would have had to knock him out or slip a sleeping pill in his coffee or something — anyway, he'd have to restrain him in some way. But once he's done, what he'd need is a guy in another silo. Once he has that, then it's open season.

Horton: So it's not just that one guy in one silo and one guy in another silo able to control those two missiles, but working together, they have access to the whole network?

Cockburn: They have access to the whole squadron, which is 50 missiles, and if one of them is in the command silo for the squadron — there are five launch control centers for each squadron — then the two of them together could launch not just missiles in that squadron, not just the missiles in that base, of which there were already several hundred, but the whole U.S. nuclear arsenal. The bombers probably would have checked and wouldn't have done it, but the rest would go. Certainly, the land-based missiles, of which there were a thousand in 1954, would go off without recall. It's pretty scary.

Horton: And there's nothing that they can do to turn off an ICBM that's on its way, right?

Cockburn: Absolutely not. There's no self-destruct mechanism on board, but come to think of it, there should be.

Horton: Although I guess from their point of view, that would make it vulnerable to hacking.

Cockburn: Yes, that's exactly what they would think, of course. A self-destruct mechanism would minimize the risk that it would continue on its way to blow up Moscow or Beijing.

Horton: This is a big revelation in Dan Ellsberg's new book *The Doomsday Machine*. He talks about how he was a nuclear war planner back in the day and how he just absolutely could not believe the nuclear war plans as they were set. When he started advising on this for the Pentagon and for the RAND Corporation, he saw how if anything happens anywhere — a conventional battle breaks out in Berlin or something — then every Chinaman and every Russian dies, all of them, with no way to turn it off. And I guess he tried to change that, but you report in here that they changed it back. Any war with Russia means we're going to go ahead and take out all of China, too.

Cockburn: Right, several things happened, but let's just deal with the China thing first. When China became our friend in the early 1980s and we normalized relations, had ambassadors and all that good stuff, they took China out of the war plans. They decided that we don't need to kill every single inhabitant of China anymore. Then, in the 1990s, under the Clinton administration, Clinton okayed a new nuclear posture — out at Omaha at Offutt Air Force Base, the home of strategic command…

Horton: This is where George Bush, Jr. ran to hide all day long on September 11, 2001, deep underground.

Cockburn: That's correct, and even deeper underground was the joint strategic targeting planning stuff. They got a document saying that Clinton changed the nuclear posture in whatever way, and they said, "Aha! This means that China's our enemy again," which it didn't really say, but they chose to interpret it that way. So they put China back in the war plan and once again, if anything broke out, up goes China in a puff of smoke.

Horton: Isn't that funny? It's just like anything else, such as when Congress passes a law and then the FCC decides to interpret it this way or that way, that these guys can do that with nuclear war plans and presidential orders.

Cockburn: Not only that — and that's a good example with the FCC — but at least Congress might know that that had happened. In this case, Clinton and the White House didn't know that this had happened. They didn't know that we were going to kill a million Chinese until Bruce Blair, my informant who told me about this, was on a trip to Omaha, and he was chatting with the general in charge who just casually mentioned that they'd put China back in the Single Integrated Operational Plan. When he got back to Washington, he called up the White House and said, "You might like to know that this has happened." They had no idea.

Horton: That's something Ellsberg talks about too, where Jack Kennedy said, "I want to see the war plan," and they told him no. Then they said, "Okay, we will," but then they showed him something that was not the war plan at all. He was the President of the United States, and they told him to go to hell and that he wasn't cleared to see this.

Cockburn: Even then, there are layers and layers, because it turned out that since Kennedy's time, the White House didn't get to see the real war plans. The Joint Chiefs over in the Pentagon had what they considered to be the real war plans, which they wouldn't show the civilian leadership, but that wasn't the real war plan either. The real war plan was the one that was being drawn up deep in the base, many layers down, at Offutt Air Force Base.

You talked about Ellsberg just now. Ellsberg was horrified. He and others were so horrified by this scheme that you outlined, which was that if anything happens — a tank backfires in Berlin — we instantly push the big red button and blow up the entire world. They were horrified by that, so they worked hard to get that plan modified. So, for various reasons, we got the counterforce plans, which included different options. For example, the president would have the option of saying, "Okay, we're just going to fire nuclear missiles at the Russian nuclear missiles and other important military bases like bomber bases, and we won't blow up their cities." That improvement might save a few hundred million or tens of millions of Russians, at least. Meanwhile, the people at Offutt didn't think too much of this scheme, so they found military targets in the city or very close to the cities. They found military targets around Moscow. Moscow itself wasn't targeted, but there were plenty of multi-megaton aim points around the outskirts of Moscow, which meant that it would have been completely obliterated. So every Soviet city — even if the president said, "We're going to withhold the city targets, we're not going to blow them up, and we're just going to go for military targets" — even then, every single Soviet city would have been obliterated.

Horton: There's an anecdote from when the USSR still existed, about when Dick Cheney was being sworn in as Secretary of Defense in 1989 under George H. W. Bush. He was briefed on the war plans, and at some point, even Dick Cheney started squirming in his seat and said, "This is insane! How many H-bombs can we hit Moscow with in a row?" I mean, what are we trying to do, dig down to the center of the Earth? There's nothing left to blow up there after the first fifty have gone off. It really makes you wonder about — and this is what you talk about in the article — how so much of this has to do with the structure. I guess I'll put off the incentive structure about the generals and the presidents and all that, but what's the incentive in the bureaucratic structure that makes them write a war plan like that, where not only do we find a loophole where we really want to drop an H-bomb on Moscow, but as you say in here, Moscow itself could be subject to 100 nuclear explosions? I mean, what is this? And I get it that the submarine officers want in on this too, but still, how do you get to 100?

Cockburn: Here's how I think it works, and the only way to understand all this is money. Let's think about the money. You have all these interests interested in producing missiles, interested in producing warheads, interested in devising plans for all this, interested in producing bombers and all that. So to keep Lockheed or General Dynamics happy and profitable, we have to place an order for 100 or 400 missiles. So now we have to find out, how do we justify this? Well, we have to find targets for them, and the basic rule was we have to be able to cover 80 percent of our assigned targets under any circumstance. That meant that you needed to have a huge excess of missile delivery vehicles and warheads. So it all starts from the need to find targets for these devices because you need to find excuses to buy them. Once you've done that, then you have to scratch around and say, "Well, we can drop a megaton bomb or more here on this entrance to the Kremlin, but maybe there's a .00001 percent chance that the other entrance would still survive, so better send another megaton bomb on that." It's all a matter of finding excuses. The whole drive of the bureaucracy is to think of excuses to find targets for these bombs.

Horton: This is why we, as Americans, have the right to do this and to hold the whole world hostage with this H-bomb Sword of Damocles over all of their necks, because we're so moral and guided by such American virtue that they'd better just bow down. They have no argument against it, even though this is the actual morality of the American empire — they'll nuke Moscow 100 times.

Cockburn: Well, right, and it's basically so that, in order for someone to make some money, we will target them, and we might end up doing it. As I mentioned, it's 100 aim points in Moscow or around Moscow alone. This is with the Cold War ended. I guess we have a new Cold War now, but the big Cold War ended in 1991 or 1990, whenever you want to date it. We said, "That's great," and the bombers were taken off alert. It looked for a moment where it seemed like peace might break out, but the central machinery of nuclear destruction carried on. There were still people sitting in the silos, and still people sitting in the bunkers drawing up target lists. This deadly machine, this mega death machine carried on even though we were, at least for a while, friends with Russia. By the way, it's absurd that there should be this obsession with Russia when the total Russian defense budget, which is roughly $61 billion, is less than the amount the U.S. defense budget goes up in a year. It's ridiculous.

Horton: Well, it's not supposed to be realistic; it's just supposed to be convincing. In fact, you wrote a great article for *Harper's* back two or three years ago about a big celebration in Crystal City, where a bunch of arms manufacturers are meeting, and they're so excited they made a lot of money bombing peasants in Afghanistan and Iraq, and this kind of thing. But the real money is in the big-ticket items — bombers, jets and all the stuff that we need for a Cold War with Russia. They all figure, "Oh, don't worry, we won't all die under fusing hydrogen atoms, it'll be perfectly cool." What a great way for them to make money without having to work.

Cockburn: Yeah, and the actual incident I described in that piece took place the day Putin took over Crimea. A friend of mine happened to be attending a breakfast. He was a lobbyist. He wasn't actually a defense lobbyist, but he happened to be there meeting with Michael Royce, who was head of the intelligence committee, and everyone else in the room was a defense industry lobbyist. I met him that day, and he said he'd been there, and he described the mood as borderline euphoric, they were just so happy.

Horton: Otherwise, they'd have to get real jobs.

Cockburn: Exactly, and I'm so glad we're talking about this because it's so hard to get across to people that this is all about money. It's all about keeping the economy afloat. Basically, what's left of the U.S. manufacturing base is mostly defense, so they think we need a cold war, we need all this to keep this show on the road.

Horton: We have a wartime economy because that's all we have left of an economy. This is what David Stockman calls the "Great Deformation," where it's a massive bubble that bends the entire economy toward militarism. Then all these other little bubbles, a dotcom bubble here, a housing bubble there, those are the little bubbles on top of the big bubble.

Cockburn: You've put it very well, that's exactly it.

Horton: Let's hope it doesn't take H-bombs to pop the damn thing.

Cockburn: I'm getting pretty scared. I've got to tell you, the hysteria that's on now has me quite frightened.

Horton: Pat Buchanan makes this point, he said the line was drawn during the Cold War that if the Russians — and in your book *The Threat* you showed what a joke this was — if the Soviets tried to roll their tanks into Western Germany across the Elbe River, then we will go to nuclear war. Don't you dare try to conquer Western Europe. But if you crush an American-provoked uprising in Hungary or in Czechoslovakia, you crush Solidarity in Poland, we're really sorry, but that's just too far outside of our sphere of influence. We're not going to go to nuclear war over Prague or over Warsaw. Yet now we've moved that line from the Elbe River all the way to the Russian border. Not that the Russians really have a plan to re-conquer Eastern Europe anyway, but it seems like we're trying to give them a motive to do so, namely to rebuild that Stalinist cushion of the borderlands, particularly Poland and the Baltic States, and to keep them as a shield to keep the West out. Now we've moved all the way to their border. You talk about the accidents and the close calls. How itchy must the Russian trigger finger be right now when we have troops right on their border?

Cockburn: It's interesting, that was the excuse over years of the Cold War. Let's call it Cold War I. The Russians were poised to invade Western Europe, and we therefore offered the guarantee that we would blow up Russia if they tried to do it. As I mentioned in the article, right after the end of Cold War I, in the brief sunshine period, the BDM Corporation got a contract from the Pentagon to go over and talk to Russian national security types about what they'd been thinking during the Cold War. There were all these very interesting interviews, including one that said, "We never had the slightest intention of invading Western Europe. That never even crossed our mind even to think of such a thing." So the justification for the whole nuclear posture was all completely spurious. There's the guy that was sort of the hero of my article, Lee Butler. He rose to the very top of the American nuclear war machine. He was head of Strategic Air Command and head of STRATCOM. Then even while he was in the job, certainly after he left, he started to think, "This is all complete madness." He has since gone around campaigning for total abolition of nuclear weapons, and he says that the whole idea of deterrence is completely stupid. Everyone says, "Well, deterrence, yes. We have to have the means to deter the Russians." Well, actually, he points out that the idea of deterrence is that if Scott Horton is going to put a nuclear bomb in my car, he has to know that I'm going to put a nuclear bomb in his car. That's deterrence. But that assumes that I know what Scott Horton is thinking, and that I know that Scott Horton knows what I'm thinking. In other words, you're prejudging everyone's reaction, and he said you could never do that. You have no idea what the other guy's thinking, so the whole premise of deterrence is false.

Horton: They say that the point of deterrence is that it's not just that we can defeat your army, but that we can evaporate your capital city. So all the people in charge of starting the war, their lives are now at risk, too. Yet we could bomb the hell out of Moscow without nukes, anyway. We've got so many daisy cutters, MOABs and all kinds of things that we could slaughter plenty of Russians without fusing atoms together. When it comes to just deterrence, we could still burn your capital city down, don't worry about that. Just ask the Vietnamese, they'll tell you.

Cockburn: Exactly, we're pretty good at that. Ask the Yemenis.

Horton: Let's talk a little bit about what you say about how they changed the structure from the STRATCOM command post to the different one, where now it's higher-ranking people who are the ones who get the intelligence and are responsible for communicating with the president. I think you're saying that, perhaps unintentionally, this actually creates the ability for the military to box the president in and force him to start a nuclear war. It changes the incentives there, compared to the way it was before. Could you explain that a little bit?

Cockburn: Sure. Initially, it was the heat-sensing satellites we have over Russia and Siberia. If they saw Russian missiles coming through the clouds from the bases in Siberia, then a minute later the early warning radars in England, Greenland and wherever else they are would confirm that the Russian missiles are on their way. Then that early warning center would report this to NORAD, the North American Air Defense Command, whose headquarters are in Cheyenne Mountain in Colorado outside of Colorado Springs. They would then call the Pentagon War Room, the National Military Command Center, which is underneath the Pentagon. Whoever was in charge there would then call the White House, wake up the president, talk to the National Security Advisor and say, "Russian missiles are on the way. You now have six minutes to decide what to do about it."

What they've done — and this happened under George W. Bush — they basically cut out NORAD. I should explain that the guy in the War Room in the National Military Command Center would normally be a colonel, so he's got to decide whether or not to wake up the president. He might take a few seconds, half a minute or even a minute to think about it. This is what I was told. They streamlined it. Now, once the satellites and the radars have confirmed that the missiles are on their way, they immediately call STRATCOM in Omaha. It almost immediately goes to the commanding officer there or his deputy, but the commanding officer is a four-star general. He then calls the president — he's a four-star general, who are like gods in the military, so he has no inhibitions about waking up the president — and says, "The missiles are on the way, and what do you propose to do about it?" Then the president might say nothing, but probably not. He's going to say, "Okay, we'll do option 2," whatever that means, maybe blow up the Russian military, and the president then gives the order. When Trump came along, people started to get nervous about the idea of Donald Trump being in the position to push the button. They asked the former head of STRATCOM, "If the president gives the wrong order, if he says blow up the world, but you think that's not a good idea, what could you do about it?" And he said, "Well, I would refuse to carry it out." Then they asked the same question at some kind of security forum of the present head of STRATCOM, General Hyten, and he said, "I would refuse to carry out that order. I would tell him what he could do." Well, that all sounded good and provided a big relief that we've got these senior military officers who would be prepared to mutiny rather than start a nuclear war for the wrong reasons. They're cool with starting a nuclear war for the right reasons, whatever they might be, but they would draw the line somewhere, and that was taken as grounds for complacency. Except that what no one said, as I point out in the piece, the president doesn't need these guys to start a nuclear war. He has the authorization, and once he gives the order using the special codes that proves he is who he says he is, the thing happens automatically. It has nothing to do with the head of STRATCOM.

Horton: What you're saying is that STRATCOM, by moving the command authority around, have made it easier for them to alarm and alert the president to try to convince him to do something. But on the other hand, if the president wants to go ahead and do something crazy without them, there's nothing that they could do to stand in his way.

Cockburn: Exactly. And furthermore, it used to be the only scenario anyone could think of that there would be Russian missiles coming through the clouds of Siberia. At least you knew how long it would take them to get here, you could figure out pretty quickly where they're all headed, and that's what gave them a degree of certainty by the whole thing. But now they're saying — and this is a good excuse for an even bigger defense budget — "Oh, all these other people have missiles now: the North Koreans, the Pakistanis, the Iranians," and they not only have missiles, but these are missiles that can maneuver. The Russians have missiles that can head off, head left and then turn right, which makes things more uncertain. So you have to alert the president earlier. And you don't even have to wait for the missiles to take off, supposing we have hard intelligence that they might be about to launch a nuclear missile, a weapon of mass destruction, then you wake up the president. So they have to light the fuse earlier, which is a lot scarier. This is all total hokum because I don't think anyone's going to launch a missile at us. I was talking to Congressman Ted Lieu of California, and I said, "What do you think about them reacting to intelligence of an imminent strike?" He said, "Well, there was good intelligence that Iraq had weapons of mass destruction, and it turned out there wasn't, but we invaded the country, killed hundreds of thousands of people, including thousands of Americans, for false intelligence." So it's quite possible to conceive of us blowing up the world on equally false intelligence.

Horton: There's the (should be much more) famous story of the Able Archer exercise in 1983, where the Soviets really thought that it was cover for a first-strike attack. They thought that Reagan was going to launch a war, and it was a traitor inside NATO who was a secret Russian spy who reported back to them, "I swear to God it's just an exercise. Don't freak out. Don't overreact." And so they didn't.

Cockburn: Right, and there was a guy called Gordievsky, he was in the KGB in London, and he went over and told the British, "Christ, back in Moscow, they are really freaked out. They think the Americans are about to launch a first strike." So that freaked out the British, and Margaret Thatcher actually called Ronald Reagan and said, "You'd better knock this off because the Russians are taking you very seriously." I think that's the same story.

Horton: Yeah, it sounds like you got it a lot more straight than I do. So they ended up doing the exercise, but at least with some reassurances or something, right?

Cockburn: Yeah, they did tone it down a bit. Remember how Ronald Reagan used to make jokes about bombing Russia?

Horton: Well, that was the thing, right? He would say, "The bombing starts in five minutes," and you can imagine a bunch of Russian counterparts to the American hysterics who might refuse to take that as a joke, even though it's just Ronald Reagan screwing around, but there's always going to be somebody who says, "Hey, that's secret code for 'He really means it.'" Look at the way the Americans are overreacting to Russia doing basically nothing right now.

Cockburn: I mentioned in the article that the Russians came up with this Dead Hand system — they called it "Perimeter" — by which if sensors detected a nuclear attack, then the Russian missiles would launch automatically, pretty much without human intervention. There was one human link in the chain, but that was it, and it was easily bypassed. We know that sensors can get things wrong all the time, so that was pretty scary, the idea that we were all dependent on a bunch of Russian sensors working properly.

Horton: Have they said that they've turned that off and it's no longer the system? Or is that still functional?

Cockburn: It's a bit fuzzy, but it looks like they have, yeah.

Horton: You know, it's funny, all this stuff about Russia now, with Donald Trump in there. He's mostly a hawk, and yet, going back to like 1986 or something, he said that he wanted Ronald Reagan to send him over there to do a nuclear deal with Gorbachev. That could be the deal of the century, right? The one that Reagan almost made in Reykjavík in 1986 to go ahead and abolish all of the American and Russian nuclear arsenals. Then we could get to work on our allies and really create a nuclear-free world. This would be the kind of thing that could get him re-elected if he could do it, and yet instead we have this entire Iraq War II-level insanity about what's going on with Russia and how they're his secret, Manchurian, brainwash masters. Anything he tries to do like that — hell, just being on TV next to Putin — is treason. So what would they call it if he signed on to the Democrats' New START Treaty? They'd call it treason, right?

Cockburn: Of course. A while back, 20 years ago, he told someone I know, someone who was involved in negotiating arms control treaties with the Russians back in the Bush administration or earlier, he said, "Here's how you negotiate with the Russians. You have a negotiating session set up, and you turn up half an hour late, so they're getting really pissed. Then you walk in and you go up to the chief Russian negotiator and say, 'Fuck you!'"

Horton: [Laughing] That'll work. The thing is, we haven't had a real war here in so long, 170 years now or something like that. The Russians, they have a little bit different experience. They tend to look at things a little bit different than the Americans when it comes to protecting themselves. I mean, when we had a war here, it was just North versus South. We hadn't been invaded since your great-great-grandfather burnt the White House in the War of 1812, which, congratulations on that, by the way.

Cockburn: I always have to say, he did it with an army of freed slaves, or at least part of his army was freed slaves.

Horton: Take that, Mr. Madison. But that really was the last time that the U.S. was attacked, and that was when the U.S. barely reached past the Appalachian Mountains. The Russians, on the other hand, they've been invaded how many times?

Cockburn: Well, let's see. With very devastating consequences, it's been three times in the last 250 years.

Horton: People who are a lot better experts than me have said that this really matters to the Russians, they're terrified. That was really what was behind Stalin continuing to occupy all of Eastern Europe after World War II, was just to get the Germans and the Americans out. To let a bunch of Poles and Estonians be the shield to protect the Russians in the event of an invasion. You could see why they'd be a little bit paranoid. Look at how paranoid the Americans are, and multiply that by however many invasions, right?

One more set of questions here, what about the new program — it was a trillion-dollar program, they're already saying $1.1 trillion, and it will surely be a $3 trillion program by the time they're done — to completely revamp the nuclear weapons arsenal, industry, weapons labs and everything in this country? I guess it's a pretty big bargaining chip to bargain away in exchange for some pretty big Russian concessions, but it doesn't really look like we're headed that way.

Cockburn: No, there's zero chance, zero possibility. Any president who tried to sign that treaty would be impeached immediately, as Trump probably will be, albeit not over any really good reasons.

Horton: If they do overthrow him, it'll be over some B.S. like this, trying to make peace with the Russians or with the North Koreans. You know what, as long as we're on that, do you really think that he might be thrown out of power before the 2020 election? I guess I'm worried that if they really try it, it could really get out of control with the reactions to that, you know?

Cockburn: I think not, but never say never. I don't think it's likely, I really don't.

Horton: Lord knows that all they got Nixon over wasn't the secret bombing of Cambodia; it was a break-in at the Watergate. As long as they have real stuff to impeach him for, they won't. They'll need some trumped-up charge instead.

Cockburn: I'm glad you mentioned the whole nuclear modernization, the complete rebuild. That's what it's all about, you know? As I mentioned in the article, Jon Wolfsthal, who was the senior nuclear guy or the weapons proliferation person in the Obama National Security Council staff, told me that the Obama White House tried to get the Pentagon to come up with a number for what our nuclear forces cost us every year, and the Pentagon refused to tell them. They said, "We don't have that figure. It'd be too hard to figure it out."

Horton: *Mother Jones* also did a review of this where they said that if you combine the military budget with the Energy Department and all the cost of the care and feeding of the nuclear weapons, and the V.A., we're at a trillion dollars a year.

Cockburn: That might be a moderate figure, because we've got the Defense budget which is north of $700 billion, then you've got the Energy Department which covers all the nuclear weapons, then you've got the VA, then you've got to include the State Department — not that it's that much, but it is certainly part of our national security apparatus — and then you've got the interest on the money that we borrowed to pay for all of this.

Horton: Well, there's the cost of the CIA, too, which is how many tens or hundreds of billions per year?

Cockburn: Yeah, well, that's hidden inside the Pentagon budget.

Horton: So yeah, imagine what it would be like if we had that money to spend making America great again here, instead of threatening the whole world.

Cockburn: Dream on!

Horton: Alright, Andrew, thanks for coming back on the show. I really appreciate it.

Cockburn: Scott, always a pleasure.

Michael Klare: The Threat of War with North Korea November 22, 2017

Scott Horton: Introducing Michael Klare. He's the author of *The Race for What's Left*. He's writing a new one called *All Hell Breaking Loose*. Also, there is this article he wrote for TomDispatch.com called "Making Nuclear Weapons Usable Again." Welcome back to the show, Michael. How are you doing?

Michael Klare: Fine.

Horton: I appreciate you joining us today. To cut right to the chase, it used to be that in order to make sure you got 'em good, you had to make a 20-megaton H-bomb. But nowadays, the accuracy of American delivery systems is such that we can dial them down to very small-yield nuclear bombs like, say, Hiroshima-sized or Nagasaki-sized nukes that they could convince themselves are perfectly usable in battle? Sir, is that right?

Klare: Oh, yes. We can deliver any kind of nuclear Armageddon you care to think about, many times over.

Horton: It really is kind of a new era, in terms of dialing down from the large multi-megaton yields to these "tactical" rather than "strategic" nuclear weapons.

Klare: There's always been a certain amount of reliance on smaller nuclear weapons, but the trend over the past 10 or 15 years since the end of the Cold War has been to move away from the understanding that having them in your arsenal made the possibility of a nuclear war more rather than less likely. That's not the argument that's made for them anymore. The argument that's made for them now is that if you have these low-yield, tactical nuclear weapons in your arsenal, the other side is going to be less likely to start a war with you, so it's good for deterrence. But many analysts believe it's the other way around: If you have these in your arsenals, one side or the other facing defeat on the battlefield would be tempted to use them, so it makes the initiation of nuclear war more likely.

The trend since the end of the Cold War has been to move away from those in order to make nuclear war less likely. That's been a blessing of the Obama administration, and we've all slept better for it, but now in the Trump administration and in the Vladimir Putin administration, there are people who feel that's tied our hands too much, and so we need to go back to having more of these tactical nuclear weapons in our arsenal so that we could threaten nuclear war more readily. That's what scares the crap out of me.

Horton: You're saying Russia feels the same way about their nukes? Because you're talking about nukes that are made for targeting Russia with, you're not saying that's what Putin is having Trump do, right?

Klare: No, this is a two-way dance.

Horton: Okay, gotcha. I'm sorry, there's so much nonsense about Trump and Putin right now, I wasn't exactly sure what you meant by that, but I'm glad that you've not gone insane.

Klare: No, this is a larger pattern of the major powers, including India and Pakistan wanting to, as they put it, lower the threshold for the initiation of nuclear combat so as to frighten the other side.

Horton: You're saying the Russians were doing this now as well? They're making smaller-and-smaller-yield nukes too?

Klare: Yes. At least that's their claim, and so our side says that we have to do the same thing.

Horton: That was what Trump himself said, right? That we've got to keep up with Russians, they're way ahead of us on nuclear technology.

Klare: Yes, that's what he said. Now, there are a number of pieces of this to discuss. One of them is: Just what exactly are the Russians doing? That's not entirely clear. Two: Is copying them and going down the same path in our best interests?

I'm saying what the Russians are doing is somewhat unclear, and we should get to the bottom of this before we do anything risky, number one. Number two, even if that is what they're doing, going down the same path as the Russians is not in our best interests because that's only going to push the Russians and the Chinese further down that path and put us at greater risk.

Horton: Of course, the obvious thing to do is open new negotiations. In fact, especially if you're Donald Trump, just invoke Ronald Reagan at Reykjavík, where he was a hair away from abolishing all nukes in the world. I mean, he is a Republican after all, right? The politics of it makes sense to me.

Klare: Well, this is where we're in this strange, surrealistic world of 2017 in Washington. On one hand, Trump is accused of being too close to Vladimir Putin for reasons that we all know about, given the last election and Russian meddling in the election. On the other hand, we need Trump to be able to negotiate with the Russians. Unfortunately, when it comes to nuclear issues, he's adamantly opposed.

Horton: Yeah, I know. All the politics, all the pressure and all this pretension that it was Putin and not the American people that elected Donald Trump last year, and that the worst thing about him is that he says that we ought to get along with Russia, means that he constantly has to prove that he's not. Or at least, that's the pressure on him. I don't know, he is kind of a wild card who does what he wants, but you know there's a story in *Buzzfeed* today about him arming Ukraine because the politics of it are that he has to prove that he's not this guy's puppet.

Klare: Yeah, but that doesn't stop Trump from giving Putin a free hand in Syria.

Horton: Yeah, well, they already won. I mean, I think that's actually the best thing that he did so far, stopping the CIA's support for the jihadists in Syria. After all, what right does the CIA have to back a bunch of jihadists in Syria? So to stop doing that, that may be giving a free hand to Russia, but the only reason Russia intervened in 2015 is that the American-backed terrorists had severed the highway between Aleppo and Damascus and were marching on Damascus. I'm not saying everything they've done there is right since then, but they're defending the internationally recognized sovereign government in Damascus against a bunch of foreign mercenaries and terrorists and murderers.

Klare: From my perspective, it's one bunch of murderers against another bunch of murderers.

Horton: Of course. I never meant to imply the opposite of that. I just mean that who started it matters, right? It's just like Saddam Hussein was a bad guy, but that doesn't mean he started the war against us.

Klare: Yes, but from my perspective, looking at this from the global geopolitical map, it looks to me like Trump is ceding to Russia control of a large piece of territory on the global chessboard. That's a question to discuss, but I think we should get back to the nuclear equation because on that, Trump is unwilling to give any ground whatsoever, and from a Russian perspective, this is really scary stuff. The Russians are bound to respond by beefing up their nuclear arsenal, and that's only going to put us in greater danger than we were before.

Horton: There's also the question of North Korea, right? I don't know how much you know about this, but I'm just guessing it doesn't seem like there's a plan for an invasion of North Korea that would not include using nuclear weapons in an attempt to take out their nuclear weapons, or the worst of their hardened artillery and other weapons that they can direct at Seoul right away. They'd have to knock out all of North Korea's offensive or defensive capabilities almost immediately, or else you're going to have hell to pay, right?

Klare: Any nuclear weapons used by the U.S. against that dug-in artillery north of the DMZ is going to create radioactive dust that's going to spill over and kill Americans and their families who are stationed right by the DMZ. So that's killing our own.

Horton: Wouldn't the theory be that you use the bunker busters so that most of all that explosion goes underground?

Klare: Yeah, I think Jim Mattis and his associates are smart enough to have told Trump that that scenario has too many risks to be plausible. I do think that Mattis & Co. have prepared a range of scenarios for Donald Trump that includes use of nuclear weapons. So I'm not saying it's not on the list of possible scenarios. I think they've given him a range of scenarios, beginning with conventional strikes of North Korean nuclear facilities and launch facilities. That would be the lowest level, a conventional limited strike, up until a full-scale use of conventional forces, and then going to limited nuclear strikes. And I'm sure no decision has been made, but I also believe that somewhere in a locked closet somewhere there is a list of these various scenarios with code names behind them waiting for the next incident to occur, which could occur at any day.

Horton: Yeah, that's my thing, too. I mean, I'm not trying to be too alarmist about it in terms of thinking that they really are going to attack. I agree especially that Mattis knows better. Mattis has said publicly that this would be the worst fighting of our lifetimes. He was being very serious about that. He was including Vietnam in that, right? He was including maybe even the previous Korean War. So he certainly was more than against it. On the other hand, like you're saying about that safe full of options, if it comes to a real regime change, I just assume it must be baked in that you'd have to use nukes in any major war, any serious war other than the slightest skirmish. If anything escalates beyond that, they've got a couple of dozen nukes, right? The fear would be that they're so crazy and irrational over there that we absolutely have to take out their nukes before they can use them. It seems like that would be the foremost priority among the Americans in the event of a war with them.

Klare: Yeah, I think that this is a topic around which other parties have also weighed in, like China, Japan and South Korea. And I do believe that when Mr. Trump was over in Asia a week or so ago, those parties made clear to him that they would rather find another way to resolve this than the ones you just mentioned, because in any of those scenarios, they would suffer catastrophic damage one way or the other. No matter what, they are likely to experience some of the fallout from all of that, literally and otherwise. So I think they're telling Trump, "Just cool it, and let us work on this," and he seems to have given them the green light. Now, I don't know how long this will last — that is to say, how much time he's willing to give China and the other countries in the region to find a non-military solution — so we may come back to what you're talking about, but I hope not.

Horton: Yeah, it sounds like he's drawn the same red line that Bush drew ever since he pushed the North Koreans into possessing nuclear weapons in the first place. I don't know if Obama ever said it this way, but he seemed to have had the same policy, which is that the nuclear issue must be resolved first, and only then can we negotiate about anything else. When, of course, that's a dead letter, that's designed not to work; it's designed to maintain the status quo.

Klare: Yeah, I agree with that. On the other hand, I think China might come up with a roundabout solution that involves some kind of concessions on North Korea's part in return for some kind of quid pro quo on South Korea's part that the U.S. will go along with. That would be the best hope.

Horton: Maybe it's all just the "art of the deal," right? You can't quite trust it, but then again, as we're discussing here, the alternative, the way that they frame it, is unthinkable, so somebody had better really prioritize this and get it right. Although, honestly — and this is a little off-topic from your piece here but it's on the same subject — for the life of me I cannot really figure out what Cheney and Bolton were thinking in breaking the Agreed Framework and more or less kicking the North Koreans out of the NPT and their Safeguards Agreement and pushing them toward a nuclear weapons program right at the time that they were preparing to invade Iraq. What did they think was going to happen other than that the North Koreans were now going to have nukes?

Klare: I think Washington has been on sleeping pills all this time. It's just been an unwillingness to face up to the consequences of not addressing North Korea more directly and doing things that are unpalatable, like meeting with the Kim family and working things out with them in a way that we may not like but will have to live with.

Horton: Well, not that Trump has the depth or would even understand this if you explained it to him, but if Nixon could go to China and shake hands with Mao, then that means that Donald Trump, the Republican skyscraper tycoon capitalist from Manhattan, could go to North Korea and shake hands with Kim.

Klare: Yeah. Put a Trump Tower in Pyongyang, for God's sake.

Horton: Sure, go over there and make a deal!

Klare: Any other outcome would be catastrophic. It's easy to say that it would only be South Koreans and North Koreans that will perish, but that's not true. There are something like 30,000 American soldiers based very close to the combat zone, and they have families there. There are tens or hundreds of thousands of American citizens living in Seoul and the surrounding area who would also be at risk.

Horton: Yeah, and if nukes start going off, then that's going to poison a lot of people in China and probably Japan and, depending on which way the wind blows, the Pacific Ocean and God knows what else.

Let me ask you about Obama, because I can think of a couple of things that Obama did right about nukes. One, he gave a pretty speech, at least the first half of the speech that he gave in Prague. I remember watching it live, and the people were so excited in the first half when he said all this metaphysical stuff. And then he got to the part where he said, "Yeah, and that's why we're going to continue the current policy," and they were all visibly, physically disappointed to hear where he was going with this. But he did, in that speech, actually back off and did not continue the Bush anti-missile program for Poland and the Czech Republic. That was really the best thing he did.*

At the same time, this trillion-dollar project to invest in a whole new revamped nuclear weapons arsenal. Donald Trump likes to pretend that this was his idea, but it was "our hero," Barack Obama. So is there something I'm missing, or was he just not quite as bad as Trump or Bush, but still pretty damn bad, on this?

Klare: That is an interesting question. Yes, President Obama let the Pentagon proceed with research and development on a new generation of ICBMs, submarine-launched ballistic missiles, and a bomber to replace the existing strategic triad, as it's called. He never authorized purchase, procurement or deployment of new weapons; he kind of left it to the next president, who I believe he assumed was going to be Hillary Clinton, and I think he believed that Hillary would be in a stronger position to slow some of this down. I can't read his mind. I don't know. There's no question that he gave authority to proceed down this path without saying we're absolutely going to go ahead with all of this.

He also said that he hoped we could negotiate with the Russians to downsize our arsenals further. He negotiated the New START Treaty, which substantially reduced the U.S. strategic arsenal, and when that expires in a couple of years, the notion is that the next stage would be a further contraction of the U.S. arsenal. Maybe he believed that through these arms control treaties, the need for the new weapons would evaporate. I don't know what he was actually thinking, but as you say, there's no question that he authorized the original work on these weapons systems, and that's deeply troubling.

---

* President Obama did back down on installing missile defense systems in Poland and radars in the Czech Republic, but then he changed his mind and went ahead anyway.

Horton: I think part of it was a compromise that he made with the Senate in order to get New START through, the idea being: "We're going to do a little bit of limiting here, but don't worry, we're going to give you just as much nuclear weapons welfare. You'll still get to cash all your checks." By the way, can you elaborate about what the limitations were in that treaty?

Klare: The New START Treaty limits the U.S. to 700 collective delivery systems between land-based ICBMs, submarine-based ballistic missiles and bombs carried by bombers.

Horton: It says here in your article, 1,550 warheads and 700 delivery systems.

Klare: Yeah, 1,550 total warheads, and the U.S. would have a choice as to how to deploy that number, as would Russia with the same number. That, under the treaty, is supposed to be completed by the beginning of February 2018, so I assume that the process is completed by now. I don't know how the final allocation was made between ICBMs, SLBMs and bombers. From a strategic point of view, the submarine ones are considered the safest and the hardest to destroy in a first strike, so therefore they are the best ones for the purpose of deterrence.

Horton: I had this guy on the show who wrote about how all of the silos, all those Minuteman missiles, those land-based ICBMs, are just meant to be targets. Those are there to be the "nuclear sponge" to attract Russian ICBMs to Colorado and North Dakota and Montana so that they can take most of the hits and hopefully spare the population centers on the coasts, while the rest of our nukes, our bomber force, and our submarines go and try to take out theirs. You can see where *Dr. Strangelove* comes from with ideas like that. Once you accept the premises about how to fight these wars, then all these things become very thinkable, and then they become reality. They really dug these holes and buried these missiles, and they have guys sitting at home with keys right now, but really, their only purpose is to be a target.

Klare: That is the most secret part of the whole thing: What are the targets, and what is the firing sequence? And I have no clue about that. I wouldn't even know where to begin. And I think you're right, there's a way in which "Dr. Strangelove thinking" takes over. That's what my article is about, how there's a new element of the Strangelove thinking that's taking over, the idea that we have to be able to make our arsenal more usable, that we're stuck with these old city-busters, and, God forbid, President Trump may have such scruples about bombing every city in Russia in the event of some incident occurring in the Ukraine or somewhere that he may hesitate to slaughter 100 million people. So, better that we give him some smaller nukes so that he won't have so many scruples about slaughtering millions of people. He could do it on a smaller scale. That's what really sickens me. That's what all this new talk is about, to make it easier for him to lessen his conscience about ordering the use of nuclear weapons.

Horton: Well, that sounds crazy, but no, that's nuclear weapons doctrine. There's a whole history of this since the end of the Second World War about when and under what circumstances we use these nukes. In fact, I have the PDF sitting on my desktop of Daniel Ellsberg's brand-new book, *The Doomsday Machine*. His job at the Defense Department was coming up with this stuff. His father actually designed the assembly line for the first generation of atom bombs, but he turned down the contract to design the assembly line for the H-bomb, saying, "I don't want anything to do with that thing." Anyway, he published an article on TruthDig one time where he reproduced a DOD chart that showed how many hundreds of millions of people are expected to die in the first seven days of a nuclear war, and you'd get to a billion by Friday.

Klare: We have a local doctor here, Dr. Ira Helfand, one of the principal figures in the International Physicians for the Prevention of Nuclear War, and he says that all these studies underestimate the level of death, because it's going to create a permanent cloud, block out the Sun, spread radiation, and kill the food supply for a large part of the human population. So what's really going to be the consequences is starvation for half the world's population in any imaginable nuclear encounter.

Horton: You've got to figure too that we're talking about mostly the Northern Hemisphere nations fighting against each other, which are most of the most developed economies in the world, so the result would be the GDP of Earth being cut by three-fourths or more. All global trade, all market distribution systems, completely turned off overnight.

Klare: Yeah, that is about the size of it. So it's not just the initial slaughter, which would be bad enough, but everybody would be affected one way or the other, and sooner or later.

Horton: Well, I sure appreciate your attention on this issue. Thanks very much, Michael.

Conn Hallinan: The Risk of War Between India and Pakistan December 13, 2016

Scott Horton: Introducing Conn Hallinan from Foreign Policy in Focus. We reprint a lot of what he writes at Antiwar.com. He recently wrote "A Global Nuclear Winter: Avoiding the Unthinkable in India and Pakistan." Welcome back, Conn. How are you?

Conn Hallinan: I'm fine, Scott. And you?

Horton: I'm doing great, and I really appreciate you joining us today. So listen, what in the world makes India and Pakistan the most dangerous conflict on the globe? Why, don't you know there are other conflicts going on?

Hallinan: There are other conflicts going on, but this is a conflict between two nuclear powers who are in the midst of a major nuclear arms race. The leaders in both countries talk very loosely about the business of using nuclear weapons. One specific thing that got me to write this article was that recently, the Pakistani military made a decision to deploy tactical/battlefield nuclear weapons and said that they intend to use them if there's a conflict with India.

The Indians don't have tactical/battlefield nuclear weapons, only strategic nuclear weapons. If the Pakistanis use their tactical nuclear weapons, the Indians are going to respond with strategic nuclear weapons, the *big* ones. The Pakistanis are then going to respond in kind, and there's going to be a real possibility of nuclear war that will not only kill at least 20 million people in both India and Pakistan, but will also deeply damage the ozone layer over the Northern Hemisphere. It would produce an enormous cloud of smoke that could lower temperatures worldwide. It would make, for instance, growing wheat in places like Canada and Russia impossible. You would have a worldwide food crisis. Yet nobody's talking about this. It just doesn't get into anyone's radar, and right now it is — not just in my opinion but in the opinion of a lot of people — the single most dangerous place on the globe, and the United States has much to answer for the current situation.

Horton: Well, let's get back to that in just one second, because America's role is a very important point. But back to the Pakistanis developing these tactical nuclear weapons, it makes sense on the face of it, right? If their land army is far overmatched by the Indian army?

Hallinan: Yeah, the Indian army is a little bit more than twice as big as the Pakistani army.

Horton: But on the other hand, as you're saying, if the Pakistanis use even one of their tactical nukes, you know that the Indians are going to have to nuke back. And not just that, but I've read that the Indians are now developing hydrogen bombs, thermonuclear weapons.

Hallinan: Yes, they are. As to whether they actually have one functioning at this point, we don't know, because those facilities are closed to international inspection. However, we do know that they're attempting to build a thermonuclear weapon.

Horton: I never taught math at MIT or anything like that, and I'm not really an expert of game theory, but it sounds like what you're talking about isn't even "calculus," as the horrible people in Washington, D.C. call it. It's just simple arithmetic: "If you guys use tactical nukes, we are going to use really big A-bombs, if not H-bombs, to wipe your cities and your civilization off of the face of the Earth." This all seems like the height of irresponsibility. Is there any kind of other counterargument that we're leaving out as to why they're pursuing this, when it's so obvious how terribly destructive this could be? One false move, and everybody dies? Why put themselves on the line like that when they weren't necessarily in that position before?

Hallinan: That's a really good question, Scott, and I'm not sure that I've ever heard a really good answer to it. Partly it is that the polls that have been taken of Indians and Pakistanis indicated an enormous ignorance about the power of nuclear weapons. It's been a long time since nuclear weapons have been used. The last time was 1945.

Horton: I remember in 1998, when they tested them, I saw footage of a Pakistani general saying, "You tell those Indians we're not afraid of their atom bombs!" [Laughing] Yeah? Well, somebody needs to explain to this guy that your relative level of fear of the enemy's atomic bombs doesn't really affect the function of those atomic bombs at all. They'll still turn you to not even dust, pal.

Hallinan: That's what's kind of amazing, the way that people talk about this. The former Indian Defense Minister said that they could easily destroy Pakistan. The Pakistanis have talked about that. One of the former heads of the Pakistani military said, "Well, you know, you could die crossing the street. You're going to die at some point anyhow, so what difference does it make how you die?" This is an indication of people really not paying attention.

Horton: It's really been since Mao Zedong that somebody who actually was in the position to do something about it talked like that about civilian casualties on their own side.

Hallinan: Yes, and the thing is, what we really are talking about is how it would be an incredible tragedy for not just Pakistan and India, but for all of the countries in the region, which would be deeply affected by nuclear fallout. But the whole question of nuclear winter has just not been talked about.

There have been studies on nuclear winter. This is not over-the-top stuff, this is pretty straightforward, and they have a good base on which to do it. What they base it on is what happens when you have these huge volcanic eruptions, like Mount Pinatubo in the Philippines. The effect of Pinatubo was that it lowered world temperatures about one and a half degrees, and they know how much ash Pinatubo put into the atmosphere. They know how much ash will be created if cities were destroyed, and the only weapons that the Indians have are city-killers. They don't have tactical nuclear weapons. They know soot, smoke and everything will be released. It's a pretty straightforward formula, and we're in trouble.

On top of all that, one of the things that radiation does is that it destroys the ozone layer. That means that you have an enormous influx of very powerful ultraviolet radiation, which is not only dangerous to people because you get skin cancer, but it also kills plants. There may be places where as much as 70 percent of the ozone layer would be destroyed in the case of a war like that, and nobody's really talking about it.

Horton: The few people who are talking about it, though, are saying the same thing as you, that hey, we kind of underestimated this before. We always just think of nuclear war as between the U.S. and the Soviet Union and the end of all of mankind; but a tiny, little bitty old nuclear war in Central Asia would be enough, as you say, at the very least, to disrupt the ecosystems around the world.

Hallinan: If you take a look at both sides, they each have in excess of 100 nuclear weapons. The Pakistanis are moving very rapidly towards 200 nuclear weapons. Eventually, they're going to pass Britain as a major nuclear power. One of the reasons why we don't talk about this certainly in this country is that both the Bush and Obama administrations share a major responsibility for the current crisis, both in terms of the danger of nuclear weapons and in terms of the issue that Pakistan and India are fighting about, which is Kashmir.

Horton: I was reading my buddy Eric Margolis and talking with him a little bit about it, and he was explaining that the Kashmiris are not really Indians or Pakistanis. On top of that, Pakistan is really four different countries held together by a military, and who knows how many countries India really is? But he was saying that Kashmir really is its own independent land with its own rules. They're pretty much their own people, and it's one of those just completely intractable problems where the Indians rule it, but the population is Muslim and lean much more toward Pakistan, and the Indians refuse to acknowledge.

Hallinan: I agree with you. Most of the polls show that if there was a referendum at this point, the Kashmiris would vote for independence.

Horton: But regardless, the Pakistanis want it anyway, whether the people of Kashmir want that or not, right?

Hallinan: Right, and in 1947, when Pakistan and India were divided, it was a terrible tragedy. There were millions of people killed on both sides when that happened. At that time, the UN promised Kashmir that it would have the right to a referendum, but the Indians blocked it. Then the Pakistani military began infiltrating terrorists into Kashmir, and it became a military situation. Ever since, Kashmir has essentially been under martial law, and has been that way since 1949, which is part of the problem. The big protests that are going on right now in Kashmir really don't have to do with the autonomy question, although that's where they want to go eventually. But it's an absolute exhaustion with decades and decades and decades of military rule, particularly this law called the Special Powers Act, which was originally designed for Northern Ireland in 1925 to keep the Northern Irish Catholics in their place.

The Special Powers Act is what the Indians use in Kashmir. Essentially, it's a law which says that you don't have to do anything; if the Indian authorities decide that they think that you might be having bad thoughts, they can arrest you, they don't have to charge you, they can put you in jail, they can keep you there for a year, and then they can continue to keep you there year after year after year, with no charges and no proof. The other thing is that it makes the Indian police forces, paramilitary forces and the military immune from both Indian law and Pakistani law. So it's a military occupation with complete impunity for the occupiers, and it's produced a level of rage that is the basis of the demonstrations that are going on right now. It's the same law that the Israelis use in the occupied territories. It's the same law that was used as the basis for the apartheid legislation in South Africa. It was the ultimate colonial law, and all these post-colonial countries decided that it's not a bad law if you want to stay in power, and so they kept it.

Horton: So the people of Kashmir basically live like Palestinians under Indian rule, or what?

Hallinan: Yeah, exactly. And there have been thousands and thousands and thousands of them that have been disappeared. The current demonstrations are really awful because the Indians have been using birdshot in the shotguns, and there are these people that they call "dead-eyes," who have birdshot in their eyes. There have been more than a thousand people who have been blinded in these demonstrations by the Indian troops. The heavy-handedness of the Indians is really a scandal, and the major reason for this is that they are completely immune to any laws. They can do whatever they want.

So here we have this situation where the native Kashmiris are demonstrating, asking for the end of the Special Powers Act, for more autonomy and for, eventually, the right to have a referendum. The Indians claim that all this agitation is the result of Pakistani infiltration, which isn't true in this case. There was a good deal to say for that in the 1980s, but it's not the case now. Now it's become between Pakistan and India, because nobody's paying any attention to the Kashmiris. So for the first time, some Pakistanis infiltrated across the border and killed a bunch of Indian soldiers. Nine Indian soldiers. For the first time, the Indians crossed the border into Pakistan, announced that they were doing so, and attacked what they called a terrorist base. It's not clear what it was, but there were some Pakistani military killed. There certainly is close coordination between the militants in Pakistan and the Pakistani army, and we know this. It's a longstanding thing. The problem is that the Pakistanis are saying that if India invades, they're not going to be able to stop the Indian army, so they're going to use tactical nuclear weapons. Then you and I are in trouble, Scott, and we live a long way from India and Pakistan.

Horton: A lot of other people are in even worse trouble. Now, of course, everything in the world is America's fault. There's no point in pretending it ain't. So explain to me about what India and Pakistan's roles are in their relationship with the world empire. They are both satellites of ours in a sense, aren't they?

Hallinan: What started the current situation was that the Bush administration back in 2003 officially changed China from a strategic competitor to a strategic opponent. It was a formal kind of thing they went through. Then they began to set up a series of alliances that were aimed at containing China. The jewel in the crown was India. If they could get India to move away from its traditional position of neutrality and join an alliance with the Americans against the Chinese, that would be an enormous plus. Eighty percent of China's energy supplies go across the Indian Ocean, so if India could essentially be part of an American alliance, that would be a big thing.

Now, the Indians and the Chinese have their own issues. They fought a border war in 1962, and the Indians are still not happy about that. They claim that the Chinese are occupying Indian territory, and I think they have a certain justification for at least part of that. In any case, what happened was that India said, "Well, what do we get in return?"

What the Bush administration said was, "Okay, if you don't sign the Nuclear Non-Proliferation Agreement" — neither Pakistan nor India signed the treaty — "you cannot buy uranium supplies on the international market." That's not a problem for Pakistan; they've got a lot of native uranium. It's a major problem for India; they have very little uranium, and they have a nuclear power industry. So the Indians have always had to try somehow to keep their meager supplies between their nuclear power industry and their nuclear weapons industry. What the Bush administration said is, "Look, we'll waive the part about you being able to buy uranium on the international market, as long as you promise that it will only be used for civilian purposes." The Indians were more than willing to do that, because then they could take their other supplies and direct them totally towards nuclear weapons.

Horton: This was a wink and a nudge thing anyway, right? Even George W. Bush understands that uranium is fungible.

Hallinan: The other thing is that Pakistan asked the U.S. for the same deal, and they said, "No, you can't have the same deal." They called it the 123 Agreement. My hope was when the Obama administration came in that they would back away from this, because it basically knocked the prop out of the Nuclear Non-Proliferation Agreement, and you really don't want to do that. The Nuclear Non-Proliferation Agreement is one of the most important agreements on the international control of nuclear weapons, and in general, it's worked pretty well. It's true that you've got India, Pakistan and North Korea that have nuclear weapons.

Horton: And Israel.

Hallinan: But there's a lot of other countries that could have built nuclear weapons very easily, but they didn't do so, so the Nuclear Non-Proliferation Agreement really has worked fairly well. This just removes the central props from it, and one of the things that Pakistan said was, "Look, this is going to start a nuclear arms race in Asia." It absolutely has. The Indians have gone on a tear to build nuclear weapons. The Pakistanis have gone on a tear to build nuclear weapons. The Chinese, fearing that most of those strategic nuclear weapons are directed at them, are on a tear, building more nuclear weapons and putting multiple warheads on their mobile ICBMs. North Korea's building nuclear weapons, too.

So here you have this decision by the Bush administration and then followed through by the Obama administration, and it's been a goddamn disaster. Now we have this powder keg, although when talking about nuclear weapons, "powder" doesn't work. Gunpowder is about as dangerous when you compare it with nuclear weapons as distilled water is to gasoline. Nuclear weapons are just in a different category of reality, and it wouldn't be hard to get something going here because tactical nuclear weapons are commanded by battlefield commanders.

Horton: Yeah, that's what I was going to ask you about next. They didn't just say that they were thinking about that; they have actually gone ahead and given colonels the power to use nukes if they think they need to?

Hallinan: That's right, you've got some Pakistani colonel on the Indian border, and he's going to make a decision as to whether or not we're going to destroy 70 percent of the Northern Hemisphere's ozone. It's really not a place we want to be, and there should be an international full-court press on stopping this, on reversing this, and it could be done. This is not an impossible situation to solve. It will require pressure on both Pakistan and India, but both Pakistan and India are countries that certainly are subject to pressure. First of all, India couldn't be continuing this situation if it wasn't for the 123 Agreement with the United States.

Horton: You're saying that the entire American strategy is the opposite of that because they're not even concerned about Pakistan, and they're just looking at China?

Hallinan: That's right.

Horton: And yet the Obama administration came into power being told, "You know, what was wrong with George Bush was that he didn't pay any attention to Pakistan. Pakistan is everything." Obama spent his whole first couple of years in office bombing the hell out of them. I don't know if he was obsessed, but he certainly was preoccupied, with Pakistan. Pakistan on his mind, reading briefing books all night long. So it's kind of funny that it seems out of sight, out of mind, or they really think, "Oh well, this is just a gamble, but it'll probably be fine."

Hallinan: I guess they think that. I don't know why they're thinking that, but to do anything else would be to move away from the Obama administration's Asia pivot. And that was to move the bulk of the American military forces into Asia. To go from 50–50 to 60–40 in terms of military force, to set up this alliance system, and to place these anti-missile systems in Japan and South Korea.

The anti-missile systems, we say, are there for North Korea. Personally, I think the only people that the North Koreans are a threat to are the North Koreans. But if you're China or Russia, those anti-missile systems could be just as well aimed at you, and the first response to anti-missile systems is to overwhelm the system. You do it by building more warheads, more launchers and multiple warheads on your launchers. That makes everything more dangerous. So the effect of the Asia pivot has been a serious rise in the level of tension in Asia.

Now, I think that the Chinese actions in the South China Sea are illegal. I think they're a violation of the international law of the sea. On the other hand, I can perfectly understand why the Chinese are doing that. I mean, to them the pivot is the United States moving close to China. It isn't Chinese ships maneuvering off of San Francisco or Seattle or San Diego or Martha's Vineyard; it's the U.S. Seventh Fleet, the U.S. Sixth Fleet, etc., maneuvering in the Maluku Straits, in the South China Sea, or the East China Sea, etc. That was the effect of the Asia pivot: "If you do this, you're going to get a response."

The thing about the Chinese is that they were humiliated by the Clinton administration. In 1996–97, there was a lot of tension between mainland China and Taiwan. The Chinese bought up a bunch of missiles, and they put them on the coast aimed at Taiwan. Nobody thought that the Chinese could invade Taiwan. They couldn't today. They don't have the ability to do that. It was saber-rattling. It was very dumb saber-rattling, but the Clinton administration put two aircraft carrier battle groups into the Taiwan Straits, and the Chinese couldn't do anything about it. They were humiliated, and they said, "That's not going to happen again." Since then, they've gone on this expansion of their military forces and have been very aggressive about control of the East and South China Seas.

Again, I'm not supporting what the Chinese are doing in the South China Sea. I think it's illegal, but I completely understand it. It wasn't the Chinese who started this problem; it was the United States that started this problem, and it scares me a lot. China and the U.S. scares me, but right now the thing we've got to be scared about most of all is Pakistan and India, because we have the least amount of control, and it's the most volatile situation.

Horton: On the China thing, as long as we're on that for a minute, I wonder what you think of, for example, what David Stockman says. He says that China is the thinnest tissue paper tiger; that despite all the fear of the rise of their economy and of their military might, they have a fascist system which means that their economy is based on political decisions, not economic ones. Therefore, they have horrible distortions all over the place, and so many corrections coming due that it'll be a million years before they really are a peer competitor with the United States.

Hallinan: I think there's truth to that, and I also think that they have an absolutely terrible ecological situation. They have pollution like you couldn't believe, and I don't know how they're going to move away from that, because the current Chinese government keeps in power by keeping the growth rate fairly high. The thing about a growth rate being high, that means you're producing a lot of stuff, and in their case that means they're producing huge amounts of pollution. A lot of their power still comes from coal. They have a water crisis which is going to get worse. They're already in conflict with a lot of other countries.

So, I think that David Stockman is right to a certain extent. I think he maybe overstresses it a little bit. The fact is that China is an enormous economic power, and people are going to have to come to terms with that fact. How we deal with it, it's not entirely clear, but I think the way that you deal with it is that you don't make it a military confrontation. I don't think that the Chinese military is a threat to us, and I don't think it's going to be a threat to us for a very long time. Nuclear weapons, however, certainly are a threat.

Horton: Right, and they have just enough of those to erase all the major cities in the western half of our country.

Hallinan: They could end civilization as we know it, and in my case, Scott, I'm on the West Coast.

Horton: Well, I'm pretty sure you know Austin, Texas is on that target list too, if it comes down to it. Thanks very much, Conn. I appreciate it.

Hallinan: Okay. Thank you for having me, Scott.

Ray McGovern: Russia's Latest Nuclear Weapons March 6, 2018

Scott Horton: Introducing the great Ray McGovern. He's a former CIA analyst for 27 years. He was actually the chief analyst of the Soviet division there for a while. He used to brief Vice President H. W. Bush back in the Reagan years. He is the co-founder of Veteran Intelligence Professionals for Sanity, and you can find all of his stuff at RayMcGovern.com, tons and tons of it at Consortiumnews.com, and, of course, at Antiwar.com as well. Welcome back to the show, my friend. How are you?

Ray McGovern: Thanks, Scott. I'm doing well, how about yourself?

Horton: I'm doing okay, except the last time we spoke last week, you said to me that this is the most dangerous time since the Cuban Missile Crisis. By the time I was sending that file — I guess it was the next night — Vladimir Putin gave this two-hour speech, and a good half hour or 40 minutes of it was debuting an entire new generation of Russian nuclear weapons. So can you just list real quick and summarize the four or five new weapons systems that they are announcing here?

McGovern: Let me just go back to half a step here. I have a personal vignette about September and October 1962, the Cuban Missile Crisis. I had been commissioned by the U.S. Army Intelligence and Infantry School, I was on my way to Fort Benning, and when I got there — which was November 3, 1962 — we found that there were no weapons there, which was sort of odd, since this was the primary weapons training program for the infantry. So I made some discreet inquiries, and I was told, "Yeah, well, two divisions came through over the last three weeks, took all our weapons, and they're down in Key West right now." It was the Cuban Missile Crisis. They were ready to go into Cuba. Three weeks later, as those divisions came back north, we got the weapons back.

But that was a tangible indicator, not only how close we came nuclear-wise, but if there was a conventional attack, the weapons were not at Fort Benning; they were in Key West. Now that's really important because we're talking about a situation where John Kennedy was president, and John Kennedy had the presence of mind in the face of these mad generals like Curtis LeMay, who, as Daniel Ellsberg has pointed out, when asked how many people would die in Russia in the event of nuclear war, said 20 or perhaps 30 million. How many in Eastern Europe? Probably another 20.

How many in the U.S.? Probably only 10 million. Give me a break. These guys were mad, and Kennedy said so.

So, what did he do? At every planning session during that crisis, he had ambassador Tommy Thompson, who knew more about Russia and the Soviet Union than just about anybody but George Kennan. He had to participate. And when these generals went off half-cocked, Tommy Thompson would say, "Well, you know, last month I had lunch with Nikita Khrushchev, and this is how he looks at it…" And when those famous two messages came from Khrushchev — one really fire-breathing, weapons-rattling one, and the other one more conciliatory saying, "Let's talk about this" — Tommy Thompson says, "Look, all you have to do is answer the conciliatory one." The other one was written with his military breathing down his neck, just like Curtis LeMay on our part. So that's what JFK did adroitly through Bobby Kennedy and a guy in Washington where they could do this discreetly. That was the way the crisis ended.

Now picture today. We have three Marine generals running the show: Mattis at Defense, Kelly as Chief of Staff of the White House, and the Chief of the Joint Chiefs of Staff, Dunford. They're all Marine generals. What do Marine generals know about nuclear warfare? They're not supposed to know a lot, and clearly they don't. So who is Trump going to turn to for some sage advice that comes from experience with Russia? Tillerson? Come on. Nobody. So that's what makes it even more labile, even more delicate, even more dangerous than it was during 1962, in my humble opinion.

Horton: Man, yeah. Now let's talk about these missiles.

McGovern: Scott, I don't have them committed to memory, so if you have the list before you, why don't you just dash off a quick rendition of the ones that he displayed in those videos?

Horton: Yeah, he describes them all, and I'll link to it at the Libertarian Institute.* He shows these cartoon videos. The first one is a new heavy ICBM that is capable of flying over the South Pole and back around, for which there's no missile defense that has even been attempted to be set up so far. That's a brand-new one.

Then he has the hypersonic glider that is, they say, maneuverable and completely controllable, can evade any missile defense system, and has an incredibly long range. Then there's the nuclear-powered underwater drones, basically a remote-controlled submarine that, he says, can deliver high-yield nuclear explosives. In other words, there goes San Francisco, and we have apparently no defense against that.

---

* Scott Horton, "Putin Announces New Generation of Nuclear Weapons, The Libertarian Institute, March 6, 2018, https://libertarianinstitute.org/blog/putin-announces-new-generation-nuclear-weapons/.

Then last, but certainly not least, was the global-range, nuclear-powered cruise missile. It has unlimited range. In other words, it can fly around the whole world for however long it can on nuclear fuel. But in the cartoon, Ray, they show it launch from northern Russia, go around Northern Europe, down catty-corner across the Atlantic Ocean, around the southern tip of South America, and then up the Pacific coast where it cuts away, but you can guess that it finally nukes Hawaii or L.A. or San Francisco.

By the way, in the cartoon when they showed the new heavy intercontinental ballistic missile that could go across the South Pole and up from the south, they show the target as being south Florida, even though it cuts away right before the detonation, but man!

McGovern: Well, I think that President Trump needs to have the Chinese president there at Mar-a-Lago all the time because I think that would be the only deterrent.

Horton: [Laughing] Yeah, "You guys want to stay the night? Go ahead. Yeah, we'll go golfing again tomorrow."

McGovern: Oh, how macabre! Well yeah, these weapons, I don't know if they're all operational. One system Putin claims is, and that, he says, is in the southern district. Now, what's the southern district of Russia? That's the one right opposite the Middle East.

The ABM Treaty was the bulwark, the guarantor of strategic stability. The ABM Treaty provided that there could be no more than two anti-ballistic missile installations in either country, the U.S. or the USSR. In three years, it was changed to one. Why just one? Well, then neither country could conceive of the notion that they could do a first strike and escape immediate retaliation and blow up their own country and most of the world.

So that was a balance, and that's what they called it: MAD, Mutual Assured Destruction. Now, that doesn't sound very good. Sounds sort of mad, doesn't it? But it was a hell of a lot better than what preceded it, which was the "balance of terror," when we never knew what mad general might be tempted to shoot off an ICBM and perpetrate a nuclear war. That Treaty was concluded in 1972, and it's one of the proudest accomplishments that my branch, the Soviet foreign policy branch, ever had. We had three people working on that, two with the delegation either in Helsinki or Vienna, and one backstopping all this information and activity at headquarters in Langley.

I got to go to the signing of the ABM Treaty. It was May 26, 1972, and when that thing was signed, my friend next to me said, "You know, Ray? You breathed an audible sigh of relief as soon as Brezhnev signed his John Hancock to that thing." And it was true! My God! That was an incredible factor for stability from 1972 until when? Until 2002, when George Bush, having announced six months before, canceled it in 2002.

Now, you can imagine the Russian reaction when they saw these little missile defense systems that were being put in Romania, Poland, the Black Sea and the Baltic Sea. Putin himself has told Western reporters and journalists, "Look, we know that these can fly 500 kilometers, but we also know that you can put a different missile in the same holes and it can fly 1,000 kilometers. We know when that will happen, and the Pentagon knows that we know. Then it will fly even farther." What does this mean? This means that our ICBM installations, most of them, are directly vulnerable to a strike, not from missile defense but from cruise missiles and other kinds of missiles in these same holes that could be fired in a first strike, if those generals at the Pentagon thought they could get away with it.

So, they're not going to get away with it. And Putin said, "Look, you know, we tried to talk to you," and indeed they tried like hell to talk to us, and nobody listened. They tried to get into the U.S. press, but they couldn't. So here's Putin saying, "Well listen now, okay? Because this is what we have on the drawing board. This is one system we have already deployed," and as I understand, the big thing here is a lot of these are nuclear-powered. I mean, they are nuclear weapons that are nuclear-capable, but they're also nuclear-powered, which enables them in a completely new art form to fly around the South Pole, North Pole, and around again until they see a hole in the defenses and then wreck our naval port in Norfolk, for example. So what he's saying here is, "Look, it was a real bad idea to revoke the ABM Treaty. We tried to talk to you to get some sort of amelioration of that, but you wouldn't listen. Now you have to listen because the balance has been restored. We have strategic parity no matter what kind of rhetoric you use against us."

Horton: Well, boy, do I feel dumb, because this whole time I've been saying, "Listen, if we keep building up these missile defenses, then that just means that they have to be more ready." In other words, if they're already on launch on warning, then that just means their trigger finger gets that much itchier. The time scales get shortened and that kind of thing. But of course, maybe they'll develop a 15-megaton H-bomb that they can deliver by nuclear submarine that you're never going to find. Or a hypersonic glider that goes Mach 20, he claims. There was a great analysis of this I read, where they talked about how this just makes our surface fleets obsolete. That's it. We can use them to threaten poor little brown countries that can't do anything, but in terms of war with Russia? Our Navy is useless.

McGovern: Well, that may be an overstatement, but not by much.

Horton: I mean, the way this guy was talking about it was almost like it was September 11th or something like that, where this is a new day in history, this is like a pivot in the history of the world. What you thought was happening, that needle just scratched off the record, and now there's this whole new reality. And what's it going to take for the Americans to even realize what just happened? Their bluff has just been called, big time.

McGovern: If you read the *New York Times*, my favorite — this is a little sarcastic — my favorite journalist there is a fellow named David Sanger. I've liked him since he started talking about weapons of mass destruction in Iraq as a flat fact usually seven times in each article, in the prelude to the war on Iraq. Now, he focuses mostly on Russian hacking as a flat fact.

Horton: And it belongs in his reputation here that he has been the single worst person on Iran's nuclear program. For years, he's been calling Iran's nuclear weapons program a flat fact; their illicit nuclear weapons program is a flat fact, without ever proving it. For years and years, he went on like that.

McGovern: He shares that distinction with this sidekick of his, William Broad, and there are other people. So it's kind of hard to single him out, but I agree, mostly. Anyway, what did he say right after Putin made his speech? He said, "You know, skepticism is in order here. Do these weapons really exist, or is Putin bluffing?" Now, here's the proof. What Sanger and this fellow MacFarquhar, who's in Moscow, next cite as evidence is this: "Analysts writing on Facebook and elsewhere lean toward the bluff theory." QED! Analysts writing on Facebook and elsewhere lean toward the bluffs, so maybe he's bluffing.

This is stage one. He may be bluffing. But stage two, which is really a hard thing to investigate, is this. Raytheon, Lockheed, and the other arms developers and merchants have a lot at stake here, and there's a question that Trump has to face: Will he accept the invitation, given the new strategic parity relationship? The invitation to talk about this stuff? Or will he not be powerful enough? Will he not be strong enough to say, "Yeah, this is the time we talk," just as back in 1972 we talked? We've got to do what Nixon did, what Reagan did, we're going to do something sensible. Or is he going to salute the Raytheons and Lockheeds of this world and say, "My God, my God! We not only have to build more anti-ballistic missile systems, but they have to be more sophisticated. We have to test them more, and then we have to build a whole new generation. Maybe we can do nuclear-powered things like the Russians claim they have."

In other words, all the defense expenditures that you and I pay in our tax money, more than half of our taxes go to these so-called defense expenditures, what will happen to the people at the bottom of the scale who are already being squeezed by all these tax measures and everything? What's going to happen to that part of America, I'd say maybe 40 percent, who can't make a decent living without doing two jobs or sometimes even three? What's going to happen? Well, they're going to suffer, and they're going to suffer badly. Who's going to profit? The profiteers. The Raytheons, the Lockheeds, General Dynamics, all these guys line their pockets with proceeds from systems that engineers say, and scientists say, will never work because they can always be easily defeated at much less cost.

The best guy on this is a fellow named Ted Postal, professor emeritus from MIT, where he got his degrees. He's in retirement now — he's almost as old as I am — and he's writing and analyzing, doing all this good stuff. He and his colleagues, the serious ones, say, "Look, the ABM System was always just a system to enrich war profiteers. It never would have worked." Edward Teller sold it to Ronald Reagan by saying, "Yeah, you can have something like an umbrella, and they can never get through the umbrella," and Ronald Reagan bought it.

When Gorbachev came to Reykjavík back in 1986, I had a front-row seat to watch this from the Washington area for Langley, and he surprised everyone by proposing a destruction of nuclear weapons over a period of 10 years. Reagan said, "My God, sure! That sounds good to me." What did he do? Well, he went to his advisers. Now, I happen to know that George Shultz would have been very optimistic about that, so long as we could verify and not just trust. But who were the other guys?

Long story short, Reagan was sabotaged by people like Bill Casey, Caspar Weinberger at Defense, Bobby Gates, who was sort of the windsock that went with the prevailing winds, and this fellow Fritz Ermarth, who was the National Intelligence Director for the USSR in those days. They put the kibosh on this. How did they do it? They said, "President Reagan, listen. You can't have your Star Wars if you agree to this. They won't let you do anything but research. You can't even test anti-ballistic missiles. You can't buy this." And Reagan said, "Oh, I can't buy this." And there perished a unique chance in 1986 to rectify this problem. Fritz Ermarth was asked later how he felt about that, and he said, "You know, I should have been less pessimistic about what Gorbachev had in mind." Well, that's very nice. Where did Ermarth work? Well, Northrop. Where did he go? Well, some of these other main arms manufacturers. The whole system is very, very corrupt. This was the first time in my experience where the intelligence apparatus, led by Casey, Gates and Ermarth, was similarly corrupted for very political and money-making reasons. It was a terrible thing to watch when that possibility was squandered. It was squandered.

Horton: I read a story and I saw at least a still picture of Reagan getting in the car when Gorbachev says to him, "Mr. President, wait! What if we walk back up those stairs, go in there, and talk about this for like five more minutes and see, man, huh?" And Reagan's like, "Sorry, out," and leaves. Like, ah man! He even had the one last extra bonus chance, and he still didn't do it. Damn.

McGovern: Well, he wasn't the smartest knife in the drawer, but I tend to be a little bit more sympathetic to him. If you look at Reagan, I think his instincts were correct. I think he was besieged by a White House, by the characters I just mentioned, and he was attracted to this notion, but he was also extremely gullible. When Edward Teller, for God's sake, comes and says, "Mr. President, you must realize that this Star Wars can give you a blanket protection against any missile from the Soviet Union." Well, Teller is a big name. So he asks Weinberger, "What do you think of this?" And Weinberger says, "Oh, Teller is directly and correctly right."

So, here he is a victim of circumstance. The big thing is the reality that that golden opportunity was squandered just as a few years later when the Berlin Wall fell and Gorbachev was deceived. We know from the cables that have been released that he was deceived into thinking that if he was able to accept a reunited Germany, if he would withdraw the 300,000 Soviet troops from East Germany, then he was promised by Secretary of State James Baker and other statesmen in Western Europe, including the chancellor of West Germany, that NATO would not be moved one inch eastward from East German territory or within East Germany.

To put that in context, anybody who knows anything about Russian history realizes that after they got rid of two centuries of domination by Genghis Khan and his Golden Hordes (from about the 1200s to almost the 1500s), they then had to contend with people coming through from Western Europe. Those were the Hanseatic League, the Lithuanians, and Poles attacking Russia from the west. Later Napoleon, and still later, it was Hitler.

Now, I just mentioned Hitler. People say, "Well, with regards to Hitler, we won the war." Give me a break. As you know, Scott, but not many Americans know, was that it was the Russians who won the war. It was the Soviets. It was the Soviets that turned the Nazis back at Stalingrad. And how many casualties, how many Russians perished in World War II? Well, Putin says 25 million. Historians say 27 million, but let's say it was only 25 million. Give me a break. Who was among them? Putin's big brother, that's who was among them. How did he die? Well, he was in Leningrad during the 900-day siege, and I don't know how he died, but I imagine that, like so many others, he hungered to death. That's the experience that Putin comes out of. That's the experience the Russians come out of. We have no, I emphasize *no*, comparable experience.

Getting back to the fall of the Berlin Wall, they're asking Gorbachev and Shevardnadze, the foreign minister, for an awful lot, but the promise was very attractive: no NATO east of East Germany. What happened? Bill Clinton came in, and he said, "Screw that. That's not written down on paper, and we'll just do it." And now we have NATO more than double the size that it was at the fall of the Berlin Wall. So what does this mean? Well, there's always been a dispute about: "Did James Baker really promise that?" Senator Bill Bradley, former senator from New Jersey — he's a Rhodes scholar and he specialized in Russian/Soviet history — went and asked Jim Baker, "Jim, it's in some of the 'memcons,' the memories of conversation: Did you promise not to move NATO one inch to the east?" Baker is a lawyer, right? So he says, "Well, I didn't sign anything like that." Well, finally we have Baker's signature, and when those documents were released about a half-year ago, nobody reported anything about them to the *New York Times*. So you go to Baker. You go to Scowcroft, who knew about this at the time and said, "Well, we didn't promise anything." Well, I know what was promised because I talked to a fellow named Kuvaldin, a professor at Moscow University, who was with Gorbachev in Moscow when these things were worked out with Baker. I said, "Mr. Kuvaldin, why is it that you didn't write that down?" He looked at me and he said, "Mr. McGovern, this was February 1990, the Warsaw Pact still existed, we didn't talk to the Germans yet. I mean, you should really talk to the Germans if you're talking about German history and German future, that's the main reason. But the other main reason, Mr. McGovern" — as he looked me straight in the eye, Scott — "We trusted you."

Trust is important in international relations. The Russians had a very, very bad experience with the expansion of NATO, and culminating all this was the Western-orchestrated coup in Kiev on the 22nd of February, 2014, and everything you've seen since then, has been a Russian reaction to what they saw as even greater danger on their doorstep, the very place where the Poles, the Lithuanians, the Hanseatic League, Napoleon and Hitler came through. So, you know, you've got to understand this stuff. The Russians have a completely different history and a different outlook, and yet they were willing to do what they did to let Germany be reunited. I mean, my God! The whole post-World War II era was informed by a joint wish, not only of the Russians, but of the West, to keep Germany divided so that it will never be a problem anymore. So we were asking Gorbachev and Shevardnadze to swallow a whole lot, and they swallowed it. Now, they're subjects of ridicule from just about everybody in Russia, including Putin, for being so naive as to trust the Americans.

Horton: To finish your point there, George W. Bush picked up where Bill Clinton left off, and then Barack Obama did the same. Trump has now continued the policy and brought in Montenegro, finishing that process there.

McGovern: You're in dangerous territory here because when Rand Paul asked innocently, "How does incorporating Montenegro enhance United States security?" John McCain immediately accused him on the Senate floor of being Putin's puppet.

Horton: Well, I have a far stronger case for John McCain's treason, so I'm not worried about that. But now listen, I want to go back to what you were saying about this most inefficient welfare program for people who are already billionaires, executive vice presidents over at Raytheon and so forth. It was Jack Matlock, the second-to-the-last ambassador to the Soviet Union in 1990 or '91, who had this conversation. I saw a clip of you talk about this before. It was Matlock who says to Putin something like, "You've got to understand, this is really just a welfare program for politically connected arms dealers in America. We're not really trying to get a first-strike capability against you to attack you and all this stuff, man." But Putin, as you put it in your email this morning, refused to let him down easily. Can you talk about that for a minute?

McGovern: Yeah, this is quite amazing. There's a discussion club, which the Russians call the Valdai Discussion Club, and here was Putin sitting on the dais with Jack Matlock, for whom I have great respect, who was actually helpful in getting me to Moscow for the SALT (the Strategic Arms Limitation Talks) signing. So Jack is a good friend, and he's very bright.

Horton: I've interviewed him one time before, too. That's in the archives, everybody.

McGovern: So he's sitting there at Valdai, and Putin says, "Well, this is a discussion club. So, for discussion, let me ask Mr. Ambassador what he thinks of the American unilateral withdrawal from the ABM Treaty." The Russians put out the text of what Matlock says, and he said, "I was personally opposed to that withdrawal, and I take your point, but I would say I don't think that any subsequent plans for the sort of deployments of ABM were or could be a threat to Russian systems. In general, I am not supporting ABM systems, but I would point out that I think the main source of that is not to threaten Russia, but to secure employment in the United States. A lot comes from the military-industrial complex and the number of people it employs." If you watch the clip, you see Putin, who was at first shocked, and then he said, "Mr. Ambassador, couldn't you create jobs with a result that would be different for all of humanity? New missile defense systems are creating jobs? Why can't you create in other areas? Why not technology, biology, high-tech industries?"

Bear in mind, this is October 23, 2015. The treaty that prohibits Iran from developing nuclear weapons, at least for 10 years, had already been signed. So Putin says, "There is no Iranian nuclear problem. We have been saying that all along, so why develop a missile defense system when there's no chance that Iran would be a threat?" In other words, this business that Matlock apparently was persuaded about, that these emplacements in Romania, Poland and the Black Sea were not against Russia. No, no, no! They just happen to be there in the periphery. They're against Iran or maybe North Korea. You know, if you have a globe, just look at the globe.

Horton: I know! I mean, just anecdotally. I don't have the footnotes and everything, Ray. But at the time, when George Bush at one point announced,

"Oh yeah, this is to protect Poland from a first strike from Iran," all the people in the room, at some international event, they all laughed. Everybody just laughed. Nobody believed that it was true for a minute. And the Russians, there was a clip of Putin or Medvedev or whoever saying, "Yeah, right. Obviously, we can see right through that."

Then, when Obama came in and said the same thing, everyone was like, "Oh yeah, that's really serious. You've got to protect Poland from Iran. [Laughing] The missiles they don't have, to deliver the nukes they're not making."

McGovern: I was reviewing this last night and looking at Putin's protestations, and how could the American people know about all this if they just read the *New York Times*? They wouldn't know. They would have no way of knowing. They would have no way of knowing that back in 2012, at a summit in Seoul, an ABC microphone was on when Medvedev, who was president at the time, said, "Mr. Obama, Vladimir asked me to ask you, when are you going to talk to us about ballistic missile defense?" And Obama says, "Oh well, give me some space. Wait 'til after the election. Just give me some space. Will you tell Vlad that?" And Medvedev very politely said, "Oh yeah, I'll tell him that." Now guess what? Obama apparently forgot to take this up after the election.

Horton: Yeah, and he got completely slammed, "Oh, look at what a two-faced backstabber this guy is." When, my God, is that not what we want him for? For him to back down and find ways to be able to get along with the Russians?

McGovern: Yeah, and you know, the next thing I came across was how Putin holds these four-hour press conferences hosted by press people from all over Russia, which stretches over seven time zones or something like that. Four hours. And during this one on St. Patrick's Day, March 17th, 2014, right after the coup in Kiev and right before, like the day before, the Crimea was annexed to Russia, what does he say? He says, "You know, the encroachment eastward of NATO was a real problem, but the threat of a missile defense was an even more important factor than keeping Crimea out of NATO hands." Well, hello? They didn't want ABM systems. They didn't want missile "defense" systems in Crimea, or more in the Black Sea, or in Ukraine. So what Putin is saying to all his people — not picked up in the Western press — is: "Look. Yeah, NATO? That was really bad, but think about the national missile defense systems that would be put in, and we couldn't let that happen in Crimea."

We mentioned the Valdai stuff, where Jack Matlock made that faux pas grande, saying it was just a jobs program. This too was after the Iran deal was signed, sealed and delivered, restraining Iran from having a nuclear weapons potential for many, many years. Here's Putin on June 17, 2016. What does he do? He recognizes that there's a whole group of Western journalists in St. Petersburg at an International Economic Forum and decides he wants to talk to these guys. He assembles about 20 of them, and he's looking at them, and he talks to them about what it means. He repeats to them, "Look, these missile holes are capable of cruise missiles, but they're also capable of missiles that reach far into Russian territory. We know the schedule for when they can knock out a lot of our ICBM force, and the Pentagon knows that we know." Then he looks at these reporters, and you know one of them is sort of falling asleep, and he loses it. He says, "How can I get through to you people? We're in danger here. It's worse than it was before."

You ought to see that clip. It's only about a minute and a half, but it's on my website, and you get a sense of how this guy is saying to himself: "They won't listen. They won't listen, and so we have to be involved in these very esoteric systems that will finally make them listen and hopefully get them to the negotiating table."

Horton: That's what he said in that speech today: "You wouldn't listen to us before, but you'll listen now," or something very close to that. I don't speak Russian or understand it or anything, but it just seems like this guy is always very calm and methodical and always means what he says. Not that he's always honest, and I'm sure he's capable of misdirection and this and that, but he seems like the type not to make a bunch of inflated claims. He doesn't need to engage in a bunch of bombast like Trump for political purposes. He doesn't have a powerful enough opposition that he has to resort to this kind of thing.

I would take him at his word on this as far as these new developments in these nuclear missiles. And, back when I compared it to 9/11 earlier, I didn't mean to say that this is an attack on America; I just meant this is a new turning point. Because it's not just that Russia has now restored the "balance of terror" after the Americans had made it so lopsided that they were threatening to cancel Mutually Assured Destruction and be able to threaten this first-strike capability, but now, they really tilted it back the other way. Not that they necessarily have first-strike capability on the U.S.; we still have 7,000 H-bombs, so not too big of a problem with that, we can still kill all of them too. But it seems like the overall balance of terror, if his claims are true, really has shifted toward them. So, looking forward, what do you think is going to be the reaction to that?

McGovern: My fear is that this will just feed the military-industrial-congressional-media-intelligence-security-services complex, and more and more money will be diverted to them. But I don't rule out the notion that somebody will get to Trump and say, "You know, you thought you could deal with Putin, but he's now sort of claimed, if not proved, that he has strategic parity."

Now what's the real danger? How can you deal with this guy? Well, number one, you can look at the flashpoints, the ones that exist now. The things that the U.S. and others are doing that could end up in a real armed dust-up with the Russians.

I'm thinking, of course, of Syria. Now, Bibi Netanyahu is in Washington this week, and I'm sure that he's trying to persuade President Trump that: "We can't tolerate Iranians or the Syrian government or Hezbollah people in Syria. We have to keep that war going. So this business about just fighting ISIS and then getting out of there, that's for the birds, Mr. President. Don't say that anymore. We've got to do more. We have to ensure the chaos in Syria persists for a couple more years." That's what he's saying.

There are a lot of Russians, and there are a lot of Iranians now, in Syria. They happen to be there at the invitation of the Syrian government. There are a lot of Americans and a lot of American-supported so-called "moderate rebels," and they're not leaving, despite what Trump says. I don't know if Trump wants them to leave. There's some doubt in my mind as to whether the security services — the CIA and the Army — would withdraw them. So that's a flashpoint.

Even more pressing is this, which has sort of dropped off the radar screen. The U.S. has decided to give lethal weapons to Ukraine: Javelin missiles, 210 of them, that can demolish Russian or any other kind of tanks and this kind of thing. It's sort of like giving SAMs (surface-to-air missiles) to the mujahideen in Afghanistan to shoot down Russian helicopters and planes, which they did with great accuracy. So what does this mean? It means that Trump has gone a lot further than even Obama would go in giving lethal weaponry to those — well, there are a lot of fascists in that regime in Kiev, and the Russians know that, and they know what fascists do, from history.

So the Russians are looking on at this, and they realized that there was some restraint on Obama three years ago when Angela Merkel, Chancellor of Germany, was in Washington. The subject of discussion was very alive at that time in Washington: Should we give weapons to Ukraine? And at a joint press conference at the end of her visit, one of the American journalists says, "Mr. President, have you decided? Are you going to give lethal weaponry to Ukraine?" And he says, "Well, you know, we have that under discussion," and Angela Merkel interrupts and says, "It's a lousy idea." Okay, wow! I'd never seen that happen before. So then Obama goes on and says, "Well, we're discussing it," and then two minutes later, she says, "It's a really lousy idea." So now what can Angela Merkel do?

Horton: Wait. Flashback to half an hour ago in this conversation, the ultimate cataclysm of humanity, at least Europe, was between the Germans and the Russians. They've been through this before, for real, in a way that Americans don't really remember, as you were saying.

McGovern: Yeah, people don't remember, and my grandchildren aren't taught how bad it was.

But I think that Merkel did stand up to him then, and I think Obama had been had, or almost had, by the neocons that wanted to blame that Sarin attack in August 2013 on Bashar al-Assad. He had been told, "These were your moderate rebels that did that." He was already chastened to disbelieve some of the stuff he was getting. So I think he was not very gung-ho to do this, and Merkel solidified that.

What's the situation now? Well, there's been a grand coalition reintroduced in Germany, except the Social Democrats have not yet officially voted to approve it. Will they approve it? Smart money says that of course they will, but you know, that doesn't happen until they vote on Sunday. I'm thinking that young Social Democrats, who are not enslaved to this alliance — that's 73 years old now, that comes out of World War II, that says you have to do everything that the Americans say — are going to vote against this thing. I don't know if there'll be enough of them, but it may not be a grand coalition. Merkel may have to rule all by herself, and that's going to cause chaos in Germany and the rest of Europe as well.

Horton: There's too many different things I want to ask you at the same time. Let's stick with Ukraine for a minute. There's still fighting going on in the east, and in the Donbass breakaway region in the east, they don't have tanks for shooting Javelin missiles at. So what is even the purpose of that? They're just going to shoot them at buildings? Or is it just to play along with the narrative that "the Russians are coming!"? Clearly, if they wanted to have absorbed eastern Ukraine, they would have done so when the eastern Ukrainians voted to ask the Russians please to absorb them into the Federation, right? So what do you think the plan is there?

McGovern: That's the key thing. This has already been suggested by the Luhansk region and the Donetsk region. And the Russians say, "Yeah, we annexed Crimea, but don't have any illusions. You guys have got to fight your own battles. We will support you, but there's no chance we're going to incorporate you into Russia proper." So that's the key there.

Now, a very ominous sign has come just this week with this Special Envoy to the Ukraine crisis, Kurt Volker, who had been U.S. Ambassador to NATO a decade or so ago. He's a neocon par excellence, and what did he say yesterday or the day before? He said, "We have to liquidate, we have to destroy Luhansk and Donetsk. Those people have to be destroyed. We need these new weapons, and we need to have countrywide elections so that these provinces will be reincorporated into Ukraine." What does that say? That says that these so-called Minsk Agreements between France, England, Russia and the Ukrainians, ostensibly at least, first and foremost provided for regional autonomy in these provinces. Not for a breakaway, but to give, under the new constitution of the Ukrainian republic, regional autonomy so that these people can manage their own affairs, and so that the fascists — I use that word advisedly — in Kiev won't be able to work their diktats out there in Luhansk and Donetsk.

So what is this? This is Volker saying, "No, no. Minsk? Those accords don't matter. We're going to proceed in a very different way, and we're going to liquidate these people." I mean, the guy sounds like he's part of the fascists. I have had personal contact with Volker. I was in Berlin just before November 2016, and I asked him a question at a fancy meeting of the NATO-supported council. I said, "What about what Trump says about reaching out to Russia?" And he looked at me and said, "Trump? Are you a Trump supporter? Trump doesn't have a chance of winning." That was one day, mind you, before the election, okay? So here's a guy who somebody told Trump would be a good guy to implement his policies toward Ukraine.

Trump famously said on the campaign trail, "I'm not sure what really happened in Kiev. I'm told there was a coup there, but I don't know." Well, that's no attitude to take. If you're going to be a neocon, you've got to make sure that these folks in Kiev, who really want to become a part of NATO, are being trained by NATO forces to use this weaponry. This kind of thing is a flashpoint, and I don't know what the Russians are going to do if Javelin missiles start destroying homes or any armor in that part of eastern Ukraine. I don't know what they're going to do, but the prospect for significant escalation exists all the more so since there are so many "American advisers" teaching these Ukrainian people how to shoot these things.

What I fear is that Putin will feel a lot of pressure. I mean, these are Russian stock. These are Russian-speaking people. This is part of the Ukraine that has a manufacturing base and still does a lot of trade with Russia. He'll be under a lot of pressure to say, "Look, you know, the first Javelin that's fired, the whole battery is going to be obliterated right quick by one of these fancy missile systems we have, and we'll see what happens next." Well, what will happen next? Chances are better than even that the McMasters, the Dunfords, the Mattises and the Kellys will say, "Mr. President, we're Marines. We don't take that stuff. We've got to retaliate." And you get an upward spiral. That's the kind of thing that is altogether possible when you sell lethal weaponry to these proto-fascists in Kiev.

Horton: Mark Perry wrote a big deal about how McMaster's whole specialty is planning for war with Russia and Eastern Europe. This is his game. In fact, part of it was the rivalry between him and Colonel Macgregor over which was the best way to do it — the form and the size of the brigades, the teams, the resupply and this kind of deal. Should we do the McMaster plan or the Macgregor plan when it comes to this? But it's sort of his whole focus.

McGovern: Mattis is a lot less delicate with his phrasing. As you know, before he was selected to be Defense Secretary, he bragged that it was a lot of fun shooting people, it was a real high, a lot of fun. If that's his attitude, then you have him kind of acquiescing when President Trump, on the basis of zero intelligence, decides to shoot 59 cruise missiles into Syria. That's okay, because McMaster will cover it up. He'll do a "government," not an "intelligence," but a "government assessment" to show what the threat was, and Mattis will hold his fire and not say anything.

Now, there's a hopeful sign out there. I think Mattis is aware of the danger of war with Iran, and the last thing he wants to do is to stoke up fires in Syria and have the Israelis or others killed by Iranians or Syrians and say, "Well, you know, it's because they're using chemical weapons." That almost worked in August and September 2013, when Obama pulled back, with the help of Vladimir Putin, by the way. But what about now? Mattis is being very cagey. The press have asked him, "What about those chemical weapons that the bad guys are using in Syria?" He says, "Chlorine? Anybody could do. Sarin is what we're looking for, and we haven't found it, but I'm still looking." Then the spokesman from the Pentagon says the same thing, "We're looking for if they're using any Sarin."

Horton: You also had some Russians — I don't know if they were mercenaries or exactly who they were, but some number of Russians — bombed by the Americans in the conflict zone between the Kurds, the Turks, the Syrian Arab Army (SAA), and God knows who else. The Russian Foreign Minister, or maybe it was the spokesman for the Ministry of Defense, said, "Well, you know, sometimes violent things happen in violent places," trying his best to play it down. I actually saw anti-Russia hawks attacking him for that, saying, "See? The Russians don't even care about their own people," when what he was really trying to say was, "Let's not fight about these hundreds of dead men right now," which was exactly what I wanted to hear. Sometimes you just have to let a couple of hundred dead men go, you know?

McGovern: That was Maria Zakharova, the former ministry spokesperson, and she said, "Yeah, we have a lot of civilian technical support there, and it wasn't hundreds." I think she said there were eight or nine. It's very unfortunate, but you're right, she played it down.

What's going to happen now? Well, Mattis, according to the *Washington Post* today, is in a kind of contest with McMaster. McMaster wants to play up the "sarin" aspects of what he claims the Syrian government is using, and Mattis is reluctant. Why is he reluctant? I think he sees a powder-keg in Syria. He knows what the Israelis want. We know what the Israelis want. They told us.

Now, back in September 2013, when Obama was saying, "Well, maybe I'll attack Syria, maybe I won't," before he went up to St. Petersburg and

Putin reminded him, "Hey Barack, don't you remember when we were at Loch Earn just in June, we set up this working group to see if we can get the Syrians to destroy all their chemical weapons?" Obama says, "Yeah…" and Putin says, "Well, we've persuaded them. Bashar al-Assad is willing to have all these very archaic chemical weapons destroyed under UN supervision on U.S. ships specifically configured to destroy chemical weapons."

Obama says, "Really?" Putin says, "Watch TV tomorrow. The Syrian Foreign Minister is going to announce that," and indeed that happened.

During this time, Jodi Rudoren, the new Chief of the Jerusalem Bureau of the *New York Times*, decided, to her credit, "Maybe I ought to talk to the top Israeli officials. They'll talk to me, and we'll find out how they look at the preferred outcome in Syria." So she goes to talk to several of them, among them former Consul General in New York Alon Pinkas, and she says, "Gentlemen, what do you think? What's your preferred outcome in Syria?" And Pinkas says, "Jodi, this doesn't sound very humanitarian, but our preferred outcome is no outcome. We look at it as a playoff game where you don't want either side to win and you don't want either side to lose. As long as the Shia and Sunni are at each other's throats in Syria, Israel has nothing to fear from Syria."

How do I know that? I know that because it was on Labor Day weekend, and the censors and all the muckety-mucks from the *New York Times* were out in East Hampton sipping martinis, and Jodi Rudoren got that on page 1. Front page of the *New York Times*, September 6, 2013. Look it up, it's right there. That is what Israel wants.

Horton: This is the one with the quote, "Let both sides hemorrhage to death," right?

McGovern: Yeah. "Let them hemorrhage to death." As long as they're doing that, we have nothing to fear for Israel. That's what the game is. Part of it has to do with the way Hezbollah in Lebanon, which is now part of the government, gets resupplied. It used to be that they'd have to come in by ship and then be transported through Syria to Lebanon. As long as there's chaos reigning in Syria, it's really, really hard to do that. Now, of course, they get it through from the east. It's just as easy. So there was a logic to what the Israelis wanted. Whether we understood that or not, who knows? But that was the preferred outcome.

Jodi Rudoren found it out, she reported it on the *New York Times* front page. Nobody else has ever said, "Who's driving U.S. policy towards Syria?" Eighty percent of that answer is Israel. Colonel Larry Wilkerson, a good friend of mine, was working for General Powell before Iraq. He rues the day when he bought the business about weapons of mass destruction and ties between Saddam Hussein and al Qaeda for that UN speech, which was a disgrace. But he also points out that there was great pressure from Israel to do that. They wanted us to do Iran first, but Wolfowitz and the others said, "No, no. Low-hanging fruit first. Iraq can't defend itself. We'll do Iraq first."

Now, why do I say that? I say that because I've always said since the invasion that there were three factors prompting U.S. policy. One was oil. I mean, hello? It's always oil. Second was Israel, and the third were the permanent military bases that we coveted in that part of the world. We were getting kicked out of Saudi Arabia, we wanted them in Iraq, and there's less evidence that's true. With the Israeli factor, they'll say that McGovern is an anti-Semite. Well, here's Larry Wilkerson, who was talking to the Israelis, and he's saying it was Israel that had a fundamental influence on George Bush, Dick Cheney and Defense Secretary Rumsfeld. And they're doing the same thing now with respect to Iran.

Horton: In fact, Wilkerson told me that out of the whole neocon network that was working for Cheney in the first Bush, Jr. term, Douglas Feith and David Wurmser, he thought, were outright agents of influence for Israel. That they worked for Sharon before they even worked for Cheney. He singled those two out as compared to just Abram Shulsky, Eric Edelman, Richard Perle, Paul Wolfowitz, and Scooter Libby, and all of those who were really acting on Israel's behalf or what they saw as being in Israel's interest.

McGovern: I would defer to Larry on that, but I look at it this way. I was the National Intelligence Officer for Western Europe at one point in the mid-'70s, and the flag went up. Everything started to fall apart in Northern Ireland. It was really serious. The British were getting bombed, the conflict was flaring, and I lived in deadly fear that Casey would come down and say, "McGovern, you know a lot about Ireland. We want you to be the principal analyst on what's going on there in Northern Ireland." I would have had to say, "Mr. Casey, no way can I do that because of all the things I heard from my grandmother about the British in Ireland and what they did. So I couldn't possibly, even if I tried manfully, I couldn't possibly be objective."

Now, translate that to people who are dual citizens, who are Israeli citizens as well as U.S. citizens. People who were brought up the same way I was, to have a very prejudiced view — about the British, in my case — about Palestinians or anybody else.

Horton: Only, no reluctance whatsoever to go ahead and act on those loyalties. That's the thing. It's not like Michael Ledeen or any of the Axis of Kristol, any of these guys at PNAC, AEI, JINSA, WINEP or whatever, that they disguise their motives whatsoever. It's all right there. And, you know, Julian Borger in *The Guardian*, James Bamford in his book *A Pretext for War*, and Robert Dreyfuss in *The Nation* have all shown that Ariel Sharon's office was outright in on ginning up the fake intelligence and "stove-piping" it to the vice president's office. They had their own Office of Special Plans in

Israel at the same time that Feith was creating the one in the policy shop at the Pentagon.

McGovern: Even more important than that — and it's hard to think of anything more important — was how George Bush wanted to play with the big guys, and he just really loved Ariel Sharon. At the first cabinet meeting in January 2001, he says to the cabinet, "Anybody know Ariel Sharon?" Powell raises his hand, "Yeah, I've dealt with him."

Bush says, "Well, I think he's a pretty good guy. I think he's got his head screwed on right. I think we have to give him his head." Powell says, "Well, sir, that could cause immense damage to our impartiality in the Middle East and a lot of violence," and Bush says, "Sometimes a little violence can do a lot of good."

How do I know that? I know that because the Secretary of the Treasury Paul O'Neill was there, and he wrote a book about it. So that's how that kind of stuff evolved. What was even more interesting? Even more interesting was that in the spring of 2004, just before the re-election of Bush, General Brent Scowcroft — you know, no wishy-washy guy, he was national security adviser to Reagan, a very principled statesman — George W. Bush appointed him head of the President's Intelligence Advisory Board. It's a pretty prestigious thing, nothing more prestigious in Washington.

What does Scowcroft do? He's receiving all this information about what's going on in the Middle East, and finally, he says, "You know? I've got to speak out about this." So he goes to the *Financial Times*, and he says, "I need to tell you that Ariel Sharon has our president wrapped around his little finger. Sharon has our president mesmerized. You need to know that."

Whoa! Did that appear in Western newspapers? A brief mention. What was Scowcroft trying to do? He was trying to say, "Look, people. If you can't believe me, you can't believe anybody about what's going on here, and our policy towards that part of the Middle East is being dictated by a guy who has our president wrapped around his little finger."

Horton: Is it worth saying too about Scowcroft that, as Bush Sr.'s national security adviser, they also co-wrote a memoir of the administration together? He was publicly known as Bush, Sr.'s best friend, and not necessarily every word out of his mouth was a message from the father, but certainly the understanding was that he would never say something that the father would have considered out-of-bounds in terms of criticizing the son, right? So when he wrote "Don't Attack Saddam" in the *Wall Street Journal* in the fall of 2002, that was interpreted to mean at least that this was okay with the father if he hadn't commissioned the damn essay himself.

McGovern: And James Baker, who had been elder Bush's Secretary of State, wrote a similar op-ed in the *New York Times*. So yeah, both of these people knew which end was up, and I think, well, George H. W. Bush, I would say he was my friend. I briefed him for years. I worked for him when he was out there at Langley. We had a correspondence after he left the presidency. At one point, when I saw what was happening, when I saw who his son was inviting to be key players in national security — Rumsfeld and Cheney — I appealed to the elder Bush. I said, "Would you just tell your son why it was that you referred to these same people as 'the crazies'? Just tell him that. Tell him why you kept them one step removed from places of real power. You know, they could be Deputy Secretary or Deputy Assistant, but you've got Wolfowitz, you've got Perle. My God. Please, tell your son."

His answer — and I'm not revealing any secrets here — his answer was: "Ray, I too was concerned initially. But my son, he's a very, very wise person. I have complete trust in him. I think he can keep the crazies at bay." Well, either he was too forgiving of his son or too trusting or didn't know about George W. Bush and how impressionable he would be before people like tough-guy Ariel Sharon and so forth. So we were reduced to letting Scowcroft or James Baker say these things. It was a terrible, terrible thing because we knew these guys. We knew what they were capable of. And although Scowcroft and Baker, as you say, both did publicly what everyone should have been doing, saying how crazy it would be to attack Iraq, we did it privately. We also published our memo saying this would lead to catastrophe. Unintended, but a very real catastrophe. Well, you know, we were right. We weren't happy that we were right, but at least people might want to listen to us now, when we're talking about Iran.

Horton: There's a couple of important things I want to add here real quick. First, in the book *Rumsfeld: His Rise, Fall, and Catastrophic Legacy* by the great Andrew Cockburn, he talks about how Dan Coats was originally the first pick for Secretary of Defense. But in his job interview, he said, "Oh, anti-ballistic missile defense? That's a boondoggle, man. We don't want to waste a bunch of money on a bunch of never-worked technology like that." And then they said, "Uh, don't call us, we'll call you." That was how he blew his job interview and didn't get the job. He didn't want to do missile defense. So then Bush turned to Cheney, and Cheney said, "Well, there's my old friend Rumsfeld, but your father hates him." And Junior said, "Alright, give him a call," because he's such a little punk.

And that's kind of the whole thing, to tie this all back together again, is that this man that ruined the 21st century for everyone with this invasion of Iraq, forcing the North Koreans out of the Non-Proliferation Treaty, provoking the Russians into creating this new generation of hypersonic this and that by tearing up the ABM Treaty. I mean, it really is a whole new day here because you're not going to uninvent this technology. It's no longer the old three-stage rockets. Now we've got this whole new thing, all these hypersonic gliders and all of this stuff, and absolutely none of this had to happen at all. The whole sectarian war that's tearing the Middle East apart, that's killed 2 million people so far, and it's all because of this sniveling little brat.

McGovern: …Who didn't like his dad.

Horton: Yeah, and then here's the other thing, along the lines of Colin Powell and on Israel-Palestine, is that Colin Powell told George Bush, "Listen, man. Your approval rating is 90-something percent right now because of the greatest failure of all chief executives ever on your watch. And so listen, now is your opportunity to ram this through and do a Palestinian state." And Bush was convinced for a time. As Walt and Mearsheimer show in their essay and in their book, *The Israel Lobby and U.S. Foreign Policy*, it was AIPAC and the neocons — it was also Tom DeLay and other right-wing Christian Republicans — who came and told George Bush, "If you want to be a one-term president, then do this. Remember how born-again Christians stayed home and didn't vote for your father over Israel-Palestine? Well, that's going to happen to you." And Junior said, "Okay," and backed down and threw Powell under the bus. This is in a very official biography by *Washington Post* reporter Karen DeYoung.

McGovern: Getting back to what we talked about before, ABM Soviet or Russian concerns. Why is it that everybody who thinks they're informed, who reads the *New York Times* page after page every day religiously, why is it that they say, "Wow! Why didn't Putin warn us?" [Laughing]

Well, as I think we've established, he tried like hell to warn us. He did warn us in any way he could, but the Western press and the *New York Times* never picked up on these things that we just mentioned, the things going back to 2012, 2014, 2015, even after the Iran "threat" evaporated, we're still building these systems. So the *New York Times*? It's not a good thing to rely on if you want to be informed. And that's part of the message here. If people don't wake up and realize that they're getting a very stilted view from people like Sanger, Broad, and all these other guys who have this great record of being wrong on Iraq, if they don't wake up, what's the hope?

What they need to do is listen to Scott Horton. What they need to do is tune into Consortiumnews.com or even RayMcGovern.com, if they want the other side of the story.

Horton: So until next time, if we don't all die in thermonuclear fusion explosions. It's great talking to you, sir. Appreciate it.

McGovern: Most welcome.

Part 2

The Nuclear-Industrial Complex

"A nuclear war cannot be won
and must never be fought."
— Ronald Reagan and
Mikhail Gorbachev, 1985

Darwin BondGraham: The New START Treaty November 2, 2010

Scott Horton: Our next guest on the show is Darwin BondGraham. He is a member of the Los Alamos Study Group, and he's written a couple of articles now, originally for Foreign Policy in Focus, that have been reprinted at Antiwar.com — that's original.antiwar.com/bondgraham — and both of them are about the New START Treaty.

Yay! A New START Treaty! It has been my contention on this show in the past that you do not have to be a patchouli-stinkin' hippie to think that our government ought to get rid of its hydrogen bombs; that they could spell the end of mankind, and that you don't want that. So, a START Treaty, right? Us and Russia, we still have thousands of thermonuclear weapons deliverable by all different means. So we want any treaty we can possibly get along those lines. Isn't that right, Darwin? And welcome to the show.

Darwin BondGraham: Hi, good to be here. That may be right, but treaties do not necessarily always equal good things. They aren't always actually about disarmament and human security. In fact, treaties, more often than not, make people more insecure, oddly enough.

Horton: Well, we'll get to how that is, and I think that the story basically is the story of the military-industrial complex. If we can skip down to the bottom of your first article here about the influence of the companies involved in America's nuclear weapons complex and their relationship with Republican and Democratic members of Congress — particularly, in this case, the U.S. Senate. The military-industrial complex is not this amorphous idea. This is a perfect portrait of it in action. "Oh, I get it! This is exactly how it works."

BondGraham: Right, that's exactly how it works. There's nothing anomalous about what's going on right now in terms of negotiations in the U.S. Senate to ratify the New START Treaty. That the Senate would use the New START Treaty as a proxy to discuss instead the enormous levels to which they want to fund corporate contractors to build new weapons systems — nuclear and non-nuclear — there's nothing out of the ordinary about this, except this is kind of a more hyperactive example than what we're used to seeing.

Horton: In the article you break this down into the New START Treaty's three major beneficiaries as a result of all the lobbying and all the riders and amendments added to the thing. That's missile defense, the Prompt Global Strike and more nuclear weapons — newly designed, improved and updated nuclear weapons. I guess part of that is new facilities, an incredible number of new facilities, for building nuclear weapons. But first, if you could, please start with the Prompt Global Strike, because I think that this is perhaps the most important part of this entire article.

BondGraham: Yeah, Prompt Global Strike is a dream that the military has had for quite a while now. What it entails is military planners' desire to be able to provide the president with the option of killing anyone anywhere on the planet, or destroying anything anywhere on the planet, within the window of about 30 minutes of making the decision to do so. So right now, if the people who run the United States wanted to carry out a military strike or attack someone or destroy something, a military operation like that actually takes a long time to piece together. It can take upwards of 24 hours to hit something with a missile or to send an expeditionary force somewhere. Prompt Global Strike will allow the president and the Pentagon to push a button and fire a missile that will then deposit some kind of warhead somewhere on Earth within about 30 minutes.

Horton: So, a non-nuclear, three-stage intercontinental ballistic missile?

BondGraham: Yes, it takes an ICBM, modifies it so that it launches, and it puts into orbit this hypersonic vehicle which will then re-enter the atmosphere at many times the speed of sound. Then that vehicle will either launch missiles or drop munitions that work purely by inertia, like tungsten rods or something. So it's entirely non-nuclear, but it's strategic like nuclear weapons. It's strategic in the sense that it's an option that the United States will have that is meant to deter or scare the crap out of anyone on the planet because they know they can be killed within 30 minutes of the decision.

Horton: It seems to me — and maybe I'm crazy — that it's probably a bad idea to go shooting off three-stage intercontinental ballistic missiles in a world where all of Europe, Russia, China, Israel and soon India will be able to respond. What if they get the wrong idea? What if we're just doing an assassination with an ICBM, and somebody gets the wrong impression that we're trying to nuke Beijing in some kind of first strike?

BondGraham: Right. During the Cold War and even in the 1990s, there were instances in which the United States, Russia or others launched missiles — not even ICBMs, not even having the same characterizations that ICBMs would have — but launched missiles which were then seen on radar screens in Moscow and elsewhere as possible first strikes or attacks, and then the warning systems all go off.

Yes, Prompt Global Strike will be very destabilizing in a very bad way. It's bad for multiple reasons. It's probably also bad to assume that the security of the United States would best be served by spending many billions of dollars creating this sort of God-like weapon to erase people from the planet.

Horton: "Fear will keep the local systems in line." That's my favorite *Star Wars* quote this week. That's how an empire rules.

BondGraham: Yes, it is.

Horton: We're talking with Darwin BondGraham. He's a member of the Los Alamos Study Group, and he's written these two pieces, "New START: Arms Affirmation Treaty" and "New START's Big Winners: U.S. Nuke Complex, Pentagon and Contractors." Both of them are at Antiwar.com. Talk to me about the building projects and the nuclear weapons factories.

BondGraham: The Obama administration is set to reinvest in the U.S. nuclear weapons complex at a level that has not been seen probably since Ronald Reagan, maybe before. They're going to spend upwards of $80 or $90 billion over the next ten years to build a transcontinental set of factories to work on nuclear weapons and their components. This perhaps constitutes the single largest coherent capital infrastructure project that the federal government is working on. That's kind of an amazing thing to think about, given the infrastructure that this nation needs right now in terms of renewable energy.

Horton: And the numbers are absolutely astronomical. You say there are 14 building projects that are costing at a minimum $20 million, and yet some of these projects make $20 million look like a joke.

BondGraham: Right. Some of these projects are in the range of several billion dollars.

Horton: ...Brand new nuclear weapons production factories.

BondGraham: Yes. And this may be a surprise to a lot of people who are under the impression that President Obama is somehow for nuclear disarmament because he won a Nobel Peace Prize and said some nice things about it. But if you actually look beyond the rhetoric, and if you look at what the administration is spending money on, they are for nuclear weapons in a major way.

Horton: And which politicians are the nuclear weapons manufacturers spending money on? It's all there at Antiwar.com.

It turns out that in order to get a START Treaty ratified, it has to leave enough loopholes in it to allow for even the increase of nuclear weapons? Tell me about the accounting of the numbers of nuclear weapons and their delivery systems in this treaty, and how ridiculously farcical they are, please.

BondGraham: Yeah, accounting methods in the treaty are very strange. What it says is that the United States and Russia will each have no more than

1,550 active deployed nuclear weapons on 700 platforms — the platforms being the ICBMS and the bomber aircraft. The biggest loophole that the treaty allows is that it counts every single bomber aircraft as one nuclear weapon even though the two types of bombers that are currently deploying nuclear weapons for the United States can hold 16–20 nuclear weapons each. So instead of 16–20, it's one. When you actually calculate that out, you realize that there's kind of a hidden capacity within the treaty so that the United States and Russia could actually, if they wanted to, deploy more than 1,550 weapons. That's only the first level at which the treaty is somewhat duplicitous and not at all about disarmament. Moreso, people need to keep in mind that a reduction of nuclear weapons from (some say) a level of 2,200 to 1,550 is rather meaningless, when 1,550 nuclear weapons is still more than enough to wipe out most life on the planet. The whole universe of nuclear strategy is a very secretive and kind of bizarre thing. I mean, why on Earth is 1,550 considered the magical number at this point? That completely escapes me. So, instead of looking toward militarily and strategically what may or may not be required to totally wipe out Russia if there was ever a nuclear exchange, I more often look toward the economic forces at work in trying to understand how nuclear weapons budgets get made and how strategy gets made. The powerful institutions that build and deploy the weapons. The corporations that have the contracts to run those institutions or facilitate their work. If you start looking at that, then you can kind of understand why there's kind of a minimum threshold which they constantly pressure for. Because if you reduce the U.S. arsenal to 400 weapons, there might not be enough work to go around for these companies like Bechtel and Lockheed Martin, but if you keep it at 1,550, maybe then there's enough work to go around.

Horton: Yeah, from reading your article, it's clear that that's the only thing at stake here. Although are there not guys in the Pentagon who have their first-strike strategy, where they want to be able to hit Russia so hard that they can't even shoot back?

BondGraham: Certainly the Dr. Strangeloves still exist, but…

Horton: Is that not the current plan for America's nuclear forces, to be able to do that? You mentioned Obama and his anti-nuke rhetoric — the best anti-nuke rhetoric he ever gave was the first half of his speech in Prague, in the Czech Republic. In the second half, he said, "Yeah, we're keeping our radar station here, and we're putting anti-missile missiles on the border of Russia and Poland."

BondGraham: Those people, the nuclear hawks, they definitely still exist. They're not quite as powerful as they used to be, and that's partly evidenced by the Prompt Global Strike Program. A lot of the generals and a lot of the national security strategists think that having a non-nuclear strategic deterrent is more useful than having a nuclear weapon. They intend to use Prompt Global Strike, so they would rather spend billions of dollars developing and deploying that than keeping this large Cold War-era nuclear arsenal around. That said, yes, there are still a lot of people in the Pentagon, the corporate contractors and the political establishment — in particular places, like New Mexico and California — who strongly support nuclear weapons as a matter of politics because they bring money to their states, to their constituencies and to people who give them campaign contributions.

Horton: It's just amazing, where the people and the companies that make nuclear weapons lobby the government to have them make more nuclear weapons all the time. I mean, never mind amorality; our country has been completely and totally hijacked by those who would kill all of humanity if they could turn a buck.

BondGraham: Yes, and there's a similarity worth pointing out here to another story that recently broke. It was reported that the law in Arizona, SB 1070 — which basically criminalized all immigrants there and turned the local law enforcement agencies into a very militarized anti-immigrant police force — that law itself was written by the for-profit prison industry which has a lot of prisons in Arizona. They see locking up immigrants as a growth area.

Horton: Yeah, it's the only one left — the police state and the overseas empire.

BondGraham: Very much so. So these private prison corporations or these corporations that run nuclear weapons facilities, they're very similar. They are very involved in the legislative process from the very beginning. They have full-time staff, and they put a lot of resources into protecting their bottom lines.

Horton: Yeah, and this is a great lead-up to our next interview with a couple of my favorite economists, Tom Woods and Charles Goyette, about David Broder's piece in the *Washington Post* about how we ought to nuke Iran because war is good for the economy.

BondGraham: Oh, my. Well, I don't know if war is good for the economy. I think the last nine years of the Bush administration's large expansive overseas wars have proven rather disastrous for the economy, and I'm not sure how anyone could actually debate that.

Horton: Garet Garrett wrote years ago that in empires, specifically in our empire, all the money and all the profit is made as the money goes out, but virtually nothing comes back. Yeah, the Saudis buy our debt and that kind of thing, but for the most part, all the money is made on the expenditure by the government on the empire itself, not in looting the booty from the countries that we invade, like a typical empire from the past. So it's clearly just an equation for bankruptcy. No wonder we're all broke — everything we spend is on nuclear bombs and locking each other up in cages. I mean, how are we supposed to produce any goods and services in an economy like that?

BondGraham: This gets to the ratification of the New START Treaty in the Senate. In some ways, some of the senators are simply opposed to diplomacy, and they want to maintain the United States as a very strong-fisted kind of empire. But a lot of people would say that this constant resort to militarism, to spending all our money on prisons, nuclear weapons budgets and so forth, that that's actually an indication of an empire that has become very weak precisely because it's lost its ability to win the struggle for hegemony in the economic and political spheres. So not being an economic power anymore, the empire constantly reaches for the only thing it has left, which is just brute force.

Horton: Carroll Quigley always said that world empire is the last stage of a civilization before it finally completely kills itself. I guess we can see why.

Listen, I really appreciate your time and your insight. I strongly urge you all to read these articles. They're both very long, in-depth, very well-written, well-researched and well worth noting.

BondGraham: Can I add one more thing really quick?

Horton: Please.

BondGraham: Everything we just talked about — listeners should not receive that as some kind of bad news. It's not that these things are just happening; people are opposing them. My organization is opposing the Obama administration and the Republican reinvestment in the nuclear weapons complex. If people want to learn about our work and support us, please visit us on the web at LASG.org.

Horton: Hey, thanks very much. I really do appreciate your time.

Kelley B. Vlahos: The Insidious Nuclear Weapons Industry July 17, 2017

Scott Horton: Introducing Kelley B. Vlahos. She is the Managing Editor of *The American Conservative* magazine at TheAmericanConservative.com. Welcome back to the show.

Kelley B. Vlahos: Oh, thanks for having me, Scott.

Horton: Very happy to have you here. So you write about the nuclear weapons industry here in *The American Conservative.* I'm glad you still have time to write stuff. "Dr. Strangelove and the Los Alamos Nuclear Fiasco." Oh, come on! There couldn't be a fiasco at anything nuclear because that's where we take the most intelligent and responsible adults in our society and have them run things to make sure that it's all good, right?

Vlahos: This is where the privatization debate really takes a bizarre and ironic turn. Many conservatives and free marketeers for years have presented a case in which privatizing certain areas of the government will save money and create efficiencies that didn't exist before in the federal bureaucracy. But in multiple cases, such as war, national security, and in this case the nuclear arsenal, privatization hasn't worked out so well. Part of that is government oversight, so there's still an issue with the government's role in this, but also when you're putting private companies in control of things that have to do with the apocalyptic power of nuclear weapons, or people's lives overseas in war, things get to be a little dicey. When greed or corporate profits get in the way, other things are lost. In the case of Los Alamos and other labs under the government's purview that have been farmed out to major contractors, like Bechtel in this case, the records have been pretty dismal in terms of accidents happening at the lab, radiation leakage, workers exposed to hazardous material, or workforce morale. You name it, it's not looking too good. What happened in the case of Los Alamos and Bechtel, it took about 10 years of these issues before they lost their contract, all while the taxpayers were footing the bill.

Horton: Now, I'm going from mostly ignorance and memory here, but I thought that all the labs at Livermore, Sandia and Los Alamos always had been run by American corporations, but as non-profits; that it was sort of like Uncle Sam had leaned on them and said, "We'd really like for you to do this for us." So is that what's changed? Is it now just straight up a costs-plus government contract like they're making tanks?

Vlahos: Yeah, it's gotten more privatized, for lack of a better word. In the case of Los Alamos, up until 10 years ago it had been run by the University of California, which had farmed out a lot of the work to private corporations, including Bechtel. They did such a dismal job of it, and they had been in charge of it for decades, that it went pure corporate. Of course, when Bechtel took over with its own consortium, University of California was also part of that partnership. So they just sort of switched deck chairs around on the *Titanic.*

It's gotten to the point where you have this rotating roster of corporations in this field of nuclear management or nuclear construction of these laboratories, and so you have a minimal amount of competition. When that happens, like in the war industry, you have very few players, they tend to get greedy, they tend to get fat and lazy, they take advantage of the government hand that's feeding them, and so that competitive edge is lost. So, as I write in my story, even though Bechtel is losing this $2-billion-per-year contract there, there isn't much confidence that the contract is going to go to any better partnership or consortium after 2018 when the contract is finished, because there are so few companies in this lane that have all worked in these labs in some configuration. They all suffer from the same symptoms of this contracting problem.

Better oversight would be first holding companies accountable. It's kind of a dry subject, but as you and I know watching how these war contracts have evolved, when there's a lack of accountability and you keep giving these same huge corporations contract after contract after contract and they *still* engage in misconduct and fraud and abuse and cost overruns, there's no incentive for them to get any better. Sure, if they're delivering food to the troops, that's one thing, and if they're overcharging the government for that, that's one thing, but when you're dealing with nuclear material you really need to start stepping up to the plate and holding some of these companies accountable.

The contract is $2 billion a year for Bechtel, but Bechtel is bringing in $32 billion a year as of 2015 for all of its contracts across the globe, so taking away this contract from them is almost a drop in the bucket. Maybe it's slightly embarrassing, but after 10 years, what the heck. They'll move on, and most likely one of their subsidiaries will be involved in the next contract.

Horton: You know, it's a funny thing, the "economics of privatization," as they call it. It really should be two different words, where "privatization" would mean that the market takes over. For instance, if you abolished all the government schools in Texas and then you just had a free market in education where people made their own companies of all different sizes and descriptions and just educated each other for prices, that would actually be privatization.

Whereas this is just the government contracting private companies to carry out government functions. So I'll agree with you and criticize some libertarians for actually promoting this kind of thing, although I don't know about the nuclear industry necessarily. But libertarianism as a philosophy doesn't prefer government contracts to private companies to carry out state functions. If anything, we might be the first to warn against that because it's easy to see the economics of how it works. You take state functions that we disapprove of, like rounding people up and locking them in prison or making hydrogen bombs or something like that, and now you just give a whole new extra incentive for people playing that game, which is taking home giant profits of billions of dollars. After all, making an H-bomb or an H-bomb factory, that's not cheap.

As we saw with the New START Treaty, Obama basically had to compromise with the Senate and say, "Okay, we'll do a trillion dollars' worth of revamping the entire nuclear weapons industry and arsenal if you guys will sign on to these minuscule limits of the number of deployed weapons." As you quote an expert in your piece, saying it's a "make-work" program at this point. It's funny because of all I've learned about government, and all of the corruption in all kinds of industries — in medicine and war and banking and all these things — I just didn't really ever learn about the nuclear weapons lobby. I always just had this childlike naivete in my mind that somehow there weren't these blatant conflicts of interest and lobbying in that industry as you do in everything else, right? But yeah, of course there are. You have people whose only job is lobbying Congress to have more hawkish policies against other nuclear weapons states to justify the production of more nuclear weapons, just the same way the alcohol industry lobbies for keeping pot illegal or whatever. The same conflicts of interest that you have in any kind of government work; only we're talking about H-bombs! The naive thinking is that it's some bald, old think tank guy who's concentrating really hard about what should be the very best nuclear weapons policy based on some intellectual game theory or some brilliant thing, but no, it's just money. It's just the same old conflicts of interest as everything else.

Vlahos: Yeah, it's a lot of money, and you're entirely right about the whole issue of real privatization versus what we're really talking about here, which is crony capitalism. We have these major corporations and favored defense contractors like Bechtel who have an army of lobbyists on the Hill. They get these contracts, and then the rules are different because they're government rules. They're more protected. When they screw up, they get a slap on the wrist. They're not fired right away. They don't lose the contract right away. They're operating under different circumstances than on the open and free market.

What happens is that when they all go up for these new contracts, they have their members of Congress whose campaigns they've been funneling money into, who are basically lobbying themselves to get these contracts or to get other people on board. When the University of California lost its contract, members of Congress from New Mexico made sure that the University of California was on the new partnership. So it's just a hand-in-glove relationship that Bernie Sanders talked about during the last election. That Trump, to a certain extent, had talked about in the last election, but it still goes on every day in Washington. It's not the free market by any means. It's just a real bastardization of "privatization."

Horton: Absolutely. I totally think so, too, that it's important to always make that distinction. I don't really see anybody making H-bombs in the free market. What good would that be?

Anyway, let's talk about some of these incidents. You kind of gave a brief overview of some bad things happening, but I think you're saying that there was at least some accountability where Congress or the Energy Department actually stripped Los Alamos of the contract to continue producing plutonium pits?

Vlahos: They shut that particular part of the facility down called the PF-4 facility. It's a plutonium handling plant or laboratory within Los Alamos, and there's been a big debate about whether we need more of these pits — that's a whole other story that I've written about — because that's part of the whole make-work theory that these labs are just in operation to sustain and justify themselves. One of the things that they want to do is make more of these nuclear cores, otherwise known as "pits," but they're not making any of those right now because of some of these accidents that have happened. So the DOE shut down that PF-4 facility, and that was several years ago, in 2013. It's still not open for business. They're not making any pits there.

Horton: Can you talk about some of those incidents? What kind of accidents have happened there?

Vlahos: From what I understand, it was a culmination of events. One of the more highly-cited incidents was where these workers were trying to demonstrate that they had reached a certain level of getting these plutonium rods assembled, and they had placed them on a table side-by-side-by-side, which apparently is a huge no-no because of the fission that could create an explosion of cataclysmic proportions by having them in such proximity to one another. You or I might not even think twice about it, but anybody with any experience in nuclear physics would know. There was a big emergency reaction from the manager of the lab, but then the other protocols weren't taken after that. The Center for Public Integrity put together a report about the various lapses that occurred at the lab. This event seems to be one that's one of the more well-known, in terms of the idiocy that we're talking about here.

There have been a few cases of transferring nuclear waste and not using the proper protocols because they wanted to cut corners or wanted to do it more quickly. There have been cases of employees being exposed to different materials because of protocols not being followed. It's a combination of cutting corners, laziness, not having the right people in there, and just a lack of oversight.

Horton: Part of me, I'm just objecting, but everyone involved in this, these have got to be the highest-level professionals, right? These are nuclear physicists, these are people who are so smart and so jealous of their own intelligence and their position, how do they tolerate this kind of thing? Forget the money and the accountability — they don't hold each other to account at all here? I mean, damn. I expect my family doctor to act on a very professional level even if the insurance company isn't really treating him right, or whatever the problem is, right?

Vlahos: I think there's probably a major difference between the managers and the scientists. A great guest for you at some point would be the guy that I quoted in my story, Greg Mello, who is extremely well-versed in the culture of Los Alamos. He's the head of the Los Alamos study group, and he and his wife are total gumshoe watchdogs over Los Alamos. She used to work there. I believe he used to work there. They know what the culture is like, and apparently, it's a dismal culture that's been beat down over a number of years. We're talking decades of management changes, poor morale, the management ethos, and this push-and-pull between corporate profits and science. All of this takes its toll over time.

So, when you don't have the right people in place and you don't have the right oversight, the government people who are supposed to be there watching, the bureaucracy, is so burdensome that the real issues end up falling through the cracks. I don't think it's all just one thing, but this is what happens when everything's done wrong. The government should take just as much blame because, on one hand, it's easy to blame Bechtel and make them out to be the bad guy, which I don't mind doing, but on the other hand, there are layers in place of the DOE and the NNSA falling down on the job.

Horton: And Congress! They're the demand for all these nuclear bombs anyway. You know, I read a thing where there was some kind of survey that said most people thought we didn't even have nukes anymore, that that was all over with the end of the Soviet Union. I mean, why would we keep nukes after the Soviet Union was gone? So people just sort of imagined that that's probably what happened, that they got rid of at least almost all of them.

Vlahos: Exactly. But then there's not only a lobby to modernize or maintain, but to modernize and increase. A lot of that is profit-driven, and the bureaucracy seeking to sustain itself by justifying why it needs to be there and in constant operation. This is all going on under the surface. Most people aren't paying attention to nuclear weapons anymore. Obama wasn't fooling around last year when he announced his huge modernization program, which would cost a trillion dollars when all is said and told. The industry was applauding and quaking because this is exactly what they wanted. Some people were looking around going, "What is he talking about? We thought he was the peace guy. We thought he was about de-escalation and disarmament." But no. There's an active industry, including parts of the air force, very influential areas of Washington, and the military and national security contractors, who want to keep this machine humming.

Horton: It's amazing when we're talking about H-bombs, where just one will kill your whole city and everybody in it and make it a permanent exile wasteland forever, never to be rebuilt. A few thousand of these are enough to get rid of all of Russian civilization and all of American civilization in a day or two. But as long as you're taking home a nice paycheck lobbying for just a little bit more…

Here's the other thing that hardly anyone mentions, but since we're both good capitalists we ought to: the opportunity costs. Just think of all those geniuses and all the goods and services that they could have helped to distribute. It's just incredible to think about or even to daydream for a moment about what might have been without all of this genius being wasted on fusing hydrogen atoms together. It's almost funny.

Vlahos: Yeah, it's almost funny, until you start thinking about how little we spend on our neediest and our meekest in society. Whether it be children or the elderly or disabled, and we're constantly debating about whether to keep them on Medicaid or have the states pay for it, or how are we going to pay for Grandma being in a nursing home? And all that is just a drop in the bucket to what we're throwing away on the war machine, whether it be nuclear weapons, like building these plutonium pits, maintaining the arsenal or throwing more money into a black hole in Afghanistan. It's amazing to me that we can't seem to have that debate, but we're perfectly fine debating whether or not we can bail some sick children out.

Horton: Let's talk about Donald Trump for a second. It's just incredible that he's actually the President of the United States right now, if you could believe it. All the Democrats — it's not even leftists, the leftists see through it, it's the liberals — are taking all the fun out of being against the president, with all of their stupid Russia nonsense. However, I spy an opportunity in it. It seems to me that Donald Trump's whole personality and persona is actually perfect for making a really great deal to abolish, let's say, three-quarters of America and Russia's nuclear arsenals, which are both pretty equivalent now, right around 7,000 warheads each. What a great way to stick it to the haters, right? All they want to do is accuse him of being a Manchurian candidate all day long and being Putin's slave all day long, so he should say, "Oh yeah?

Well, I'll tell you what, I'm going to abolish three-quarters of the nuclear weapons in the arsenal, and I'm going to get them to do the same thing and let you attack me over that. Let's see how the Democrat rank-and-file voters of America like the Democratic Party attacking me for getting rid of nuclear weapons."

There's an interview with him from like 1986 where he talks about his brilliant uncle at MIT who taught him all about nuclear weapons and how he wanted Ronald Reagan to appoint him to be the special negotiator to go over there and negotiate with the Soviets to get rid of the nukes. That would have been a disaster, don't get me wrong, but I'm just saying you could see his mindset even back then. Thirty years ago, he was saying, "This is crazy. Why should we have these H-bombs that could wipe out everything with just one wrong move? Let's do something to solve it." You don't have to have a lot of vision to agree with that. Apparently, it was easily within Donald Trump's range. It seems like maybe this could somehow be put out there, as this would be a great way for him to get revenge against the Democrats for all their stupid Russia conspiracy theorizing. He could do some huge Nixon-esque move to shake hands with Mao or do *détente* with the Soviets and just take all the wind out of their sails.

Vlahos: Yeah, I can sympathize with that, but I don't know if he's even in that mindset anymore. He seems to be totally on board with Reagan's "peace through strength" mantra. I could see Trump today saying, "Well, we need to have these nuclear weapons as a deterrence because we have to show strength to Russia," and I can see Putin mirroring that on the other side. It's almost like it will never happen, even though I would love to see the Democrats eat it. But I just don't think Trump is there anymore. Where's the Trump that talked about scaling back our operations overseas? He seems to be gone. He's hiding somewhere because I don't see that happening either. As much as I'd like to see your scenario unfold, I don't think you can trust that Trump would even pursue something like that, sadly.

Horton: It's really amazing, the fact that you don't have to have much vision to see how Donald Trump ought to be handling this whole Russia thing. It's to find a way to stick it back in their face, right? What's he going to do? He's just going to keep retreating and retreating and tweeting out that it's fake? Go ahead and go to Moscow. Go to Moscow, sign some giant treaty, and stick it to them! What else can he do at this point, except hit them with it, right? That's my thing. Anyway, thanks for coming back on my show. I really appreciate it.

Vlahos: Thanks for having me, Scott.

Len Ackland: Obama Breaking His Pledge Not to Build Nukes August 3, 2015

Scott Horton: Next up today is Len Ackland. He is a former newspaper reporter and is the editor of the *Bulletin of the Atomic Scientists* magazine, and he wrote a book, *Making a Real Killing*, about the Rocky Flats nuclear bomb factory near Denver. That factory actually plays a part in this article, which can be found at RevealNews.org, and it's titled: "Obama Pledged to Reduce the Nuclear Arsenal, Then Came This Weapon." Welcome to the show, Len. How are you?

Len Ackland: I'm fine, Scott, thank you.

Horton: I'm very happy to have you on the show here. What is the B61-12?

Ackland: The B61-12 is a guided nuclear bomb. It will be the first guided nuclear bomb in the U.S. arsenal, and it is currently being developed and tested at the Sandia National Laboratories in Albuquerque, New Mexico. The B61-12 is just one U.S. nuclear weapon, but its development really illustrates a much larger issue, which is the power and influence of the nuclear weapons industry, which has rebranded itself in recent years as the "nuclear enterprise."

You'll recall that President Eisenhower warned the nation in his farewell address about the military-industrial complex, and the story of the B61-12 really talks about the military-industrial-congressional complex.

Horton: Alright, well, a lot of things to go over there already. First of all, that seems to me a big blunder on their part, even admitting they exist under any name, especially admitting to this concept of a military-industrial complex; but giving it the name "enterprise" instead as some kind of chintzy euphemism. Whereas before, at least in my mind, it took a long time to really get through to me that the nuclear weapons lobby in America acts in exactly the same way as tobacco, Big Pharma, the banks and insurance companies or any other arms dealers. As far as they're concerned, they would like to sell one million hydrogen bombs, and they will push to create a nuclear weapons policy friendly toward that end, just like any other industry would. There's no limit on how many nukes they would want to sell, and they don't look at it from any other point of view. That's their job, selling nukes. They don't look at it from a "Well, you know, humans only need so many nukes" kind of point of view.

What I'm trying to get to is that I think most Americans are like me. They never even really imagine that you would have the most cynical lawyer lobbyists up there lobbying for and pushing to sell H-bombs. They figure that that is a demand-side equation only, where the military says exactly how many they need for defense purposes, and that there are no other cynical conflicts of interest. So for them to say, "Well, okay, but we're not a complex, we're just an enterprise," to me, is a huge admission that such a thing is even real. To me, it's a great talking point to point to for people to imagine what probably was previously unimaginable to them.

Ackland: Well, there's a lot unimaginable when it comes to nuclear weapons. You know, this week is the 70th anniversary of the United States' having dropped two nuclear weapons of mass destruction on Japan. Two bombs, and more than 200,000 people died in the short term. Many more died from radiation sickness after that. So it's a good time to reflect on all of this, in particular after President Obama and Cold Warriors, including Henry Kissinger, George Shultz and William Perry, have all called for the elimination of nuclear weapons because of the high risk to the world. And yet, the United States plans to spend some $348 billion, just for starters, over the next 10 years to modernize U.S. nuclear weapons.

Horton: Back to the B61-12 for just a second. You guys were talking about how they've kind of exploited a loophole in the policy here, where they basically added a satellite-guided tail kit to what was previously a dumb ballistic bomb. But what that means, though, is that it's a giant shift, as you explain, in the usability of this weapon. You say it even has a dial on it, where you can dial down the yield to an acceptable level for a politician to go ahead and hit the button to launch it, is that right?

Ackland: Well, that's the concern. This bomb, it's part of what's called a life extension program, and it's supposed to be part of the notion that as long as nuclear weapons exist, we have to make sure that they are reliable. There are nine countries with nuclear weapons, and they rationalize it by saying that you don't want to encourage any of those countries to use such weapons. So at its maximum, the explosive force of this bomb is 50,000 tons of TNT equivalent, but it can be dialed down and reduced to 300 tons.

I'm just a journalist reporting on this. I'm not an expert. The concern from the experts is that if the military argues that here is a 300 ton equivalent with much less collateral damage and much less fallout, then, in a conflict, they might be able to use this bomb against a target without having radioactive fallout that would fall on our allies and friends. That could encourage the military indeed to recommend to a president that nuclear weapons should be used. Once that were to happen, there's great concern among everyone who's studied nuclear weapons that if nuclear weapons are again used, no one knows what the outcome is going to be. Is that going to start a regional nuclear war? A global nuclear war?

There are 15,000 weapons out there, which is a good drop from the peak of 64,000 that we hit back in the 1980s, and we should be happy about that drop. But still, those 15,000 weapons have the explosive equivalent of well more than 100,000 Hiroshima bombs. So we're not out of danger when it comes to nuclear weapons. The Cold War ended, but the nuclear weapons age didn't.

Horton: I just interviewed Chris Woods about the air war in Iraq, and previously he had done all this work on the drone wars. Part of that whole discussion is the lessons of Iraq War II, which is that Americans really don't want to lose a bunch of infantry and Marines. But air war? Sure. So I worry that that argument you just made about the usability of these tactical nukes would have that much more play in that same post-Iraq War II era where no one really wants to do a Bush-style invasion. I'm afraid someone could say, "Well, here's an alternative, Mr. President."

I love this quote you have in your article. It's straight out of Henry Hazlitt. "There's no question that the labs are a major portion of the economy, especially in Albuquerque in northern New Mexico. They employ thousands of people." That's a little taste of the seen and not a question of the unseen there.

We're talking about Uncle Sam holding humanity hostage here with a gun to its head. I want to get to the revolving door stuff here in a second. You guys bring up what Daniel Ellsberg has talked about on this show, that even a very limited nuclear war between India and Pakistan could mean what for humanity?

Ackland: Well, it could mean disaster. There are scientists who worked on nuclear winter back in the 1980s who have done calculations. Brian Toon at the University of Colorado has worked with colleagues, and their analysis shows that if India and Pakistan, for example, were to fire 100 nuclear weapons at each other, the explosions would send so much dust and particles and crud into the air that you would have a nuclear winter kind of outcome for the Earth. That's with just a relatively "small" nuclear war.

Just the idea that they're "weapons" is such a misnomer. These are devices that are made for mass destruction. They're not weapons in any common sense of the word. This is something that people forget about, don't think about, and yet, it's the greatest short-term risk to the planet.

Back in the 1980s, we used to talk about fast death and slow death. People who worked on nuclear weapons issues worked on fast death. In those days, we thought about 15 or 20 minutes before launch time, so that, for my generation, was the constant nuclear fear. Slow death was environmental degradation. These days, I think a lot of people pay attention to climate change, global warming and environmental problems, but they have forgotten that, in the absence of attention, the nuclear weapons enterprise has gone along its merry way, making weapons and profit.

Horton: As you mentioned, the term "weapons of mass destruction" is always kind of a silly way to try to conflate mustard gas with an H-bomb that could kill all of Houston. Some of these larger multi-tens-of-megatons H-bombs that the U.S. has and that our would-be adversaries like Russia and China have could just wipe our major cities completely off the map. As you say, it's just kind of an issue in the background. Hardly anyone ever even talks about it, and they go on their merry way. Nobody ever even questions the policy except for, it seems, the worst people in the world, like Henry Kissinger, George Shultz and all these former senior-level State Department and Defense Department officials. They're coming together and saying, in a nutshell, "Hey, we actually know about these things, and before we die, we want to try to see if we can provide a little wisdom now that it's too late, and now that we don't have power anymore. Let's advise a severe reduction in nuclear weapons." But their cries are falling on deaf ears. Nobody even seems to notice them.

Ackland: Well, that's the big concern, absolutely. How do you wake people up? People have so many concerns, and, of course, in a mass-consumption society, people have many trivial concerns as well. How do you get them and the policymakers to address these issues?

One of the concerns that we found with the debate about the B61-12 bomb was whether it was really a full debate, or it was skewed by the power of the defense contractors and their allies so that questions weren't raised. If you look at the language, the idea that there was to be no new nuclear weapons with new military capabilities, but then you take a bomb that's a freefall gravity bomb and suddenly you make it a guided, accurate bomb, how is that not a new military capability?

Horton: Sounds like a new capability. But again, they can do what they want. Laws are for the little people, not for how the government is to operate.

Ackland: Well, obviously, when you get down to the low numbers of nuclear weapons, 15,000 is, I guess, something of a low number if you compare it to the 64,000 that were there. But of those 15,000, some 90–93 percent of those are in the hands of the United States and Russia. So clearly, the U.S. and Russia are the countries that people should be focusing on to reduce the arsenals as well as trying to keep other nations like Iran from joining the nuclear club. The problem is that if the United States, one of the two nuclear superpowers, and arguably the one economic superpower left, is developing bombs that look like they are valuable and might be possibly considered in a conflict, then how are other countries — nuclear nation states as well as those who aspire to obtain nuclear weapons — going to react? If you say,

"Hey, this is so valuable that we have to make it," then why not everybody? That's a big concern.

Horton: Have they decided whether they've achieved what they think is first-strike capability, meaning the nullification of the Mutually Assured Destruction doctrine? Because it's always argued that as soon as we think we've achieved first-strike capability, that means that the Russians have to launch on warning, they have to have an even lighter hair-trigger than before if they even think something is wrong because of the balance being tipped so far in that favor. Are we there yet? Is that right?

Ackland: Well, I hope not. One of the great unresolved issues is: How many of the nuclear weapons in the U.S. arsenal are on alert? And on alert for what? We've seen instances where, for instance, 25 years ago there was the launch of a missile in Norway that the Russians mistook for the beginning of a missile attack. They were very close to calling a retaliatory strike.

The world is too precious to have these kinds of risks there. If you care about a sustainable Earth, then you have to say, "We've got to get rid of nuclear weapons," and the best way to start is to reduce the arsenals. But, of course, that goes against the interests and the profit sheets of the defense contractors, and the bureaucratic interests of the Air Force and Navy. So there are a lot of countervailing forces that really have to be met by public opinion. If the public is silent, then the military-industrial-congressional complex is simply going to proceed, and the world is going to become ever riskier.

Horton: Could you give us an example or two of a couple of the more obvious examples of the complex at work? That revolving door or iron triangle, where politicians become lobbyists and all that kind of thing in this area?

Ackland: My colleague, Burt Hubbard, did a really great job. If your listeners go to the RevealNews.org website, there's an interactive graphic that takes just six figures who are part of the nuclear weapons enterprise and shows all the various connections between the positions that they've held in the defense industry as well as in the public sphere. For example, you have a Congressionally appointed 12-member panel appointed by Senator Tom Udall of New Mexico — well, not appointed by him, but he was the person who, along with Jon Kyl of Arizona, proposed that this panel be instituted. Half of the members of this panel have connections with the weapons lab. The chairman of the panel was the former chairman of Lockheed Martin.

Their report, which was titled, interestingly enough, "A New Foundation for the Nuclear Enterprise," essentially criticizes the National Nuclear Security Administration that's under the Department of Energy for being, in their words, dysfunctional because it doesn't have proven management practices. They basically are calling for less oversight of the nuclear enterprise. That is good for the contractors, but what's good for the contractors isn't necessarily good for the nation.

Horton: Yeah, and in this case, it couldn't possibly be worse. Even if you were just making it up. I mean, we're talking about H-bombs. It's just incredible, and the fact of how alone you and your co-author are in doing this kind of journalism is almost as shocking as what you report.

It's not even a matter of opinion, right? It's almost a mathematical fact that, assuming that the future of humanity survival is important, then this is the most important issue facing all of us all day, every day, until they're all gone. It's as simple as that. What else could be more important?

Ackland: There are other issues that are very important, and I applaud people who are engaged in so many socio-economic issues. But when we're talking about the survival of the planet, we're really talking about nuclear weapons.

Horton: Hey, listen, thanks very much for your time on the show. I sure appreciate it, Len.

Ackland: Thank you.

Michael Klare: The Catastrophic Consequences of Nuclear AI December 19, 2018

Scott Horton: Introducing Michael T. Klare. He is a regular at Tom Engelhardt's great website TomDispatch.com, and therefore, we run pretty much everything he writes at Antiwar.com as well. He is the Five College Professor Emeritus of Peace and World Security Studies at Hampshire College and a Senior Visiting Fellow at the Arms Control Association. His most recent book is *The Race for What's Left.* Welcome back to the show. How are you doing, Michael?

Michael Klare: Always a pleasure.

Horton: Very happy to have you here, and what a great piece you wrote, scaring the hell out of me. We've all seen this coming since *Terminator* came out in 1984, but we're really getting there! The title is "Alexa, Launch Our Nukes! Artificial Intelligence and the Future of War." What exactly is at issue here? We've known for a while that we have drones piloted by humans, and yet there have been stories over the years about autonomous weapons systems where the artificial intelligence will be able to pick targets and kill them. It sounds like this is rapidly advancing, and the story itself is changing to what degree American military forces or other military forces could be turned over to artificial intelligence for almost full command and control.

Klare: That's the way things appear to be moving. Now bear in mind, when drone technology was developed in the past 10 to 15 years or so, the U.S. was fighting small, isolated wars in Africa and the Middle East, and drones were used to help out our ground soldiers by spying out enemy hideouts, tracking down isolated bands of violent extremists in desert locations. For the most part, there was no opposition to this. These were really intended for counterinsurgency and counterterrorism operations.

What the Pentagon is thinking about now is something very different. They're thinking about high-intensity warfare with Russia and/or China, where both sides are going to have armed robots and drones, and the intensity of fighting is going to be a thousand or a million times greater than the kind of fighting we see in Iraq or Syria or Somalia. It's going to be thousands and thousands of weapons all firing at once. In these conditions, no battlefield commander is expected to be able to keep track of all the moving pieces on the battlefield. So increasingly, they expect that machines are going to have to take over the tracking and the monitoring of all these forces and the decision-making, because humans won't be able to keep up with it all. At least this is the way the Pentagon sees things.

Horton: I'm just thinking immediately of Neil Postman and his great book, *Technopoly*, about the outsourcing of decision-making, and you talk a bit about this in the article, particularly in the context of nuclear weapons, where a human American could decide that it'd be better to lose a war against China rather than to go ahead and break out the H-bombs. Whereas a computer doesn't have feelings; it might not think about things that way. It's just garbage-in, garbage-out logic according to the programmer, and the computers very well could get to a position where — never mind H-bombs, necessarily — they start making decisions that humans would never make. There's this paradox, right? Where you have these silicon chips and the flawless logic of the way computers work, compared to the kinds of mistakes that mushy human brains can make, but mushy human brains can also conceive of things. Computers can't conceive of anything at all. Computers can't feel, can't know; all they can do is process data. They can't even really understand what information is, other than just pushing it through an algorithm.

Yet humans seem to have this huge incentive to promote this sort of thing in order to diffuse responsibility away from themselves and automate the decision-making. It just seems like absolutely a recipe for disaster where, like you're saying, a human battlefield commander might not be able to keep up with all the different things happening on the battlefield in a way that a computer could, but he also could understand context in a way that a computer never could, no matter how smart it is.

Klare: I think that's the essence of the problem. You also used the words "mushy humans" and "flawless machines," and one of the things we learn about machines is they're not flawless. They contain all kinds of programming errors that are not obvious until something goes wrong. A lot of the algorithms that would be used in making decisions in warfare are akin, similar to, of the same nature as, the algorithms that are used in self-driving cars. And we know that self-driving cars have been put on the road, and sure, they work fine a lot of the time, but we also know that they are capable of making mistakes. They're capable of misinterpreting road signs and traffic conditions and killing people, as has occurred.

So, imagine the same thing when you're trusting machines with decisions to launch nuclear weapons or not, or to escalate to nuclear war or not. We might assume that they're flawless in their decision-making as compared to mushy human beings, but that's not the case, because they've been programmed by humans, and they reflect the biases of humans. The mistakes that we make, we put into them, and we can't see those mistakes until it's too late.

Horton: That's the real joke, right? A computer will flawlessly execute your bad code. I guess I was thinking more that that's the conception, right? That a human brain is wet and makes terrible mistakes? I mean, I one time called for Glenn Greenwald to be executed, and then I only realized later that I meant Paul Wolfowitz, the guy that Greenwald's article was about. I didn't even notice my big mistake, where a computer wouldn't make a mistake like that. That's the idea, right? Computers are here to prevent the kinds of careless mistakes that mushy human brains might make. But then, as you're saying, they still open up entire new and different categories of mistakes to make instead.

Klare: There's that, and talking about the human brain, the event in my life that was so decisive was the Cuban Missile Crisis of 1962. I could remember the headlines because it made such a big impression on me as a young person. I thought my days were coming to an end. That was an occasion where the generals were ready to take action, were ready to put nuclear weapons on alert, they were ready to go all the way. And from the declassified documents acquired later, we know that the Russian generals were the same way; they were prepared to push the button, to move to nuclear launch conditions. The political leaders in this case, in particular President Kennedy and Khrushchev, took a look at this, had second thoughts and said, "Wait a minute, wait a minute. Maybe this is a bad idea."

You want somebody in the loop who's saying, "Wait a minute, wait a minute. Maybe this is a bad idea." You don't want this to happen automatically without anybody giving it a second thought.

Horton: We see this all the time, right? Where it's not even necessarily computers, but say, for example, "zero tolerance" policies in school that say that "a weapon is a weapon" and "a drug is a drug." So, a girl gives an aspirin to another girl? They're both expelled. The valedictorian has a butter knife in the back seat of her car parked out in the parking lot? Expelled. Zero tolerance. No thinking allowed, no judgment, no discrimination, no decision-making by humans. We go with what the words say on the paper, and the rule must be implemented.

We see that all the time, right? Like, even when it comes down to expelling the valedictorian over a butter knife, they will do that rather than stop and say, "No, pardon. This is where our humanity interrupts and intervenes and we make decisions instead and do something different." So if we're talking about a computer, it's the same difference, right? The computer goes and implements the algorithm whether it makes sense or not.

Klare: I'm not familiar with that valedictorian story, but if it's true, that's pretty upsetting.

Horton: That was just one. There was one where the girls give each other Tylenol for menstrual cramps or something, and they get thrown out as though they're heroin dealers, this kind of thing.

Klare: That's why we have to look in particular at this issue of AI and decision-making in wartime. Scott, the worry here is time, the compression of time where the pace of warfare is accelerating or is expected to accelerate in the future, where you're not going to have as much time as was once the case to ponder these decisions and say, "Wait a minute, let's think about this," because the pace of action is going to be so rapid that thinking will be pushed aside and it's going to be go/no-go, where you have three minutes to make a decision. This is what terrifies me. We have to figure out how to slow this process down so that human minds can take a look at the possible implications of what's about to happen.

Horton: You talk in here about the paradox where the more that we automate it, the faster all the decisions happen, and on the side of our enemies (from the Pentagon's point of view), Russia and China will be using AI, too. And they're going to be making battlefield decisions at such a rapid rate that, "We can't have an 'AI Gap'!" So it kind of keeps escalating and escalating, and now they've created this situation where they're basically forcing themselves. They painted themselves into a corner, where now they have to outsource, as you show at the end of this argument, virtually all of their decisions and even strategic decisions to the computers in order to keep up.

Klare: One interesting thing that I've discovered about all of this is that among the people raising the most alarm about this trend are people within the industry itself. It's not so much people like you and me necessarily on the outside saying, "Nah, nah, nah! This is bad." I'm learning about this mainly from people within the industry who are saying, "Gee, what have we unleashed here? Maybe we should give it some thought."

You see this in the revolt of people at Google who opposed the use of their intellectual property and their intellectual creativity by the Pentagon to develop AI systems for killing people overseas. There was a huge revolt, and the company finally agreed not to work on that project. It's called Project Maven. It's about using image identification algorithms for identifying potential targets for assassination. Google employees said that this is not what they signed up for; they don't want their work being used for military purposes and are speaking out about the dangers of what might happen when this technology is turned over for military use without any kind of regulation or thought given to the possible legal, moral and ethical consequences of this.

Horton: That's so important what you're saying about that Google revolt, because it shows they didn't all resign over it. They said, "We're staying, but we don't want to do this." I guess they signed petitions and this and that, but Google actually, at least as far as we know, backed down on that and isn't going to do that now. So that really shows that resistance is possible, but it's not automatic. People have to make the effort to go ahead and fight.

I mentioned that book by Neil Postman, *Technopoly*. The subtitle is, "The Surrender of Culture to Technology," and he's saying that anything that can be invented will be invented, and that as soon as it's cheap enough, it will be implemented; and there's no set of beliefs on the part of the population of this society that can possibly stop it. So this is kind of a last-ditch effort, but that's still pretty big. It goes to show that if people of whichever different political persuasions and ideologies, if they're really desperate to hang on to liberty; if we really want to live in a society where our sheriff's department is not flying drones over our heads, or where the traffic cameras aren't tracking and keeping tabs on all of our license plates; or where the NSA isn't keeping data on every place our cell phone has been for the last five years, then we have to make it that way. We have to insist, because otherwise this stuff is a tsunami. Especially since it's all subsidized. All the worst surveillance tech and drone swarms and God-knows-what they're doing is all on the taxpayer's dime. It's the Pentagon and the National Security State that are pushing technology specifically in these directions.

But we can resist, and we can stop it by writing articles like you've written here, by using their internet to push back, like we're doing with this show. And, like you're saying, with Google and Maven there. Now, Amazon has said that they don't care, they're going to keep up with all their CIA and military contracts. I guess there'll always be someone, but it's sort of like with BDS (Boycott, Divestment and Sanctions) in that what matters is making it an issue, forcing people to argue about it and shedding light on the problem.

Klare: If people are concerned about government intrusion through electronics, what you were saying about AI snooping, this is happening much faster than any of us think. That's what I'm being told, that the technology of intercepting our communications and our imagery is happening at an exceedingly rapid pace. As you say, if this is a concern for people — preserving the privacy of their information and location and all that — don't wait to raise objections. The time is now.

Horton: What a weird position to be in, right? Where the government is pushing the invention of fruit fly-sized drones, sensors the size of grains of dust and these kinds of things. Then all we can do, essentially, is petition that they will only use these technologies against helpless peasants in third-world countries in their horrible wars overseas and not use them against us, because otherwise they can. If there's not a law that says that the Pentagon is forbidden from giving these kinds of sensors out to local police departments, then they will. Like we're seeing with the stingrays intercepting our phone calls, we've got our local sheriff's department acting like the NSA in the neighborhood because that's where they've got the technology.

Klare: [Laughing] The more you talk, the more anxious I get. Yes, all this is true. My concern and my research is on the implications of all this for the future battlefield. When I look at that, that's where I get really anxious because of course it's not just the U.S.; it's also Russia, China, Israel and South Korea all moving to develop, to sell and to export this technology. So it's going to become a worldwide phenomenon in the not-too-distant future unless we move quickly to impose some kind of limits. Right now, there doesn't seem to be much standing in the way of this.

Horton: Let me ask you about the subject matter of Russia and China being the enemies we supposedly have to fight. The thing about Earth, I guess the problem for the military-industrial complex, is they've run out of countries and continents where they can find bad guys to fight. China's our number-two trading partner after Canada, or they're tied for first place with Canada. The Soviet Union ceased to exist back in 1991. We don't really have that much to fight about, but they're focusing on the old Cold War enemies there.

When I was a kid in the Reagan years, it was brinksmanship and the height of the Cold War right before it all ended. It seemed like all the talk was, and the idea was universally understood, that if America and the Soviet Union ever went to war, it would be a full-scale H-bomb strategic nuclear war, every major city in both countries would be completely obliterated off the face of the Earth, and it would be the war to end all wars for real this time.

Yet now there seems to be all this talk about these weapons systems, like in your article here, and it sounds like they really think that they could fight a conventional war against Russia and China that would not go nuclear. And that's what they're really planning on. They're acting like H-bombs just aren't in the picture here, and if we fought a big conventional war with Russia and/or China, here's how we would do it. The battlefield would be Latvia or whatever they imagine, and I guess they think it'll be fine. So I wonder if that seems really different to you as well, that there's this idea that we could do this without losing D.C. and Houston.

Klare: I don't know what goes through the minds of the people who are planning these war games and these war scenarios, so I can't speak for them. The impression I have is that the current contemporary military leadership, the people who are now the generals and admirals, have spent the past 15 or 20 years fighting losing wars — senseless, pointless wars. What they see are pointless wars in Afghanistan, Iraq and Syria that have gained the U.S. no advantage whatsoever, have drained and depleted their forces, and have been nightmares for them, while Russia and China have not had to do that and have been able to invest comparable sums in modernizing their own forces. Now they think, "We have to catch up with Russia and China in this conventional war-fighting capacity, and move beyond Iraq and Syria and Afghanistan, and be able to fight Russia and China on the conventional battlefield."

Do they think that they can keep it a conventional war? I really don't know. I think they do. I think they believe in their war-fighting skills. They see themselves as warriors, as people who are trained and equipped to fight and win these high-tech wars of the future. I think they're totally wrong about their ability to contain it at the non-nuclear level, because whoever faces defeat in one of these high-tech, all-or-nothing, zero-sum conventional wars is not going to say, "Oh, I give up"; they're going to escalate. At that point, we're talking about nuclear weapons. So I think there's a fallacy in all of this. That's what scares the crap out of me.

Horton: It seems too like the people who are interested in nuclear warfare game theory and this kind of thing are a bunch of insane lunatics, and so they come up with these policies where they say, "You know what we'll do? We'll escalate to de-escalate," which means, "We'll set off some nukes in order to warn the other guys that they'd better back down now, or else we'll set off even more," when no, obviously, the reaction is going to be for the other side to use nukes, too. Escalate to escalate to escalate. There's no de-escalation once you're setting off H-bombs.

Klare: Yeah, once the decision has been made, all hell breaks loose, and there's no control whatsoever.

Horton: Only geniuses could be that stupid, right? Only these people who are so good at math that they'll just blow you away, you could never understand what they're even talking about. But then only they could be so wrong, lost that deep in their navel-gazing that they just ignore the plain, common-sense realities of the world we live in that cancel out all their stupid algorithms.

Klare: This speaks to what I was saying. It speaks to the notion that you can fight a conventional war to win or defeat, and the other side will accept defeat and will do so without raising the stakes. I can't imagine that that's possible. If the U.S. attacks Russia or China with conventional weapons and destroys the hard core of their military — all their ships, all their planes and all their radars — with our superior conventional weaponry, and we assume that they'll just wave the white flag and say, "We surrender! Do to us what you wish." I think that's where the fallacy lies. And the same would be true of us if it were the other way around. One side or the other will say, "We're not going to bow our knees, we're going to escalate." As long as you have the potential for escalation, there's always the likelihood that that will be the outcome.

Horton: That's definitely the way to think about it, right? What would American politicians do in that position that they think they can put Russia in? They would use H-bombs.

People don't like mentioning this, so I think it's worth bringing up extra: that the whole problem with Russia is all America's fault. George H. W. Bush promised not to expand NATO, and then Bill Clinton broke that promise. Then George Bush's son doubled the breaking of that promise. Obama tripled it. Even Trump now has put Macedonia and Montenegro into NATO. They've expanded all the way up to the Baltic states, right on Russia's border, 1,500 miles east of where the line used to be, halfway across Germany.

It'd be like if the Russians had Warsaw Pact, Russian and other Soviet troops stationed in Mexico on our border. That would be a tripwire. It would be a really bad idea for them to do that, wouldn't it? And that's exactly the position that we put the Russians in. Then, just like George Kennan predicted back in 1998, when he told Thomas Friedman: "I'll tell you exactly what's going to happen here. We're going to expand NATO like this, and everybody who is saying, 'It's fine, the Russians won't mind because this isn't about threatening Russia. It's totally cool,' when the Russians react, those same people are going to say, 'See? That's why we have to do this, to contain Russian aggression.'"

Klare: From the Russian perspective, they see this pattern. They feel they've been betrayed and lied to by American leaders going back to the first Bush. So they think they're the victims of Western aggression, as it were. Or NATO expansionism. Now, I don't think that excuses the aggressive behavior on Putin's part. I think he's taken advantage of the legitimate anger that ordinary Russians feel, and they feel they've been taken advantage of by the West. He's exploited that legitimate feeling to perpetuate his own crony regime by whipping up nationalist sentiment and using that to perpetuate his regime while engaging in bad behavior of his own. So that puts all of us in a worse situation.

Horton: I totally agree with that, and if you look at what Russia has done in eastern Ukraine and in Syria, especially in Syria, a lot of innocent people have been killed. But in both of those cases, those are in direct reaction to American policy. Our government had overthrown the government of Ukraine twice in 10 years. Then the new coup government in 2014 was threatening to kick the Russians out of the Sevastopol Naval Base in Crimea where they've had a base since we were under the Articles of Confederation here. Since before Alabama was a state in the union. So we really put them in a corner and up against the wall, and basically, the Obama government forced them to intervene in Crimea and then in the east of Ukraine, in the way that they did. I'm not excusing all their behavior of what they've done.

Innocent people have died, and it's been horrible, but the Russians had no need to intervene in Syria until American support for al Qaeda was getting so bad that the regime there was really threatened and could actually fall. And that's right in Russia's neighborhood. That threat's still far away from here, but that is a huge deal to the Russians. It was only in the end of 2015, after four years of U.S. support for the terrorists in Syria, that the Russians finally came. And again, they've killed tens of thousands of people with their bombing there, just like America did in Iraq War III as well as in Syria and Iraq.

Klare: My interpretation of events would be different. I think that Putin is playing a geopolitical game like other imperial powers, and Syria was a ripe area for Russia to reassert itself as a Middle Eastern power after being pushed out at the end of the Cold War era. Russia, then the Soviet Union, had been a major international actor on the world stage, and they lost that with the end of the Cold War.

Horton: Yeah, it was one of their only foreign bases, their naval base in Syria there. I don't disagree with you; I'm just saying that it still was the Americans who created the entire crisis for the Russians to take advantage of. So I don't mind hearing you complain about Russia. I just don't want to hear the policymakers complaining about Russia. You know what I mean?

Klare: Agreed. Let's agree with it right there.

Horton: Because yeah, you know I'm no apologist for that regime in any sense other than just there's four fingers pointing back at our side every time we accuse them.

Klare: My equal worry is in the South China Sea because that's where I think World War III could break out. Here, once again, the blame can be equally distributed. I think the U.S. believes that it is entitled to remain the imperial power in Asia as it was for the past century in the Western Pacific. China, I think legitimately, says that era has come to an end, and China is the rising power in Asia. That creates an irreconcilable clash, and it's incumbent on leaders on both sides to say that we have to step back from the brink and figure out how to manage this world's historical event of a change in the power balance in the region without creating global annihilation. You can't turn the clock back and say China doesn't exist.

But China is also behaving in an atrocious way, I believe, by building up these islands in the South China Sea and militarizing them. And the U.S. is behaving in an atrocious way by trying to provoke the Chinese by sending ships into harm's way next to those islands. Sooner or later, somebody is going to get killed, and that could set off an escalatory spiral, leading to World War III. This worries me as much as, if not more than, the U.S.-Russia conflict in the Baltic States. So wherever you look, there are these geopolitical pressures that are pushing us to the brink of war.

Horton: Another major pressure involved in all of this is the corruption of the military-industrial complex. Not just the military-industrial complex, but the whole "Money first, M-16s that fire when you pull the trigger second" kind of attitude. When the Russians are developing their new fighter jets, they're really working hard to make the best fighter jet that they can for the price, like you would expect, but in America, it's all just about stealing as much as possible. So the F-35 is a complete piece of crap. We have 1970s model F-16s that could blow the F-35 out of the sky. None of its equipment works, and we don't have to go down the litany of everything that doesn't work about the F-35.

It's the same thing with our Navy ships, too. These littoral combat ships are basically worthless. I would assume that all the great high-tech stuff on most of those battleships and carriers doesn't really quite work right the way it's supposed to, and that if we get into even a little skirmish with the Chinese on the high seas, a ship or two gets sunk. Or imagine if an aircraft carrier had a ballistic missile fall on its deck and cut it in half, and it sank to the bottom of the sea. You're talking about thousands killed there. That's the kind of thing that no politician at home can survive, and they would rather start a war than resign, you know?

Klare: Yes, this is what wakes me up in the middle of the night, that scenario that you just described. And this is hardly unlikely. This could happen tomorrow, something like that. Maybe not quite tomorrow, but we're moving towards a world like that where all these powers feel an obligation to flex their muscles all the time to show they're not going to back down, and we're not going to let the other side walk all over us. That's probably how they talk to each other, and one side or the other is going to push too far.

Horton: Back to my old question about whether they really think that they can fight just a conventional war. I'm not trying to ask you to read all the minds of the wonks, but I just mean in the sense that you're exposed to the media the same way I am and whatever events you go to, or what have you. But talking to people that you talk to, the experts, it seems to me like this narrative about the rise of China and what America is going to do about it, people are so married to this idea. I hear very little talk of: "Who's scared of China? The richer China gets, the better. If we have any problems with the Chinese, we'd better find more things to cooperate with them on then, in order to make things better."

Same kind of thing with Russia, where we can't fight these countries. Everybody's got H-bombs, so no fighting. So the other option is to always do our best to work together. Here's a country of a billion-plus people that abandoned communism in the 1970s, essentially for more of a fascist pseudo-capitalist-type system. There are huge amounts of new wealth being generated in their rapidly growing economy. At some point, it will be equal in economic output to the United States. But then, so what? That's just more Chinamen to trade with, and more wealth for everyone on Earth to benefit from. It seems like I don't hear that narrative at all anywhere. You turn on NPR, read *Commentary* magazine or whatever you've got, and the idea is, "Oh, no! China! China! China! How are we going to prevent their rise? How are we going to contain them?" Everything begins with the presumption that America has everything to lose as China comes up from Maoism. That's the really worrying part to me, that it doesn't seem like any reasonable people are part of the debate at all.

Klare: You're right. I'm in Washington, D.C. right now, and that is the drumbeat in Washington, exactly what you're saying. How can we get ahead of China? How much more money can we spend? How many more billions can we spend to ensure that we have permanent supremacy against China? And, you know, that's all that Senator Inhofe, the new Chair of the Senate Armed Services Committee, talks about. And he has plenty of other friends who share that point of view: "Let's throw money at the military-industrial complex so we can be certain that we'll never fall behind China, and the Chinese will never catch up with us." That's the point. And this is endless until we're utterly bankrupt, because that's the only way it could go.

Horton: Well, I'll let you go at this point. I've kept you over time here, but I sure appreciate it, Michael. It's always great to talk to you.

William Hartung: The Existential Threat of the Nuclear Weapons Lobby
October 22, 2021

Scott Horton: Introducing William Hartung, Director of the Arms and Security program at the Center for International Policy and author of "Inside the ICBM Lobby." And this is a press release, a very informative one, at the Institute for Public Accuracy, that's Accuracy.org: "How the ICBM Lobby Is Threatening Armageddon." Welcome back to the show. How are you doing?

William Hartung: Good, good. Thanks for having me again.

Horton: Happy to have you here. This is one of the goofiest, ironic, crazy things of our times, that there's such a thing at all as the ICBM lobby. And I know I'm a broken record on this, but it just goes without saying that somehow the military decides how many nukes they need and then they make them or order them or something; it just doesn't really come up that you have businessmen who are trying to get rid of some H-bombs. They will do anything that they can to bribe the Congress to make sure that they can keep selling to their captive market, the Energy Department and the Pentagon, in preparation for the destruction of all of humankind, as long as they can make a quick buck right now. Just the same as any other racket. Just the same as a company that sells tanks or shoelaces or combat boots or whatever to the military. We have an ICBM lobby that you say is threatening Armageddon. So please do tell, sir.

Hartung: As you're pointing out, there's a lot at stake here. It's not like overcharging for boots. The ICBM is particularly dangerous. All nukes are dangerous, but because the president has a few minutes to decide whether to launch them in a crisis, there's a danger of an accidental nuclear war due to a false alarm. And there have been false alarms before. So if we were going to get the nuclear problem under control, the first thing you want to do is get rid of ICBMs. Yet you've got senators from the states where the ICBM bases are, or where they develop and work on the ABM, who have blocked almost every single attempt to spend less on ICBMs, to reduce the numbers, to even study alternatives. When they negotiated the New START Treaty, they got rid of 50 of them. There's 400 left. So they said, "Well, let's destroy those 50 silos."

The lobby said, "Oh, no. Maybe we'll need them in the future if we rebuild up again." So literally any little change that you might think about this lobby — which is senators from relatively small states, Wyoming, North

Dakota, Montana and Utah — have had kind of a stranglehold on policy on this. Backed up, of course, by the corporations. Northrop Grumman has a $13 billion no-bid contract to build a new ICBM. It's not bad enough that we have them in the ground, but now they want to spend over $250 billion building and operating a whole new one. Of course, in wonderful Pentagon parlance, they call it the "Ground Based Strategic Deterrent," which doesn't sound so bad. Not like it's something that's going to end life as we know it. But that wouldn't be a good name, "The Missile to End Life as We Know It." So they stuck with GBSD as their acronym.

Anyway, yes, because of what's at stake, the fact that a relatively small number of senators and corporations are keeping us down this road is one of the most dangerous and outrageous examples of influence-peddling that we have.

Horton: Yeah, it's completely crazy. Can I ask you? Do you know off the top of your head how many megatons we are talking about in the missiles that we're discussing?

Hartung: I don't know. I know it's like many, many multiples of the power of the bombs that killed a few hundred thousand at Hiroshima and Nagasaki. It's a huge, huge explosive power.

Horton: Do you know if these have multiple re-entry vehicles per missile? That's the ones we're talking about here, right? The big ones?

Hartung: Yeah, exactly.

Horton: So you could have one rocket that actually takes out four or five cities, something like that?

Hartung: Exactly. The U.S. has 1,550 deployed nuclear weapons, and about four or five thousand in the active stockpile. A handful of those, if you had an exchange of even a hundred of those thousands and thousands, you would probably end life on the planet because you'd interfere with heat getting in, you'd have famines, you'd have all kinds of horrific consequences. All this talk about how we need more because China is building a few missiles, or we don't have enough to take out their nuclear weapons in a conflict, once you start using nuclear weapons, essentially life is over. So a lot of the arguments used in Washington for why we need a tweak here or a more reliable weapon there, a different capacity here, really is irrelevant to our survival. In fact, it's counter to our survival.

Horton: Well, you're such a hippie, how could anyone be opposed to the power to destroy a city in one shot? After all, as Bill Kristol might say, it's kept the peace for the last 75 years. Never mind the Koreans or the Vietnamese or the Iraqis or whatever people like that, they don't count. But between the major powers, we haven't had major wars because atom bombs are too big to use. So thank goodness, everything's great. What do you say to that?

Hartung: Yeah, I think it does make the country think twice about having an all-out war with another nuclear-armed power. But as you said, most of the wars have been in states not only that don't have nuclear forces, but don't have major conventional forces either. There's been great suffering under this umbrella of what they call "deterrence." And deterrence is not guaranteed. It depends who's got their finger on the button. It depends on other developments.

You've now got all this clamoring for better war plans against China. What if they overreach? I mean, China is a nuclear-armed power, and yet they're making all these plans for striking deep into Chinese territory, a naval battle over Taiwan and all these things that kind of imply that we would. In the National Defense Strategy, they're saying, "We have to be able to win a war with Russia or China." What on Earth does that mean when you've got nuclear-armed powers? How do you win a war against somebody that can destroy your society just as you can destroy theirs? I think it has shifted the focus of conflict away from great power conflict towards fighting in the rest of the world to the detriment of millions and millions of people, but I think even that the notion that it's going to hold the peace is not guaranteed, and we're in dangerous waters at the moment because of the demonization of China.

Horton: Yeah, you look at the past 75 years, that's nothing, right? That's a blink of an eye. So the fact that Mutually Assured Destruction has worked so far is no guarantee at all. And boy, what a failure, if it ever fails. You're talking about losing cities at a minimum, maybe losing entire civilizations off of the face of the Earth, setting humanity back a thousand years, billions of people starving from famine.

Hartung: It's a risk that's not worth taking at any level.

Horton: Part of the context here, though, is that there's pressure to get rid of these Minutemen, right? Some reasonable people inside the military establishment even think that we could just rely on our subs and air power, and we don't really need the nuclear sponge in the middle of the country. That's the other crazy ironic thing about all this, that all these Minutemen are there mostly to distract Russian forces so that they have to waste all their nukes nuking the middle of the country, which is supposed to somehow spare the coasts, which is ridiculous, anyway. We'd lose New York and D.C. and L.A. and San Francisco in a nuclear war with Russia. But there's this nuclear sponge idea, and the reason we're talking about this is that some people are objecting to that and saying maybe we could get rid of these things. So you're really telling the story of the lobby pushing back and saying,

"No, you can't do that. We've got to keep these, no matter what. In fact, replace them with brand-new ones, etc."

Hartung: Yeah, and as too often happens, people from the military and the Pentagon don't really tell you what they think until they get out of office. They're not in there necessarily fighting for this when they've got more power to do something about it, but nonetheless, there's a whole long list of retired generals, heads of the Strategic Command. Former Clinton administration Defense Secretary William Perry has been particularly vocal about getting rid of ICBMs. Dan Ellsberg now is speaking out as this being the biggest nuclear danger. He just did a piece with Norman Solomon for *The Nation* that makes that case. There's a lot of arguments and a lot of people who know their way around this issue saying we don't need these things, but not enough people on the inside. Of course, their decision-making isn't always based on what makes us safer; it's that we don't want to lose a Senate seat in Montana, we don't want to lose these campaign contributions, or we don't want to be perceived as weak, that blanket argument for spending more and more in the military.

Horton: I always loved that line: "We're *afraid* that someone will call us cowards." Yeah, that would be a real problem.

Hartung: Yeah, if you really had the courage of your convictions, that wouldn't bother you. If you're such a tough guy, stand up for what you believe in.

Horton: I'm sure you probably saw this. They're talking about giving these nuclear-powered submarines to the Australians. I'm pretty sure they said that the time it's going to take to transfer this technology to the Australians, prepare the infrastructure and train up all their people to be able to use it, it'll be 2050. It's just all baked in there, unsaid, that we do not foresee an era of diplomacy with Russia and China; we foresee the status quo holding. If we're lucky, we'll still have a nuclear standoff with these two major nuclear weapons powers into the indefinite future, and that's the framework that they look at everything.

Hartung: The new ICBM would last until 2075 if they build it. So yeah, this is supposed to be in perpetuity, in their view. Using that time to actually do arms control or disarmament or change the calculus or have a diplomatic opening to any of these countries is not even in the scenario.

Horton: Yep. It's crazy. I really appreciate you making a little bit of time for us here today, Bill.

Hartung: Always great to talk.

Part 3

About Those Nuclear "Rogue States"

"The best estimates at this time
place Iran between three and
five years away from possessing
the prerequisites required for
the independent production
of nuclear weapons."
— Benjamin Netanyahu, 1995

Grant F. Smith: The Niger Uranium Forgeries May 24, 2019

Scott Horton: Hey, guys, on the line I've got my buddy Grant F. Smith. He is the founder and director of the Institute for Research, Middle Eastern Policy. That's IRmep, IRmep.org. He wrote all of these books about the Israel lobby, like 8 or 10 of them, something like that. The most recent is *Big Israel*. Get it? Like Big Tobacco or Big Pharma or Big JDAM Bombs or this kind of thing, and boy, he's got a doozy of a new piece here "Lesson from FBI's 'Niger Uranium Forgeries' File."

Oh, did I mention? Here's what you find at IRmep, you find PDF files of documents that Grant sued the government for, got and posted. So if you go to Google and type "site:IRmep.org .pdf," I don't know how many results you'd get, but I know that it would be a very high quantity, all of very high quality, of things that Grant has discovered and reported over the years, including this doozy: 640 pages of FBI investigatory records obtained by IRmep under FOIA about the FBI's investigation into the Niger uranium forgeries of 2001 through 2003. Welcome back to the show, my friend. How are you doing?

Grant F. Smith: What a setup, Scott. Thank you so much for having me again.

Horton: Hell yeah. So tell us all — wait. Hey, kids! Back in 2001, America got attacked by al Qaeda. Immediately after that, the Bush government decided that they were going to lie us into a war with Iraq by making your mom afraid that Saddam might give some chemical weapons to Osama bin Laden, since they were buddies and Saddam had all these weapons.

Of course, the linchpin, or the major part of it all, was that the Iraqis had a nuclear weapons program. As the vice president said numerous times, "They are reconstituting their nuclear weapons," which, of course, they never had in the first place. A big part of that was the story that Iraq had imported a bunch of yellow cake, semi-refined uranium ore from the African country of Niger. George W. Bush mentioned this — the famous 16 words — in the State of the Union address of 2002. This became, of course, "Mustard gas is mustard gas, but *nuclear bombs are nuclear bombs*," and this became really the linchpin of the case for starting the war against Iraq in 2003. So that's what we're talking about, the Niger uranium forgeries, that all sides now agree were forgeries that were the basis of the so-called intelligence that was the basis for that accusation by the president in the State of the Union.

So, that's my setup for the conversation today. A lot of young people listening, they were kids when this happened, so they don't know the story, but it's a big story. And now you've got a lot more of it, if not the whole story. So do tell.

Smith: I think it's important to remember that, for a long time, there was a lot of mystery surrounding the allegations of uranium sales direct from Niger with no participation in the consortium that's running it with the Niger government from the French, who use fuel for their nuclear energy production. There was a lot of speculation swirling, particularly after George W. Bush made that allegation in his State of the Union address, about whether these allegations were true, whether they were completely derived from something somewhat true with added false information, or whether they were just outright false. Getting these FBI files has been a long-time pursuit of ours. We've gone through multiple appeals. We really started going hard after these in 2012, if you can believe that, and it was all to find out what the FBI discovered, because they were charged, as part of this controversy, with investigating a number of things.

One of the things that I think it's good to start with is that on March 14, 2003, on behalf of the Senate Select Committee on Intelligence, Senator John D. Rockefeller IV sent a letter saying, "We want a broad investigation of the documents, who created these documents, and we want to know for sure that the U.S. Intelligence Committee or other elements of the U.S. government didn't actually create these forgeries in order to build support for the administration's policies."

Now, this letter was tucked into the FBI file. It was addressed to Robert S. Mueller III, the director at the time, from Rockefeller. This shows the level of skepticism and suspicion. He was obviously the minority guy on this committee, but right after Bush made that allegation and right after the IAEA finally, after multiple delays by the U.S. government, got their hands on this dossier of forgeries, they almost immediately were able to debunk it as a fraud. So what the Senate Select Committee on Intelligence thought at the time was that, in addition to using these documents to justify an invasion of Iraq, the U.S. government or some of its tentacles may have been involved in creating the forgeries. It's always been obviously difficult for the FBI to investigate Washington officials, although they've done it successfully. I just want to note that they didn't seem to have any problems investigating other types of fraud committed by high levels of government, even going after powerful entities that seemed like government agencies, like AIPAC, which they did just a few years later. But they had a great deal of trouble in this investigation doing anything but scuffling around in Europe; particularly the Rome embassy, where a reporter brought these forgeries to the attention of the U.S. embassy and asked whether they were authentic or not, and reintroduced them into the stream of intelligence in the U.S. inadvertently because they were sent to the State Department.

So the FBI, just to cut short the story, which is called "Lesson from FBI's 'Niger Uranium Forgeries' File," the FBI spent most of their time and resources investigating the forgeries. They were getting information from Elisabetta Burba, who was the magazine writer trying to authenticate these forgeries and not having much luck doing that. She also traveled to Niger. The FBI spent its time looking at her and looking at this Rocco Martino guy, who had connections to Italian intelligence, quickly concluding that these were forgeries by June 2003. What I think is the key point of this whole file release is that the FBI spent 551 days on its investigation being essentially limited to the provenance of the forgeries; but not until they asked for and received the full letter from Senator Rockefeller from FBI headquarters (i.e., Robert Mueller's office), they had no idea that the mandate to investigate was broad, that it wasn't just about finding out who forged the documents and why, but that they were also supposed to find out whether the U.S. intelligence community had a hand in either creating or amplifying the reaction to these documents and the misuse of these documents.

Horton: So it's not redacted, it's just that you have all these files and none of them are about the neocons in the Pentagon, or Stephen Hadley working for Condoleezza Rice in the White House.

Smith: Correct. Nor even looking at the fact that the administration desperately wanted these sorts of documents to fit their preconceived narrative.

Horton: Do they talk at all about Michael Ledeen and all the Iran-Contra II? Back in the day, Josh Marshall wasn't such a hack, and he wrote this piece "Iran-Contra II," about all those meetings in Rome with Ledeen and Ghorbanifar and all them — is that in there anywhere?

Smith: Yeah, it is. It's in the investigation by the Senate that Ledeen and Iran-Contra figures were trying to turn attention toward Iran, and that Ledeen's connection to *Panorama* magazine and all sorts of things were investigated by the Senate, but by the time they got around to that in 2007, it was far, far too late. So I think there's still major questions about the provenance of the documents. Anyone who looks at the FBI file — if they know anything about it, like you and many others do — isn't going to come away from it satisfied that there was a complete investigation and that everything is known about who created these documents and why.

Horton: Now, this investigation, is it technically — legally, how it works, it's at the behest of the Senate, so it's not necessarily a criminal investigation or a counterintelligence investigation — it is just sort of a custom-made investigation at the request of a powerful senator? Or how does that work?

Smith: It was never a criminal investigation, correct. It was an investigation at the behest of the Senate Intelligence Committee. Whenever they're interviewing anybody, they always made it clear that it was not a criminal investigation. However, the directive and the investigatory scope did flow through the counterintelligence division in FBI headquarters, so they were conducting it kind of like a counterintelligence investigation. Yet, not really, because they were able to release the FBI file — and I can tell you from experience that they wouldn't have released a complete counterintelligence investigation if it had in fact actually been that — so they were more doing "light information-gathering."

The FBI's abilities to investigate anything in Europe are extremely limited, and as mentioned, the scope was supposed to be really broad. I mean, ideally, there would have been interviews with Ledeen tucked into this file, in these special forms that they used later to charge and prosecute people. They probably could have followed up with the Office of Special Plans, but they left that until later in the investigation because there was disagreement in the Senate Intelligence Committee. They didn't really want to handle the part of: "Why was the Bush administration so easily (allegedly) misled by the documents?" That was too hot for them, and although they did investigate Ledeen and crew, it was a very light investigation in 2007, which was very damning to the extent that this American Enterprise Institute figure and his cohorts really could move the government and get a lot of cooperation even though he's really not a government official. But they never did anything about that, and they never did anything in the Senate investigation on what the administration was saying and what the available intelligence could support. If they had done that, I think we might be in a different world.

Horton: Here's the thing, too. This is old news now, but back then, it was a big deal. I still remember that the CIA had rejected this intelligence, I think 14 times, and the administration for a long time, the different government actors, were sort of forbidden from bringing this up. The CIA had said, "Take this off all the talking points. Take this out of the speeches. We're not using it. This is not true." And then, on the 14th time, the neocons were successful in getting it into the speech, and it was the State of the Union. For whatever reason, George Tenet rolled over, and, if I remember right, he claimed, "Ah, well, I was asleep, and it was late at night, and someone else decided, so it's not my fault," or something like that kind of excuse.

But that's a big deal, right? The CIA says, "No, no, no!" over and over again, and the agenda is that somehow we've got to get it past them and into the speech. All that counts is to tell your mom she might get nuked in her jammies in the middle of the night if she doesn't let us do this.

Smith: Right, exactly. And the CIA did change its National Intelligence Estimate under pressure from the administration to include this idea that this dodgy allegation might possibly be true. So they're at fault there. But you're right, they managed to keep it out of speeches.

Horton: Oh, it was in the NIE, right? I had forgotten that.

Smith: Yeah, so they were complicit. But to your point, they did manage to keep it out of various speeches and allegations. But you had people like John Bolton, Dick Cheney, and others constantly trying to push this into the top official speeches. You had Condoleezza Rice talking about it. You had Bolton rejecting Iraq's claims of innocence by citing Niger specifically.

Horton: John Hannah and Harold Rhode in the vice president's office, too.

Smith: Right. My opinion, based on what I've read, is that they knew it was garbage, but they didn't care. And now we know, from Richard Clarke's book *Against All Enemies*, that President Bush on September 12, 2001, was already asking for links between Iraq and the 9/11 attacks. The more time that passes, the more evidence there is that the administration, from the get-go, really wanted to include Iraq in their response to 9/11, the facts be damned. My personal assessment, after reading all of this, is that they knew it was garbage, and they pushed it through the system anyway, because, as some have said, this was something the administration could all agree would be compelling to the American people as a justification for war.

Horton: Right, that's Paul Wolfowitz's own words.

Smith: Right, it's what they went with. It's like they were bringing a bad product to market, like this was a flawed pharmaceutical, but they were going to push it through to accomplish some quarterly earnings target.

Horton: This is probably a silly question, but could you ever get the CIA documents where they discuss why they don't like these forgeries and why they shouldn't be in the president's speeches?

Smith: The CIA is extremely difficult to FOIA. They have a million exemptions that they can apply, and their response to anything at that level is usually no response or that, if they did exist, they wouldn't be releasable. The so-called "Glomar response." The other exemption that they would cite would be deliberative, or that they're not going to show any sort of internal dissent. Because, of course, the plebs out there in the hinterland couldn't take the fact that the government isn't unified on any given issue. They really are difficult to FOIA. I don't think that the actual twisting that went on inside the CIA is ever going to come out. They managed to keep a lot of stuff under wraps by citing, accurately or falsely, a bunch of FOIA exemptions, so I don't think we're going to get that.

This is one of the older FOIA files that we've been nagging the FBI for. It's disappointing that it takes seven or eight years to get something as basic as: "Well, the FBI investigated. What did they find?" And here, we finally have some information, but not much. Why? Because the scope was so limited, and they're not very good at going up against high political appointees in any environment, but especially not this kind of environment. So I think the conclusion that we could very easily repeat this with Iran right now is a valid one, and I think that the lesson learned for administration officials who like twisting intelligence is: Don't ever let any source files leak out, especially stuff that's as easily debunked as these Niger uranium forgeries. We could very well be in a situation of going to war with Iran where they keep everything as "foreign source, but compelling" and never present a solid case. In fact, that's what it looks like is happening right now.

Horton: That's why Assange is in prison. We can't be having all this truth getting in the way of our agenda. It could be a problem!

Hey, here's some stuff I'd like to see leaked or FOIAed or something, which is perfectly in line with all of this: the investigation into Ahmed Chalabi. And how did Ahmed Chalabi know that America had broken Iran's codes in order that he might commit the treason of letting them know that? Which then led to the investigation — which was a joint CIA-DIA investigation on one side, and a criminal investigation on the FBI side — where the CIA and the DIA later decided that the Iranians had really helped Chalabi to lie us into war against Saddam, and that the neocons were his dupes, just as everybody else — on the right, anyway — were the neocons' dupes in that war.

But the real point is that on the FBI side of that investigation, which may be easier to FOIA, they did interview the Pentagon neocons, because I remember reading Douglas Feith complaining severely about, "Oh, it was just the worst Gestapo kind of activity, the way they came and persecuted us all." I think his quote was — I'm almost positive it was Douglas Feith who said this — "Oh sure, they just questioned every Jew at the Pentagon," and framed it in that way. So let's see those, Grant!

Smith: Okay, Scott, I'll put that into the old FOIA docket. But I do think that's an interesting defense, because one of the people who was meeting with Ledeen and some other figures, Ghorbanifar, et al., trying to turn the whole attention toward Iran in the context of this investigation, was Lawrence Franklin. Lawrence Franklin was working at the Pentagon, leaking information to Israel and Israel's U.S. lobby, AIPAC. When he was finally indicted, that's the sort of thing that, with the aid of mainstream media, he was charging the FBI with — an "anti-Semitic investigation" — and it's a very powerful charge. In looking through FBI files, whenever they were looking into things like the activities of Jonathan Pollard, in two previous incidents about AIPAC, in arms smuggling cases, that's the charge that always comes up, and they really don't like facing that sort of charge. It's a lot easier to just let the investigation go away because it's an extremely toxic charge.

Horton: Although at the time it wasn't getting that much attention, because everyone else was doing them the favor of ignoring the issue. So the smear part of it didn't get very far, either. To me, the only part that was important about it was that apparently the FBI was doing a pretty thorough interview of the members of the Office of Special Plans about what they may or may not have told Chalabi about America's access to Iranian secrets, which is a big deal.

Smith: They were dealing with a known fraudster in that case, and the real interesting thing is that Chalabi had a record of criminal behavior. It's amazing that anyone would have been able to get government funds flowing to such a character. The fact that he turned out to be even worse than was known? It's so predictable, but yeah, that would be an excellent thing to come out.

I just think that, as we keep hearing of dented oil tankers and 10,000 troops being sent to Iran, it's really worth revisiting all of the fabrications that came out of the White House in this previous era, where there was such a disdain for any sort of truthfulness toward the American people about what had actually happened. It's really valuable to revisit in detail every single one of these incidents, because people have forgotten. For most people, this is ancient history, this is like the Pentagon Papers and Daniel Ellsberg. This investigation started back in 2003. That was forever ago, so why does it matter?

Horton: Yeah. Well, it matters, of course, because of all the dead people, and all the actions taken based on these lies. This is just one example of how badly we know they knew they were lying, manipulating the American people to believe Saddam was an ally of Osama bin Laden, that he was making nuclear bombs, and of course that he would attack North America somehow at his first opportunity. These are ridiculous lies.

Smith: They are really ridiculous allegations. Anybody who knew anything about the logistics of moving this stuff, about the oversight of such things, the fact that Iraq already had that much yellow cake in their possession but it was under IAEA seal. The ridiculousness of the argument and the fact that the *New York Times*, the *Washington Post*, and all of these so-called credible newspapers were just all over this stuff, spinning out yarn after speculative yarn. And nothing has changed, absolutely nothing has changed, in the intervening decades.

Horton: Did you ever see that *Chappelle's Show* skit, where it's "Black Bush," and Dave Chappelle is George W. Bush lying us into war, only…

Smith: [Laughing] Of course! Who hasn't seen that?

Horton: Everyone should. Where he's got the yellow cake: "Oh, don't drop that!" Of course, also the aluminum tubes: "Got aluminum tubes! Aluminum tubes!" Like you're supposed to use your imagination about how that tube is supposed to be a threat to you. Then the funny part is — well, all of it's funny — but the reality of it, from even back then, was that you're missing a real big couple of steps between yellow cake and your centrifuges that Netanyahu said were the size of washing machines, hidden all over Iraq, enriching this uranium. You don't just put yellow cake in a centrifuge and spin it. That's not how it works. You've got to refine that and transform it into uranium hexafluoride gas first, which takes a giant conversion facility that obviously Iraq did not, could not, have had. As you say, what they did have, though, was a big storage locker full of yellow cake uranium from back when they did have a nuclear program in the 1980s, and it was just sitting there. So even if they wanted to take some yellow cake and begin to try to do something with it, they didn't need to go to Niger for that.

Smith: All you have to do is look at a picture of the Oakridge facility to see that it's a lot more complicated getting from yellow cake to highly enriched uranium. The funny thing about the Dave Chappelle skit — and which adds to the ridiculousness of the whole thing — was that they actually had some yellow cake, like edible yellow cake, and they're saying, "Watch out for this stuff!" It was brilliant humor, but we've become so deferential as a nation that you would never see a skit like that on *Saturday Night Live* or someplace that had a broader audience, because they know they'll get into trouble. I mean, questioning authority like that? Ridiculing some of these fraudsters who put this out with such a straight face? You still can't do that. Look at the esteem with which George W. Bush and Dick Cheney are held in many circles. Whatever they did is just irrelevant to most of the mainstream, who don't see it as their job to debunk this sort of thing or even follow up. Why is it that we're the ones to get the FBI Niger uranium investigation? It's very simple: Nobody else cares.

Horton: Yeah, certainly not in the mainstream media, and yeah, you're right, in all of D.C., their whole bubble, the consensus is: "Oh yeah, Iraq? I guess we probably shouldn't have done that, but we don't like to talk about that anyway… So, how's the local sports team doing?"

Smith: That's what you get when someone says "faulty intelligence." This was not a case of faulty intelligence.

Horton: "Certainly not anybody's fault, and it's certainly not mine," is what they'll say. I remember back then, when MSNBC, trying so hard to keep up with Fox, had counters full of toy airplanes that they had gotten from Toys "R" Us — and this one sort of looks like an F-15, and this one sort of looks like some kind of long-range bomber. The hosts of the show were actually flying them around in the air like little kids as the generals are explaining, for 40 minutes or something, "This is all the cool, high-tech gear that we'll be deploying," and all of these things. All the major questions were already answered about how necessary this all is. We're only waiting for it to begin for the excitement to start. It was essentially a Lockheed promo video, and boy, were they into it, every one of them! Anyone who was involved in that, there's hardly enough sorries for that. The only one I could think of was Peter Beinart, who wrote two different books about how sorry he was for mongering that war, but the rest of them? I can't think of many who have even tried to say they're sorry for being that bad and helping to mislead so many good people, too. Think about how many fathers and daughters decided they hated each other's guts because George Bush attacked Iraq and they couldn't get along over that issue. Never mind all the dead Iraqis. That hurt American society so deeply, launching that war for no reason. The reverberations, you can't even begin to somehow add them up. All the opportunity costs, all the broken relationships and all the everything.

Smith: Just think of what should have happened. There should have been an entire generation of reporters and government officials drummed out of office and drummed out of their so-called media organizations. The fact that it never happened, and the fact that we're beginning to repeat history and that there's no indication that it will be any different, that's the problem. That's the issue. It wouldn't have been worth it, but it would have been redeemable, if we had learned anything from it and if there had been any consequences. But to this date, other than who, Lawrence Wilkerson? Who's the highest-ranking official who's contrite and doing some form of absolution for the fraud that they perpetrated on this country? I can't think of anyone.

Horton: Yeah, Wilkerson. Certainly, none of the neocons.

Smith: No. There's a whole "Never say you're sorry" mentality with them, and I think that guarantees their survivability. And although I've heard on your show that John Bolton is not a neocon, and I accept the argument that he's just an ardent hyper-nationalist, he should be paying a price for the Niger uranium hoax. He's not. He's in fact got an incredible amount of credibility and power now, thanks to his patronage network. There's a lot of bad press that he's generated, but nowhere near enough, in my view, revisiting some of the things he did to perpetrate the last major war.

Horton: And like you say, it's because there never was accountability for any of these goons and the ones on TV, too. They can't point a finger at him when they regurgitated all of his lies. Which is kind of funny because Bolton does have a reputation of being crazy old John Bolton. He doesn't represent the consensus; he's to the right of even Robert Kagan sometimes. Yet, boy, when he says, "Who needs to get killed?" — down in Venezuela or Iran or wherever — the entire media establishment line up and salute, click their heels and declare in unison that finally, Donald Trump is acting presidential. In other words, instead of being a TV goofball, he's taking control of the U.S. government, and thinking of big, horrible things that he can do with it, like a good leader should. So they're the most grateful for Bolton, and they're willing to repeat any lie he would have them believe. "Oh, apparently the whole government and military in Venezuela is ready to switch sides this morning, everybody, because John Bolton says so," according to MSNBC and CNN and the rest of them, and then they're completely surprised when it doesn't work out.

Smith: That's what happens when there's this national inability to maintain any of this relevant history. So we're stuck. We're very stuck.

Horton: It's not just that now they're doing Russiagate; it's that this whole time, they were part of the "why we have to stay in Iraq and we can never ever leave or else you're a traitor who loves terrorists," "why we have to double and triple and quadruple the war in Afghanistan and surge in to save the people from themselves over there," "why we have to go and save the people of Libya from their horrible dictator Gaddafi and give them a nice permanent civil war instead," and also "why we need to go back al-Zawahiri's suicide bombers in Syria" — because somehow Assad, the guy in the three-piece suit, is worse than al Qaeda; in fact, al Qaeda are moderate heroes compared to him, because he's friends with the Iranians, not that they ever did anything to us. Then, of course, the secret plot between Vladimir Putin and his secret spy compromised agent Donald Trump to usurp Hillary Clinton's rightful throne and seize power in this treacherous and traitorous act from the rightful rulers of our society. My God, man! These people are so lost, they're so upside-down on all of what they really think is true.

You know, I've seen this interview with Donald Trump where he talks about, "Yeah, yeah, yeah. I know, I am anti-war. Except Iran, Iran, Iran. The generals tell me, 'Iran, Iran, Iran.'" He didn't say, but he meant: "The Israelis also tell me, 'Iran, Iran, Iran.' They're going to make a nuclear weapon. They're going to take over the Middle East. I have to do this because they tell me there's a good reason for this."

Meanwhile, it's true that everyone that Donald Trump talks to tells him that that's right. Nobody will tell him that this is all a bunch of crap. The whole narrative is a lie, just like what George Bush would tell you. It's not true, none of it. The Ayatollah, he ain't so bad. Our enemies are the guys that knocked our towers down, the suicide bombers, the bin Laden-ites, and the best way to kill them is to ignore them to death; certainly, don't back them.

To blame all that on the Ayatollah is just crazy, but there's no one available, other than, I guess, Colonel Macgregor on Tucker Carlson's show, to tell Trump a different point of view on any of this stuff. Again, it's because of the same lack of accountability from the last time and the time before that.

Smith: Right. This is our system, and it's not going to produce any other results. I really do think that you're completely right about him existing in this bubble of people who really aren't able to give him any information that isn't predetermined. The complete lack of any sort of alternative view or assessment that's being put out is just terrifying. The position of people who are against this? They're marginalized because they're going to be positioned as Iranian government lovers, people who hate the will of the people over there, or Ayatollah supporters. It's just the "Iraq redux," where everyone who opposed the war at that point was positioned as supporting Saddam Hussein, and who can support him? There's very little nuance. Absolutely no unique opinions are allowed in the media.

So I guess the job at hand is to continue putting out some research and information so that people can at least get it from somewhere, and I hope people read "Lessons from FBI's 'Niger Uranium Forgeries' File" at Antiwar.com because it pretty much brings things up to date, drawing in some other insights as well. In particular, there's a pretty good book from a couple of reporters called *The Italian Letter: How the Bush Administration Used a Fake Letter to Build the Case for War in Iraq*, which was a book I wasn't aware of until recently. It's just extremely thorough and damning and never got the attention it merited in terms of doing all of the groundwork to show the timeline of admissions, reports to the press, what they were gathering and discrediting until they revitalized it. It should be required reading in every foreign policy or international relations course, but of course it won't be.

Horton: You say in your article that Rocco Martino was a former cop recruited by Italian intelligence to come up with this. Did anybody ask the question of whether Americans had been meeting with them recently, and that this was either definitely or apparently at the Americans' behest in the first place? Like, "Find us somebody to cook up some of this stuff. We're shopping for a bill of goods, and we need your help."

Smith: No. If the FBI had been doing its job, it would have thrown Michael Ledeen up against the wall. Ledeen had connections with both SISMI and *Panorama.* As far as I could tell, they never actually interviewed Rocco Martino either. He's a guy who changed his stories multiple times.

Horton: When was the first Rome meeting? Do you remember that?

Smith: You're talking about the first Rome meeting between Martino and Ledeen? I believe that was 2000. It was only one meeting, but it was Ghorbanifar, Ledeen, some SISMI intelligence officials and Lawrence

Franklin, the convicted spy for Israel. That meeting was the subject, finally, of one of the last Senate investigations, but it didn't benefit from any participation of hardened FBI officials who were actually doing a criminal investigation; it was merely based on some interviews in which he denied anything. That meeting, Justin Raimondo wrote about it, and a lot of other people wrote about it, as being the possible genesis of the forgeries. That was the point of the meeting, but they were pretty much exonerated in the Senate investigation of that particular meeting. There's a 60-page document from the Senate Select Committee homing in on that. I'm sorry, it was December 2001 when that happened.

Horton: Ledeen has always wanted to hit Iran much more than Iraq, and he was more on the Ariel Sharon side of the argument, whereas Netanyahu's guys preferred to hit Iraq first, I guess. You know, my wife did this report, where this was one of the plots that Ghorbanifar was in on, where they were trying to frame up Iran and Iraq both for having some uranium. Although the thing fell apart, I guess. But that's one of the things that they were working on, trying to frame Iran for selling some uranium to Iraq, or something like that.

Smith: Yep, and they had Pentagon heavy-hitter Harold Rhode from the Office of Net Assessment. That was an extremely suspicious meeting because they were there to take advantage of the 9/11 attack, but it didn't precede the actual base event that was used to gin up this idea that Iraq was shopping for uranium in Niger. That was already happening shortly after this Wissam al-Zahawie meeting in Niger on February 1, 1999. They were starting to spin that in the year 2000, even before 9/11. So that sort of work was not going on, as far as the investigations have determined, as a result of trying to spin the 9/11 attacks. That was occurring and being built, but it did get a major kicker. It became something that they could really try to use.

The book that I used for some of the background because it was very up-to-date was *The Italian Letter: How the Bush Administration Used a Fake Letter to Build the Case for War in Iraq*, and when they talk about the Italian letter, they talk about a completely fabricated letter between the president of Niger and Iraq talking about a numbered bill of sale and ensuing shipments of uranium. It was completely fake, and yet it was the single piece of so-called "hard intelligence" that the Bush administration and the Brits were saying was indisputable proof that the nuclear program was being reactivated. That came out in 2007. It came out from a minor publisher called Rodale Books that generates magazines like *Men's Health* and *Gardening Today*. I think it was the kind of book that you couldn't get a mainstream publisher to go with — which is an entire tale in and of itself. And these are two reporters who were mainstream reporters, and they just wanted to get to the bottom of it, Knut

Royce and Pete Eisner. It's just a fascinating book, and I don't think it ever received the attention it deserved.

Horton: In that book, when they go up the chain of cause and effect there, are they stumped or they are satisfied when they stop at the SISMI agent who recruited Martino? They don't have any indication either way as to whether the Israelis put him up to it, the Americans, Ahmed Chalabi's group, or anybody else?

Smith: They don't. They didn't get higher up than that. They didn't actually interview the SISMI agent. They didn't.

Horton: Even chronologically speaking, there's essentially, as far as anybody knows, nothing really to report from further up the chain just in the timeline?

Smith: Yeah, not from their timeline. And their timeline in the appendix of the book is extremely lengthy and detailed. Their story, essentially, is that SISMI was setting up Martino to spin this connection, and there's no motive given whatsoever. Martino was always in it for the money, their plant at the Niger embassy was in it for the money as well and could fax things back and forth and get letterhead. Then they reported a burglary, which you'll remember. But there's nothing in there saying, "And here's the case for why Italian intelligence…" By the way, no Western intelligence agencies really rely on Italian intelligence, because they're so inept and incompetent and compromised. No one explains why they were the generator of this operation. It's very unsatisfying, because you would think that a bona fide, high-priority investigation would have left no doubts. But I still have doubts, and I don't think anyone who has examined everything wouldn't still have doubts about why a third-rate intelligence agency would, before 9/11, be rushing to inject intelligence. It's based on truth. The kernel of truth is the visit by the Iraqis to Niger, and they're continuing to spin it as this credible evidence of reconstituting a nuclear program. There's no satisfaction to be had.

Horton: What a crazy thing, to think that the American people were somehow convinced that we had to attack Iraq before they attacked us first. Maybe with nukes! It is, especially in hindsight — you've almost got to be kidding me that this worked — that they said, "Look, our policy is regime change, and we're going to lie to you and your family for the next year and a half until you're afraid enough to let us start a war," and then that worked instead of not working.

Smith: Yeah, something like 90 percent of Americans believed that Iraq was developing nuclear weapons. Even after the inspectors didn't find anything, a significant number still believed it. That's the problem. Once you've made all these connections, and it's coming from everywhere — the government and media — it really has an impact, and it only slowly dissipates..

Joe Cirincione: Syria, North Korea, Pakistan, Iran and the Bomb
October 8, 2008

Scott Horton: Alright, folks. Introducing Joe Cirincione. He's the president of the Ploughshares Fund and is a nuclear proliferation expert. You can also find what he writes at the *Huffington Post.* He's a pretty well-known policy wonk there in D.C., formerly of the Center for American Progress and the Carnegie Endowment for International Peace. He's the author of the books *Bomb Scare: The History and Future of Nuclear Weapons* and *Deadly Arsenals: Nuclear, Biological and Chemical Threats.* Welcome back to the show, Joe.

Joe Cirincione: It's a pleasure to be on, thanks for having me.

Horton: It's very good to have your expertise here, and I really appreciate you lending them to the radio show today. The last time we spoke, the primary topic that we were interested in covering was the bombing, a little more than a year ago now, of a so-called "reactor" in Syria by the Israeli government. It was a very strange thing because the Syrians didn't really complain about it. Everybody was sort of quiet, and the accusation from the War Party has been that there's no doubt that this was meant to be a weapons-grade plutonium-producing nuclear reactor and that this so-called "preemptive strike" was justified. I don't really know all the details, but I'm sure a lot more evidence has come in since then. I know that the IAEA has gone and done an inspection, and supposedly they were expecting to find graphite spread around everywhere and found none. I just wonder if you can update us on what we all need to know about what that building was. Was it a nuclear reactor? Did the North Koreans help them build it? Were they giving weapons-grade plutonium to the Iranians? What's going on there?

Cirincione: I'm happy to. You're very well informed. You summarized it very accurately. At the time of this mystery strike in the desert — this was Israeli F-15 fighters hitting a target not too far from their borders — against a mystery building that Syria had constructed in the desert and had gone to great pains to hide, was blown up by the Israelis. It gradually made its way into the paper. It wasn't that the Israelis or the U.S. were eager to talk about it, and the Syrians, as you say, didn't complain about the strike.

At the time, I was extremely skeptical. Syria did not have a serious nuclear weapons program. They don't have the financial or industrial or technical basis for maintaining a program. So I thought this was crazy that it would be a reactor. Well, just because something's crazy doesn't mean the state won't do it, and there is now pretty strong circumstantial evidence that something nuclear-related was going on there. That's based primarily on photos that the CIA released showing what looks like a reactor vessel. This is a very distinctive kind of structure. We don't know for sure whether those pictures are genuine, but they appear to be. It does appear that something was going on there, and if Syria were doing this, the likely supplier of this technology was North Korea. But nothing is known for sure about this.

As you say, the IAEA inspectors went to the site, and I just spoke to an IAEA official this week about this, and they were looking for signs of a reactor. There wouldn't be any radioactive material, but there would be graphite. Graphite is like the stuff in a pencil that is used as a moderator. It's put in between the fuel rods of a reactor, and some of that would have been left there after the destruction. They couldn't find anything like that. So this remains a mystery. I would say that the evidence is circumstantial but strong that something nuclear-related was going on there, but we still don't know. Whatever it was, it was very far away from any bomb capability. This was a preliminary facility, not something that posed an immediate threat.

Horton: A couple of things to follow up on there. First of all, how convincing were those pictures? Because I thought there was a whole scandal, and that people were laughing at these pictures in the *New York Times*. Some of them were clearly Photoshopped. I mean, they were as bogus as could possibly be — at least a couple of them, right?

Cirincione: There was a scene of a Korean official next to a Syrian official, and this was part of the evidence that was presented that did look funny.

Horton: I was thinking of the pictures of the buildings, too. No?

Cirincione: This is controversial, and as I said, we don't really know. So I don't dismiss anybody's questions about these photos, and they do look a little funny. I've talked to several reporters who have dug into this on both sides, reporters who were convinced that it really was a nuclear facility, and there's no question about it, and they've spoken to senior officials in the administration that have no axe to grind here that are not of a neocon persuasion. These officials are convinced that there was a nuclear facility. I've talked to other reporters much more skeptical about this, who still insist that there are people in the intelligence community who think this is a fraud and a hoax. I'm just telling you what I know, and that is that we don't know. The jury's still out on this.

Horton: Satisfy my curiosity on this point if you could, please. Aren't the North Koreans completely inept here? I mean, they're harvesting plutonium from a reactor that the Soviet Union built for them back when there was such a thing as a Soviet Union. How can it be that the North Koreans are helping the Syrians build a nuclear reactor? I mean, seriously, come on.

Cirincione: That's right, and it's actually a British design. This is very old technology, but so are their missiles. Their so-called Nodong missiles are based on Scud designs that are based on Nazi Germany designs. This is very primitive but workable technology. So even though it's old and we wouldn't be doing it, we wouldn't be building it, that doesn't mean that they're not doing it. We do know there were North Koreans that were there. The North Koreans have a quirky country, but they do trade in missile technology. There's some indication that they've traded in nuclear technology.

Here's the key issue for us. This has become an issue in the U.S.-North Korean relationship because we're engaged in talks to have them end their nuclear program, and the U.S. officials have confronted the North Koreans about this, and what they say is, "We won't do it again." They don't admit having ever done it, they don't acknowledge any trade with the Syrians on this, but they said they won't do it again. I just spoke with former Secretary of Defense, Bill Perry, at a seminar we sponsored up at the University of Maryland, and he said, "Well, that would be good enough for me." One of the problems we're having is that some in the U.S. government are insisting on a full confession from the North Koreans, or else they don't want to go ahead with the negotiations. This is the kind of thing that we will learn about eventually, but it's going to take a lot of steps. It's going to take a new climate of trust between the two countries before the North Koreans will come clean on the whole history.

Horton: With the limited information available, do you know of any evidence that the North Koreans actually imported uranium enrichment centrifuge technology from the A.Q. Khan Network? Then secondly, is there an atom of evidence anywhere in the world that the North Koreans actually used any of that uranium enrichment equipment? Because supposedly this was how they broke the Agreed Framework with the United States before the United States broke it with them. But I believe it became a scientifically proven fact that when they tested their nuclear weapon a couple of years ago it was made out of plutonium, not uranium. It was harvested from their old Soviet reactor, right?

Cirincione: Absolutely. So there's two ways to make a bomb, and in World War II, we used both of them. You can use highly enriched uranium, or you can use plutonium. Highly enriched uranium these days is made through centrifuges, which is what Iran is trying to do. You spin uranium gas around and you enrich it so it's of purer and purer quality until you have a quantity that's good enough for a bomb.

The other way is to build a reactor, and this is what the North Koreans did. You build a reactor, you radiate fuel rods, and in that process, some of it is turned into plutonium. You then can extract the plutonium from the fuel rod. If you have somewhere between four and eight kilograms, you could make a bomb. That's the North Korean route, and you're absolutely right, the bomb they detonated in 2006 was a plutonium bomb, and all our estimates of their capabilities are based on how much plutonium we think they've produced. So we think they have enough for between six and ten bombs. The evidence is pretty clear that they did buy some uranium enrichment technology from the A.Q. Khan Network. We know this because parts of the Khan Network told us they did this. They told this to the Pakistanis who told it to U.S. officials, and when confronted with this evidence back in 2002, they admitted it.

Horton: Is that a fact that they admitted it? Or is that what John Bolton said that they admitted it?

Cirincione: It's a little unclear because there's a real question of translation, and now the North Koreans say they never admitted it, but I've actually spoken to some of the officials engaged in this. The people I've spoken to, they're not neoconservatives, they don't have an axe to grind, and they felt that they were being told yes. But they were being told: "Yes, we do have that, and let's talk about it. Let's put that on the table." So the North Koreans were willing to make a deal.

What happened was that the Bush administration in 2001 used this as an excuse to break a deal they never liked. They didn't want to be negotiating with North Korea. That was against their whole ideology, their whole foreign policy. It's still one of the big divides in U.S. foreign policy between, for example, Senator Obama and Senator McCain. It's summed up by Vice President Cheney who in 2004 said, "We don't negotiate with evil; we defeat it." They didn't want to be negotiating with Iraq, Iran, or North Korea; they wanted to topple those regimes.

Horton: But on that particular point, it wasn't just John Bolton who claimed some anonymous munchkin told him that the North Koreans told him at a cocktail party. Is there actually some real indication besides John Bolton's assertion?

Cirincione: Yeah, there really is. What happened was that Assistant Secretary of State Jim Kelly, who was then the negotiator with North Korea, flew to Pyongyang and he had this information on the uranium enrichment technology. This was in 2002. He confronted the North Koreans with it, and there was a general hubbub. There was a recess in the talks. The North Koreans came back, and they admitted it to Kelly. That was unexpected.

Kelly was on a very tight leash, and there were very strict instructions about what he was going to do, which were given to him by the National Security Council where, at the time, the neoconservatives were in control. He didn't have talking points for what to say if the North Koreans admitted it. They thought they were going to deny it. He wasn't authorized to continue the conversation. He had to leave, come back to the United States, report what happened, and that's when the whole deal unraveled and the whole thing fell apart. The North Koreans started processing more plutonium, they kept taking step after step, and the whole thing got worse and worse until it led to the detonation of a nuclear bomb in 2006. Only then did the administration start to reverse course. By that time, Donald Rumsfeld was out, the administration had lost the 2006 congressional elections, and the whole balance of power in the Congress had shifted. There was tremendous pressure to start negotiating again with the North Koreans. They sent a new guy in, Assistant Secretary Chris Hill, who did a great job and was able to start the talks up again and got us back to where we were when the Bush administration came into power — that is, a freeze in the North Korean nuclear program. Then he went even further and actually got them to start dismantling the facilities. That was progressing pretty good up until the most recent hiccup, and now it's starting to unravel again.

Horton: What I want to get to is the order of events here, and hopefully I can get you to help perfect my understanding. If I remember this right, I'm not exactly sure what steps have been taken by the North Koreans, but they did something right, and so then Bush announced that he was going to remove them from the state sponsor of terrorism list. They said, "Great," and tore down their cooling tower. Then Bush said, "We're not taking you off the terrorism list," and, as best I understand, nothing had happened in between then that the North Koreans had done that gave him an excuse for that. Then the North Korean response was, "Oh yeah, well, we'll rebuild our cooling tower, and we don't want to be off of your state-sponsored terrorism list anyway, pal." Is that right?

Cirincione: You're exactly right, Scott.

Horton: [Laughing] Ah, man! You were supposed to say, "No, no, no, Scott, you missed an important point there."

Cirincione: The *New York Times* might put it a little differently, but your facts are exactly right. The way to understand this is that you have to understand that there's always been a struggle within the administration between the pragmatists and the hardliners. The pragmatists like Colin Powell or Condoleezza Rice want to make a deal. If you let them do what they can do, they will get us a deal, and indeed they have gotten a deal. The hardliners don't want any part of it. So the struggle continues, and every time we get a deal, the hardliners are trying to kill it. And by the way, there's similar struggles going on inside the North Korean regime.

So this is a very difficult set of circumstances. Here we are, cruising along with this deal, they're blowing up their cooling tower, U.S. inspectors are all over the North Korean nuclear facility, we're poking holes in the reactor, we're building up relationships, and we're starting to get a history of everything they did. They hand over 17,000 pages of information on their nuclear program. Now there's a lot of information there, but not all of it. We know they're not telling us everything. It's at that point where we're supposed to take them off the terrorism list, which is just a piece of paper called the "State Sponsors of Terrorism" list. There's only 4 or 5 countries on it: Cuba, Syria, Iran, North Korea.

We're supposed to take them off a piece of paper, that's all. We don't do it. The administration comes back, led by acting Undersecretary John Rood of the State Department, with a list of demands to verify what they told us in the declaration. The North Koreans freak. They say, "We're not Iraq. We're not going to open up our country to dozens and dozens of U.S. inspectors coming in. This was not in the original agreement. This is a new demand that you're introducing. We won't do it."

Eventually, they came back and they actually agreed to about 70 percent of what we want. That's not good enough for the administration, the whole deal grinds to a halt, and the North Koreans are now step by step starting to go back to rebuilding and re-opening some of the facilities that they'd been shutting down. Still, the way they're doing it indicates to me that they want to make a deal. They're not going pedal to the metal. They're not going all-out. They're doing it step by step. They want to make a deal, and I actually think there's a chance in the next couple of weeks to get this deal back on track because George Bush himself, as president, wants to leave office with foreign policy successes. North Korea might be the one foreign policy success he's going to be able to claim. I think that pressure might push us back to the negotiating table.

Horton: I hope that's true. It's clear he can't claim Libya because they had been sucking up to the West for years trying to normalize relations. Bush refused to allow them to until after the Iraq war so that he could pretend to give the Iraq war credit for that. So that doesn't count as a victory at all, and really, since everything was fine when he took office with regards to the North Korean issue, the best he could do is give them a better welfare deal than they had when he took power. So even that would be a failure at best, right?

Cirincione: The deal that we're making now is the one that was there in January 2001, when he came into office. If he had let Colin Powell do what he wanted to do, which was continue the negotiations that had been started under President Clinton, we wouldn't have a North Korean nuclear program now. They completely screwed it up, and now they're trying to walk it back and repair the damage. So you're right, North Korea is a failure. Iran is a failure. Both North Korean and Iran have made more progress in their nuclear programs this decade than they made in the previous two combined.

Horton: Speaking of Iran, I noticed you called it a nuclear program and not necessarily a nuclear weapons program. Was that a deliberate omission?

Cirincione: Yes, that's my evaluation. This is actually now the National Intelligence Estimate (NIE) on Iran. Our intelligence agencies agree that Iran does not have a dedicated nuclear weapons program at this point. They do have a civilian nuclear program by which they're acquiring the technologies and knowledge that would allow them to build a nuclear weapon sometime in the future, should they decide to do so. I think what the Iranians are doing is trying to do the whole thing legally and openly: acquire the nuclear technology, build the centrifuges, build a facility to enrich fuel, and then decide down the road if they want to turn that facility into a bomb program.

Horton: This is basically the position the Japanese are in right now, for example. They could begin to make a bomb and have everything ready in six months or so, but they haven't done that.

Cirincione: That's exactly right. Japan is the only other country in the world that has done what Iran is doing, and the Iranians point to Japan as the only non-nuclear weapon country that has the capabilities to build a nuclear weapon. All they would have to do is try to do so.

Horton: First of all, did I understand the NIE right? Secondly, was the NIE right that if they perfect their enrichment capabilities — under the presence of the IAEA inspectors, they're currently only enriching to low electricity-grade rather than above-90 percent pure uranium-235 — it is a matter of six months, or a year, or two years before they would have enough to make the first atomic bomb?

Cirincione: It's a question of two things. The way this works is that these factories are built to make fuel. And as I say, besides Japan, every country that has a factory like this also has nuclear weapons, So Iran would become the second state outside the nuclear weapons states to get one. These factories enrich uranium to low levels for fuel, between three and five percent enrichment. The trouble is that the very same machines could just keep going with some slight reconfiguration and enrich it to very high levels, to 90 percent purity, which is what you need for a bomb. So it's a question of trust, and one of the techniques that have been set up over the years is that you have inspectors in the facilities sampling the gas, watching what's going on, looking at the configuration. If Iran were to use the existing facilities now, we would know it because we have inspectors there. The problem is what happens, say, a year from now, when the whole thing is finished, they kick the inspectors out, declare they're leaving the Non-Proliferation Treaty, and they go to build a bomb? They might have enriched enough low enriched material that they could just feed that back into the machines, and somewhere between six months and a year of doing that, they could have enough material for a bomb. Nobody knows for sure, but they could be somewhere between two and five years away from having a bomb.

Horton: That's still, right there, you said it. The worst-case scenario is that they announce to the world, "Inspectors, get the hell out. We are now about to start making bombs. Give us about a year or so, and we'll get back to you." Is that the worst thing that could happen here?

Cirincione: That is the most likely worst case. There is some suspicion that they might have a secret facility someplace. This is highly unlikely. We have no indication that they do have a secret facility, and it would be extremely hard for them, with their limited capabilities, to have a duplicate facility someplace else. But I wouldn't completely rule it out. So, for all practical purposes, the worst-case scenario is that they continue to build their enrichment facility at Natanz and then, at some point in the future, kick the inspectors out and leave the treaty. We would know what they were up to then, and the countries of the world could decide what kind of action they wanted to take.

Horton: Has there been any information uncovered by the inspectors that would lead you to believe that the Iranians have pursued a plutonium bomb of any kind? Because this goes back to the Syrian accusations that they were going to make plutonium and then hand it over to the Iranians to make a bomb out of it — again, a totally different kind of bomb than enriched uranium.

Cirincione: The Iranians are constructing a reactor capable of producing plutonium. It is at a place called Arak. So this is another path to a bomb. This reactor does have civilian purposes, and they say that's what they're using it for, but this looks awfully similar to what you would do if you wanted to build bombs. That's why most of us are suspicious of Iranian intentions. There are factions inside Iran, and there are some who want to go build a bomb. President Ahmadinejad is one of them. I don't think the government has actually decided to go do that, and so I believe we still have time to make a deal with Iran. We won't know whether such a deal would work unless we actually try, and we just haven't tried.

Horton: A couple of things there. First of all, it's much more difficult to detonate a plutonium bomb, isn't it? That's why the North Koreans' bomb only kind of half-fizzled, right?

Cirincione: It is hard. It's a tougher technology. The bomb we dropped on Hiroshima was a uranium bomb. It was what they call a "gun-assembly device." You put one chunk of uranium at one end, you put another chunk of uranium inside a tube that's six feet long at the other end, and you accelerate one chunk at the other. It's a very simple design, probably the design a terrorist would use. It's sort of a first-generation bomb.

A plutonium bomb is much more difficult. You have to shape the plutonium into a sphere surrounded by conventional explosives that have to be detonated with exquisite timing in order to compress that sphere into a smaller sphere. Once it hits critical mass, it detonates. It's like turning a basketball into a baseball with explosives. It's much harder to do. The North Korean design didn't work very well. That's why they got a much smaller explosion than they expected. So having the material is the hardest and most difficult step, but it's not the only difficult step. You then have to get a design that works and test it.

Horton: You brought up the possibility of a deal. Of course, there's a new report out again — we seem to learn all these things from Flynt Leverett and Hillary Mann Leverett from the National Security Council — that the Iranians were cooperating with us as much as they could on al Qaeda. They've been sending, I believe they said, 3,000-something Arabs back to their home countries. And of course, we know that there was what Gareth Porter called the "Burnt Offering" in 2003, where the Iranians offered to put everything on the table. They were willing to negotiate the nuclear deal, their relationship with Hamas, Hezbollah, Iraq and everything else.

How likely do you think that deal is? And secondarily to that, if they are absolutely crazy, if they refuse to deal with us in any way, and if they are hell-bent on obtaining nuclear weapons, would you think that that's so objectionable that America ought to launch a preemptive war to avert their nuclear program?

Cirincione: Okay, well, let's walk through it. First, the deal was real. I have talked to the Iranian official who wrote the deal. I've talked to the State Department's Flint Leverett who knew of the deal. I have also talked to the Swiss Ambassador who transmitted the deal. These were real people who aren't making this up. This was in April 2003, right after we invaded Iraq. The Iranians, then with the reformist president, Mr. Khatami, offered to talk about everything: the nuclear program, relationship with Israel, and support for Hezbollah and Hamas, both of whom were much less powerful than they are now.

This was one of the great missed opportunities after 9/11. It was a chance when we could have had everything. We could have made a deal with these guys, fundamentally changed the relationship, and there would've been no nuclear program. We weren't interested, we didn't even answer, because we thought we were on a roll. John Bolton, Dick Cheney and Don Rumsfeld thought that the overthrow of Saddam Hussein was just the beginning. They thought we would have serial regime change in the region. There were some who were even talking about overthrowing the North Korean regime. This is the way we were going to remake the world. It was a miserable failure, a complete failure of judgment. The Iranians felt insulted, spurned and ignored, and it sort of solved the debate inside Iran. They said, "Okay, there's no dealing with these guys. There's no pleasing them. We have to go ahead," and they started accelerating the nuclear technology program.

Could we still make a deal? Yes, I think we could, but the price has gone up. They're getting stronger while we're getting weaker. I think we could have shut it down to no centrifuges back in 2003, a few centrifuges back in 2007, but now we're going to have to have some kind of compromise that allows them to continue some kind of centrifuge activity. We would want to keep it as limited as possible until we can improve the entire relationship, and maybe then walk the program back or shut it down. But it's going to be a difficult set of negotiations, probably the most difficult that we have to conduct.

Assuming that we get a president who actually wants to talk to them, if those negotiations fail and they go ahead and get the bomb, I think attacking Iran at that point would be suicidal for us. It would put a third war in the region, completely alienate broad sectors of the Muslim populations around the world and jeopardize U.S. strategic interests for generations. I think you'd have to fall back to a policy of containment, trying to contain a nuclear Iran the way we've contained a nuclear Soviet Union, a nuclear Communist China, or for that matter, a nuclear Pakistan. It's not preferable, I don't want to be in that position, but that would be the unpleasant choice you would have then.

Horton: On the deal, wasn't there just news in the last week or two that the Iranians actually indicated to some degree that perhaps they actually would be willing to negotiate a situation where they import low enriched uranium from outside the country? I mean, this is the sticking point of all sticking points, right, Joe?

Cirincione: It is, and at the Ploughshares Fund, we fund people who do work on areas like this. We fund several efforts at what's called "Track II diplomacy," which is people outside the government talking to Iranians, some of whom are in the government and some of whom are outside, exploring negotiated compromises. Those sets of private, off-the-record negotiations have led many of us to believe that there are openings there, that there are pragmatists in the Iranian government who want to make a deal. The right combination of pressure, keeping up the sanctions, threatening more sanctions, and incentives. Incentives such as offering them a way for them to be part of the game in the Middle East, to sit down at the negotiating table, to be part of a new security arrangement in the Middle East, recognizing them, and leading a process that would lead to mutual recognition between Iran and the U.S., which would lead to the return of Western investments in Iran. That combination of pressure and incentives could work to convince Iran to give up its nuclear program. It is certainly worth a try now.

Horton: Forgive me, because I know this isn't your area of expertise, but you obviously have very sound judgment on these issues. I wonder whether any of the people that you look to as sources of credible information and insight also believe that you can't deal with these people because of the eschatology of the Shiite 12th Imam? Does anybody take that seriously where you live? Or is that simply propaganda for the rubes out there?

Cirincione: Iran is a big country with lots of political factions, and there's no question that there are radicals in the government — I would include Ahmadinejad as one of them — who don't want to make a deal, who do see Iran as having a messianic destiny to lead the rise of a Shia nation and are heavily religious, who see all this as just the earthly dimension of a greater spiritual struggle. Yes, those people exist. I wouldn't actually include the Supreme Leader, Khamenei, in that. Even though he's a religious leader of the country, he's much more practical, much more pragmatic, than some of the secular radicals like Ahmadinejad.

So, just like you have in this country, just like you have in other countries, you have zealots, and you have fanatics. The question is finding the people you can make a deal with, and doing what you can to strengthen them and limit the fanatics.

Horton: And of course, when it comes to foreign policy, military policy, diplomacy, and everything else, it's the Ayatollah who decides, not Ahmadinejad.

Cirincione: That's exactly right, and it's not even clear that Ahmadinejad is going to survive the March elections next year. He's very unpopular inside Iran, where youth unemployment has hit 50 percent, inflation is 30 percent. I can't imagine what this global recession we're entering into is going to do to Iran. Things are tough inside Iran, and Ahmadinejad is getting the blame for a lot of it.

Horton: Before I ask you about the India deal and Russia policy, if I can fit those topics in at the end of this interview, I want to ask you something else. I read somewhere that you, at least for a time if not currently, were advising Barack Obama on some of these issues, and I wonder if you've taught him that there's such a thing in the world as the IAEA and reports that they've issued that say, for example, that they can "continue to verify the non-diversion of any nuclear material to a military or other special purpose"? Because right about a year ago, in a debate he said, "They're making nuclear weapons, nobody thinks they're not making nuclear weapons," and Dennis Kucinich, in the debate, said, "You're wrong. You must have never heard of Mohamed ElBaradei or the IAEA before, but there's no evidence whatsoever that they're making nuclear weapons, and you don't know what you're talking about." So Joe, does he know what he's talking about yet? Because when I turn on TV and he talks about Iran, he sounds to me as though he's at least pretending he's as well informed as John McCain on this issue.

Cirincione: In my personal capacity, outside of my Ploughshares duty, I have given advice to the campaign of Barack Obama, but I've never met the Senator. I'm not a senior adviser by any means, and he has hundreds of people advising him, and the non-proliferation team is made up of some of the top experts in the field that are advising the senator. Everything that I've heard him say is quite well informed, and he does get this issue. You'll notice when he talks about it, he talks about the changes that we need in the world, especially when he's talking like he did during the foreign policy debate and brings up the issue of eliminating nuclear weapons. So this is part of his view of the change that we need, and one of the encouraging things that I've seen in this election is that John McCain has also adopted very progressive positions on nuclear policy.

Horton: Really?

Cirincione: Absolutely. You can go and see his stuff. He made a speech out here in Los Angeles on March 26th, where he said that as president, he would have the United States lead a campaign for global nuclear disarmament. Those were his words, "nuclear disarmament." I can't remember the last time a Republican candidate for president used those words, let alone thought they might help him get elected. He then followed that up a month later with a very comprehensive speech where he broke with the Bush administration on several policies, including negotiating deep reductions in U.S. and Russian arsenals. So both the candidates recognize that the existing strategy has failed and that something new has got to be tried.

There are questions of degree about how far each would go and how much of a priority each would make it. I've been in panel discussions where they had representatives from both campaigns, and I would say there's much more agreement on what the next nuclear policy should be than there is disagreement. One of the big areas of disagreement is over this issue of negotiating with adversaries like Iran, North Korea, etc.

Horton: Well, forgive me if I fail to see the nuance in their explanations of the situation between America and Iran that I hear in your statements on the issue, sir. I just don't hear that at all.

Cirincione: There are certainly issues that get too glossed over, but people can look at the two campaigns to make up their own mind.

Horton: So you don't hear them speaking as though it's an assumed truth that there's a nuclear weapons program in Iran?

Cirincione: There are many U.S. officials who talk about the nuclear weapons program and Iran, that's for sure. I don't remember if Senator Obama talked about it that way, but I must admit, I wouldn't be surprised.

Democrats are often very skittish about defense issues during campaigns and don't want to appear weak, so they usually err on the side of talking tough. I wouldn't be surprised that Obama talked that way.

Horton: Well, Joe Biden in his one debate with Sarah Palin didn't get into particulars, but he did say, "We have some time now. They're not about to have a nuclear weapon in a year, like you might read at *Newsmax*." So we'll at least give Biden a couple of points for that.

I want to ask you about Russia, and this will be a two-part question. There was the Nunn-Lugar effort to get rid of any excess weapons-grade nuclear fuel in old Soviet states. The Russians, at least supposedly, got ahold of all their actual nuclear weapons and have those back in Russia, but any other old nuclear fuel from the former Soviet satellites, America was going to buy all that stuff up and keep it out of the hands of anybody else, proliferators of any description. I wonder if you can comment on that.

Then also, I wonder if you can comment on the evolution of our military posture. Apparently, from the point of view of the posture writers, we're in a position where we could have a nuclear first strike on Russia and not risk Mutual Assured Destruction. I believe part of that is the construction of a new generation of so-called usable nuclear weapons for use in a situation like that. Am I anywhere near the ballpark?

Cirincione: I teach a course at Georgetown, and I think I devote a class to each one of the subjects. These programs are called, collectively, the "cooperative threat reduction programs," and we've had them with Russia for 15 years. They have been some of the most successful national security programs we've ever run. It's unprecedented, where a former adversary opens up their doors and allows us to come in and dismantle the weapons that threatened us for decades. Yet that's what the Russians agreed to do.

We ponied up most of the money, at a cost of about a billion dollars a year, and the good news is that we've dismantled and secured about half the material in the states of the former Soviet Union. The bad news is that we haven't gotten to the other half yet, and that's the flaw in these programs, they've been moving much too slowly. Now, as U.S.-Russian relations deteriorate, the Russians are cooperating less and less, and they're much more suspicious about our intentions. So we really need to fundamentally fix the U.S.-Russian relations. Henry Kissinger and George Shultz have an editorial about this in the *Washington Post* talking about not having a renewed Cold War with the Russians and how we need to have a working relationship with them for so many issues. Iran, for example. Energy, for example. But particularly for solving the nuclear problem.

They still have about 2,000 warheads ready to launch at us on 15 minutes' notice, and we have about 2,000 warheads on missiles ready to launch at them on 15 minutes' notice. We've got to change that posture. The Cold

War's been over for 15 years, but the Cold War weapons remain. This has got to be at the top of the next president's agenda, taking those weapons off hair-trigger alert, accelerating the programs to eliminate the material not just in the states of the former Soviet Union, but everywhere. There are about forty countries that have material that they used for civilian purposes that Osama bin Laden could get and turn into a bomb. You've got to accelerate those programs, prevent the terrorists from getting the weapons before it's too late. You've had this remarkable development in the last couple of years where Henry Kissinger and George Shultz, two Republican stalwarts, rock-solid conservatives, have joined with Bill Perry, former Secretary of Defense, and Sam Nunn, former Chairman of the Senate Armed Services Committee, two conservative Democrats, to call for a world free of nuclear weapons, to urge their colleagues of both parties to make this a top priority, and to be serious about eliminating nuclear weapons. This is the number-one threat to the United States, whether it's on top of a missile or in a terrorist's backpack. We need to just eliminate these weapons before any of them are used. They acknowledge that it's going to take a long time to do this, but they outlined a series of practical steps that can start us on that road, all of which would make us safer as we take them, all of which would help convince other countries to cooperate with us to prevent that ultimate disaster of a nuclear terrorist assault.

I'm actually very optimistic that the next president will take steps in this direction. I think Barack Obama would do it quicker and faster and fuller, but John McCain also would take a number of steps. So I think we're on the verge of seeing a fundamental transformation in U.S. nuclear policy. It won't happen just by the politicians doing it; it requires an active and informed citizenry. It requires us to be demanding that whoever is elected implement the promises he made during the campaign. That's why I'm the president of Ploughshares Fund. That's why I'm out there raising money to help fund some of the groups all over the country that are doing this kind of work. I think this is one of the most important threats to life on this planet, and I want to do whatever I can to reduce that threat while I'm still alive.

Horton: Pat Buchanan on the show recently characterized the relationship between America and Russia as the single most important thing on Earth. Period. There's really nothing else that can quantitatively compare to the importance of our ability to get along with these guys.

Can I ask you to speak to the first-strike doctrine that seems to have been evolving in these Bush-Cheney years?

Cirincione: During the Cold War, both sides developed postures where they could think about developing a first strike that'd be so devastating that it would eliminate the other side's ability to strike back. It became pretty clear that you couldn't really do that. Each side just ended up building more and more weapons, so it was impossible to guarantee. What the Russians are now worried about is that we're starting to build conventional weapons that could actually strike their nuclear facilities. Precision-guided, conventionally armed cruise missiles, for example, or putting conventional warheads on former nuclear missiles. I've been in Moscow twice in the last 12 months and have had conversations. They are concerned about this, as well as NATO expansion, as well as the deployment of missile bases in Poland and the Czech Republic. So any new discussions we would have with them about reducing nuclear arsenals have got to bring in those issues as well.

Horton: Alright, I can't tell you how much I appreciate your time on the show today.

Grant F. Smith:
Israeli Theft of U.S. Nuclear Material
April 14, 2010

Scott Horton: Grant Smith runs IRmep, the Institute for Research, Middle Eastern Policy, which you can find at IRmep.org. He's the author of a bunch of great books about America's relationship with Israel, the latest being *Spy Trade: How Israel's Lobby Undermines America's Economy*. You can find what he writes at original.antiwar.com/smith-grant. Welcome back to the show, Grant. How are you doing?

Grant F. Smith: Great to be back, thanks for having me on.

Horton: I appreciate you joining us this morning. We've got an opportunity for a big history lesson here, because it's in the news that Obama did a big nuclear summit thing. Although, for some reason, Benjamin Netanyahu, virtually alone among world leaders, other than the mullahs of Iran who weren't invited, didn't show up. Pretty much everybody else showed up for this thing except the Israelis, right?

Smith: Well, no. Actually, the Israelis sent Dan Meridor, but he obviously wasn't of the same profile or same level as most of the other leaders who attended the summit. So it was notable for that reason that they sent their intelligence guy to a high-level nuclear summit.

Horton: Okay, thanks for that clarification. That is interesting. Now, the nuclear summit is something that Bill Clinton and George W. Bush both completely failed to follow up on — the Nunn-Lugar plan. Senator Lugar is still pushing this. Apparently, he's got a little bit of a mentor relationship with President Obama, and Lugar's first order of business for years and years has been to buy up and do whatever we can to get rid of any loose, fissionable, weapons-grade nuclear material left over from the Cold War on this planet.

Smith: That's right, especially in Russia, Ukraine and Belarus, and to decommission it, get it out of the reach of anybody who might misuse it.

Horton: Why wouldn't Benjamin Netanyahu show up at something like this? Sounds like great PR for Israel, and they sure could use some good PR this week, right?

Smith: Of course. Any country wants to keep the PR spotlight shining on the right things, and by sending Meridor, the bid was to avoid exposing Israel's own clandestine nuclear arsenal, and also to keep up the drumbeat of attention on Iran's nuclear program. You and I would call it a civilian nuclear program, but they would say it's a clandestine weapons program. So by sending Meridor, it was kind of a way to keep any "ambush" from occurring, particularly by some of the Arab states, to shine a light on Israel's clandestine arsenal and put that front and center.

Horton: The U.S. has this weird relationship with Israel about their nuclear weapons, where we pretend to be ignorant. The West pretends not to know that Israel has nuclear weapons. But the deal is they're not members of the Non-Proliferation Treaty, and they don't openly declare that they're a nuclear weapons state. It's a "strategic ambiguity," that's their diplomatic term for it.

Smith: Right, that's their policy, and it's because of the power of the lobby and Israel's desire to keep this kind of thing "away from the kids" that we are complicit in that ambiguity, and it makes a lot of other moving parts of the U.S.-Israel relationship possible. If the U.S. were to come out and openly say that Israel is a nuclear weapons state, immediately, the Symington and Glenn Amendments to the foreign aid law would kick in, and the U.S. would no longer be able to deliver any taxpayer-funded military or civilian aid to Israel. Those amendments created restrictions prohibiting aid to any country found proliferating or managing a clandestine nuclear weapons program. In fact, for Pakistan, the president has to sign a waiver every single year, saying that although it is not in compliance with the Symington and Glenn Amendments, the U.S. nevertheless finds it in its national interest to support Pakistan in some ways.

So, there are a lot of moving parts in the U.S.-Israel relationship, and by being complicit in strategic ambiguity, the U.S. is actually thwarting some of its own laws.

Horton: That's the way it works. The law is for them to use against us, but when they don't want to obey, they don't have to. Scalia got up there and said, "Look, the Eighth Amendment bans torture for punishment, not for interrogation." They can do whatever they want.

Anyway, teach me a history lesson. The article on Antiwar.com today is called "America's Loose Nukes in Israel." What do you mean by that?

Smith: There's this whole catchphrase that everyone is buzzing around here in Washington about "loose nukes" and the need to get all these loose nukes in the Lugar spirit, as you just mentioned. If we're going to do that, then we're going to have to gather up a significant quantity of material from Israel. Because the history lesson, in short, is that some major Israel lobbyists with very tight connections to Israeli intelligence set up a nuclear reprocessing facility in Pennsylvania, called NUMEC.

Back in the late 1950s and early 1960s, this facility was visited by Israel's top economic espionage agent, Rafi Ethan. Subsequent Atomic Energy Commission audits found that close to 600 pounds of highly enriched uranium were no longer present at the plant. This was a plant run by a very meticulous person, Zalman Shapiro, who is still around today, who seemed to exhibit extraordinary financial acumen, but he just couldn't seem to keep track of the uranium, according to the audit. The CIA-FBI consensus opinion as early as 1968 is that "NUMEC material had been diverted by the Israelis and used in fabricating weapons." So if we're going to be gathering up loose nukes, this is one of the outstanding things that's got to be on the list.

Horton: That really is amazing. It's like the USS *Liberty*, where the majority of people who just heard that term out of my mouth don't know what that means. They'd probably guess it's a ship because it sounds like a ship, but they don't know the story of Israel attacking the USS *Liberty*. Here you're telling me they stole at least a few bombs' worth of weapons-grade, highly enriched uranium for their nuclear weapons. I guess I've heard rumors of that, and I haven't read your books, but it seems like that would be a big enough deal that everybody would know about that one time Israel stole several nuclear weapons' worth of highly enriched uranium from Pennsylvania.

Smith: What's more is that this is only one component in a longstanding flow diversion from the United States. A lot of the people who were top-level scientists in the Manhattan Project had strong ties to Israel and wanted to help it out, and they were uniquely positioned to do that through scientific collaboration. So the uranium and plutonium issue is just one small piece of what has been a longstanding position of the Israel lobby, in contrast to the official position of the U.S. government, which is that Israel is entitled to this support, that it is entitled to have its own nuclear arsenal, and that it's vital to its survival. I think a lot of people would contest that and say that their conventional deterrence is quite adequate.

Horton: We're talking about during the 1950s and the Cold War here with the Soviet Union and China as enemies, and it was around this time that China was detonating atom bombs and then even fusion hydrogen bombs. Why wouldn't Ike Eisenhower and John Kennedy just say, "Yeah, Israel is our ally in the Middle East, let's arm them up." Why wouldn't we just give or sell them nuclear weapons?

Smith: Because they're a loose cannon. That was the problem, for example, in the 1967 War. The Johnson administration referred to them as a tiger and something that had to be kept on a leash. They were really trying to negotiate a climb-down and get the Egyptians and Israelis together, when Israel went and preemptively attacked Egypt and all of the other neighboring countries.

Horton: That was Eisenhower's view, too?

Smith: Eisenhower was very concerned that his foreign policy position was being usurped by all of these groups sending tax-deductible donations overseas and that some of that was then being recycled back into the U.S.

for lobbying. He really felt that that needed to be controlled and even made some moves toward the Treasury to get that under control. That is something that Kennedy inherited and, as I mentioned in the article, 1963 was a critical year because Kennedy also wanted to reclaim sovereignty over such vital issues. He not only was sending ultimatums that Israel's Dimona nuclear weapons facility had to be put under U.S. inspection, but he was also saying that they were going to regulate AIPAC's parent organization, the American Zionist Council, as an Israeli foreign agent. They were fed up with being constantly preempted and end-run by an organization with strong ties to a foreign government. They wanted to recover the sovereign power of being able to make decisions about nuclear weapons within the U.S. government.

They failed on both counts. Israel proceeded with its nuclear program after Kennedy's assassination, and the American Zionist Council just staged the shell company reorganization, and it's still with us today, only now it's known as AIPAC.

Horton: As long as we have nation states in this world, it seems to me like America ought to be friends with Israel from here on out. Free trade and open relationships. But this alliance doesn't seem to be working out very well, and it's not even really an official alliance, is it?

Smith: No, it's really not. If you take most of the positions the lobby advances, that Israel is a strong ally of the United States, or that it's a vital trading partner, the opposite is actually true. Israel couldn't really exist without massive U.S. aid. It couldn't exist without the U.S. giving it trade preferences that allow it to export to the United States while shutting out U.S. exports to Israel. Just about every theory advanced, particularly the one about shared values, the U.S. doesn't have very many shared values with Israel anymore because we've been struggling to overcome some of the Jim Crow-type discrimination. We've been striving to overcome our own legacy of long-term territorial seizures. We've kind of moved beyond the stage that Israel finds itself in.

So, I don't call it the "special relationship"; I call it the "e-special relationship" — it's *e*specially costly, it's *e*specially troublesome, and it *e*specially sucks a lot of air out of other diplomatic initiatives. If you think about the amount of time and resources that government officials are forced to spend on one country with 7 million people, while they neglect much more important, closer and bigger economies in countries, it's an especially troubling relationship.

Horton: What did you think of what Barack Obama said in his news conference yesterday?

Smith: Well, every headline from the newspapers should say something like, "Obama Orders Israel into the Non-Proliferation Treaty," because that announcement, which came late yesterday, is a fundamental alteration and challenge to the idea of this hush-hush treatment of the Israeli nuclear program. Back in early spring, they began to chip away at ambiguity when some of Obama's lower-level State Department officials said that they'd like to see everybody, including Israel, in the Nuclear Non-Proliferation Treaty, which would involve inspections and responsibilities. When Obama says that, he doesn't say anything about the Israeli nuclear weapons because they want to maintain their flexibility.

Horton: They couldn't let them in as a non-nuclear weapons state. Besides, Netanyahu's going to buck and not do it anyway. So I'm not sure what it means. Why would Obama even say that?

Smith: This is the first time a U.S. President has said that we want Israel in the Nuclear Non-Proliferation Treaty, and that is a major deal.

Horton: He also complained or expressed frustration, as they would report it in the news, about the Israeli-Palestinian conflict, too. He referred to the Israel-Palestinian conflict as not being just some issue for our hearts, but it affects our national interests, which was seemed to be a veiled reference to the Petraeus Doctrine, that our guys get shot because of what Israel does in Palestine.

Smith: Well, that's not new. If you look at the Secretary of State George Marshall of the famous Marshall Plan, he had pretty much the exact same sentiments that the U.S. was going to be dragged into a lot of confrontations with Israel's neighbors if it were allowed to have a state. So what Petraeus is saying is something that's been voiced a long time. You and I might spend a lot of time looking at things like the Office of Special Plans, but at the top levels of the Pentagon, a lot of honest thinkers have always viewed the relationship as restrictive and problematic. They've seen Congressional and interest groups insist that the U.S. transfer lots of military technology starting back in the late 1970s and early '80s, which really have been binding the Pentagon to the Israeli defense military-industrial complex.

The U.S. did not need Israel as the regional power to confront the Soviet Union or its client states. During the Cold War, Israel in fact needed the United States and access to all sorts of high-end weapons because it wanted to become a major exporter of high-tech military industrial goods. At the Pentagon, they're aware of all of this. A lot of them would prefer that all of the fighter planes and high-tech military assets remained under direct U.S. control rather than turning them over to Israel, where they can do things like bomb the daylights out of Lebanon at a moment's notice or create events in the region that will preempt any sort of stability that peace negotiations could bring.

Horton: In Mearsheimer and Walt's book, *The Israel Lobby*, they talk about how America has all this military equipment prepositioned in Israel, knowing that it's an illusion that, "Yeah, right. Like they're a useful ally if we get into a war in the Middle East." We've had two wars with Iraq, and our governments have had to bend over backwards to keep Israel *from* intervening in them, especially in 1991 when they wanted to keep all the Arab states in the coalition. But then they talk about how all that prepositioned military equipment, it's just there for the Israelis. It's just another uncounted welfare payment from the American people to the government of Israel. They talk about the Lebanon War in 2006. That's exactly what they did. They raided that so-called prepositioned military equipment, and they used it without even asking.

Smith: That was Steven J. Rosen's entire project while he was at the RAND Corporation, getting prepositioned military equipment into Israel.

Horton: To be clear, the excuse to "preposition" this equipment was so that the American military can use them in the event of some war, right?

Smith: Yeah, Mearsheimer used to talk about that. He wrote a book called *Conventional Deterrence*, in which he mentioned POMCUS units (Prepositioned Equipment Configured to Unit Sets) in Europe, which was prepositioned equipment to be used by the United States. The U.S. Army would train to go and get it and be able to use it to deter the Soviet Union in Eastern Europe, if there was any sort of massive conventional attack.

But what you said is exactly right. This prepositioned equipment has been entirely misused. Back during the Iran-Contra affair, we know that the Israelis were raiding U.S. taxpayer-funded TOW (Tube-launched, Optically-tracked, Wire-guided) missiles to send those to Iran because they wanted to topple the government and go back to the gravy days before 1979 when they were major military contractors and able to get oil exports from the friendly Shah of Iran. So these things have tended to be misused in the case of Israel, whereas I'm not aware of any European government that ever raided the POMCUS units and started selling ammunition or diverting it for their own political ends. The argument, in this case by Steven J. Rosen, was that prepositioning was helping American security, but the polar opposite was true. It tended to exacerbate U.S. problems in the region, and it tended to be a misuse of American resources.

Horton: I want to share a couple news stories with you here and get your comment. But first of all, I want to set this up correctly by explaining that, at least from what I understand, Obama's push against Israel is virtually all talk. It's just hot air. He hasn't insisted that they stop expanding settlements. He's said, "Well, we'd kind of like for you to slow down the rate of growth," like a Republican talking about a spending program. The whole spat between Netanyahu and Obama seems like just a personal one, but when it comes to policy, there hasn't really been much of a change or much of a push about East Jerusalem, the West Bank, Gaza or anything else.

So here's the two headlines. First is "76 Senators Sign onto Israel Letter," which is them telling the Obama administration they'd better back off; it's 76 senators taking the side of Israel over the President of the United States, and that's in the Democratic-controlled Congress. Then there's this one from *Haaretz*: "73 percent of U.S. Jews Approve of Obama's Approach to Israel."

Smith: It's amazing, because there's definitely dissonance between evolving public opinion, what Obama is trying to do — which is polling well among many different influential groups — and what the Congress wants. The Congress is used to, since the mid-1970s, climbing onboard virtually anything AIPAC wants because they are key donors to most senators and congresspeople, who want to get re-elected. So it's really not surprising that Congress can be signing a letter so quickly. Some AIPAC lobbyist used to say, "Hey, we can get them to sign a napkin in a number of hours. We can get all of the senators to sign anything."

But in this case, image is reality, because there is no public session where Netanyahu and Obama are photographed shaking hands with key members of the lobby and talking about Iran, because that's not happening. There is a crisis, and it is not revolutionary but evolutionary that this has gone on so long without Obama collapsing and following the AIPAC program, which would be to de-emphasize peace with Palestinians and to de-emphasize focus on illegal settlements, simply moving ahead on Israel's hit list while providing full diplomatic cover and consequence-free, responsibility-free aid. So this is a big deal. There's even been some talk, but it hasn't been confirmed, that some Israeli nuclear scientists are having problems renewing visas so that they can go and learn things at America's nuclear labs. Some news sites have said that's true, and others have said it's not true. But the fact that there is this push against maintaining the fiction of nuclear ambiguity, the firm statement that settlements have to stop would be very tough maneuvering in this relationship. I wouldn't expect any loan guarantee cuts or aid cuts or any moves against these tax-exempt charities in the U.S. that funnel close to a billion dollars to Israel.

Horton: That's Grant Smith. He's from the Institute for Research, Middle Eastern Policy. Thanks for joining me on the show.

Grant F. Smith: U.S. Violating Its Own Laws For Israeli Nukes July 2, 2018

Scott Horton: Alright, you guys. It's Grant F. Smith on the line. He's our good friend from the Institute for Research, Middle Eastern Policy. Welcome to the show. How you doin'?

Grant F. Smith: Very well, Scott. Very well. Thanks for having me on.

Horton: Good to have you on. Hey, you wrote *Big Israel*, about the Israel lobby, and also you wrote *Divert!*, about the Israeli government spies stealing weapons-grade uranium from the United States. And you also wrote all about the trade wars. You've written 6 to 8 books about the Israel lobby and their legal and illegal antics in the United States, correct?

Smith: That is correct, and if I could write another one, it would be called *Project Pinto*, which was the Arnon Milchan-Benjamin Netanyahu smuggling ring that got ahold of a bunch of nuclear triggers back in the day.

Horton: Yeah, what a great story that is. Isn't it funny, the ratio between what a big story that is and what a big story it wasn't?

Smith: Yeah, it certainly is. I sent a tweet to Anne Applebaum of the *Washington Post* because she was saying something about it being certain that North Korea's spinning up their centrifuges, and I said, "By the way, have you ever heard of Netanyahu and the krytron heist, and why is it the *Washington Post* has never written word number one about that?" I'm still waiting for an answer.

Horton: You mean she didn't immediately give you the due respect and honest answer you deserve about what was going on?

Smith: No! That's what I expected, of course. I expected that the leading newspaper here in Washington would leap all over that because it's such a great story. It did go up on Antiwar.com in 2012, and I have to credit National Public Radio of all places for at least mentioning it a couple of months ago, but it didn't get any traction there, either.

Horton: That's interesting, though, that they would. That's the kind of thing where it only comes out if they don't really realize the politics. Whatever news editor is working that day is not up-to-date on what you're never supposed to let through, and it accidentally does.

Smith: It was a throwaway line about Milchan and the Netanyahu pink champagne controversies, so you're right about that.

Horton: So let's talk about this. I had no idea about this, and it is shocking. The Israelis make the American presidents or their national security staff put it in writing that they will not talk about Israeli nuclear weapons?

Smith: I've got to hand it to *The New Yorker* and Adam Entous. He is an amazing writer. I linked right at the top to one of his articles in my article "Four Presidents Conspired to Give $100 Billion to Israel" that's at Antiwar.com, and his articles on this subject are absolutely astounding. They're all about these secret commitments and all of these antics that have been taking place all the way back to Bill Clinton. Apparently, these papers that Israel gets American presidents to sign make it very clear that the U.S. will not do anything to jeopardize Israel's strategic deterrence capabilities. They're filled with a lot of really weaselly words for saying, "We're not going to mention your nuclear weapons."

Horton: There's this old anecdote about Bill Clinton, who had already been the president for four years at this point. It was the first time he ever met Netanyahu, and Bill Clinton reportedly after the meeting said, "Who in the f— does this guy think he is? Jesus Christ! I can't believe what just happened here!" The story goes that Netanyahu just came and started reading Bill Clinton the Riot Act like, "Here are your marching orders, punk," and Bill was like, "Hey, I'm the emperor of the world, dude, you're just the prime minister of a little Maryland-sized nothing. So let me tell you about the way things are for a minute." He just couldn't believe it. I wonder if that's what this was about, if it was Netanyahu basically saying, "Sign it, bitch!"

Smith: It is! And it's amazing because that same reaction came from Trump's staff who said to Ron Dermer, "This is our f—ing house. What do you think you're doing here? You can't demand this," and he was making all of these…

Horton: Wait, wait. Tell that story in a little more detail there, as Entous reports it here, if you could please?

Smith: Well, the Israelis came in at a very sensitive time when they're just about to dispatch Michael Flynn, the national security adviser. Flynn was about to hand in his resignation letter, and Ron Dermer, Israeli ambassador to the U.S., came in and said, "I need to have Flynn in the room, I need this guy, I need that guy…" He obviously didn't know that Flynn was on his way out; otherwise, he probably wouldn't have made that demand. But that just made everyone snap. They said, "Look, you don't just come in here with a letter or a set of points that you want to make us sign and demand the people you want in the room. That's not how it works here." But he got the letter signed anyway.

And from doing Freedom of Information Act requests, it's known that there's virtually no handoff of National Security Council information from one administration to the other, and all of the presidential libraries are like locked boxes where the National Security Council stuff is under lock and key, never to be released. Americans think that there's a lot of continuity on the NSC, but there's not. So nobody knew about these letters except the Israelis, who had all four of them.

Horton: That's so funny because we're talking about Jared Kushner here, right? Ron Dermer and Jared Kushner, they don't get along? Dermer couldn't have said, "Oh, hey, Jared. There's this thing we'd like to talk to you about"? In other words, if you're the Israeli side in this, why the need for the imperious attitude? These are your friends here. Obviously, they're going to sign whatever you need them to sign, but they can't even be cool about it. They have to be so demanding that they get pushback from Benjamin Netanyahu's godson, Jared Kushner, for God's sake.

Smith: Right, the guy who gave up his bedroom, as we all have heard ad nauseam, so that Bibi could sleep there. I think it's a matter of trajectory and a doctrine within the Israeli side, that there's never enough that you can do. The attitude is: "You can never do enough for us, so why aren't you jumping higher or moving faster when we come in to bestow our latest request upon you?"

It's terrible, and the great piece that Adam Entous wrote in *The New Yorker*: "Donald Trump's New World Order," really goes into detail about all these requests and that there is contention because they've returned to this horrible policy of "no daylight," something that Michael Oren, another former American ambassador to the U.S. from Israel, always insisted upon, wherein the U.S. has always got to present this public face that there's never any disagreement with the Israelis, that the U.S. and Israel have these inherent interests that are exactly the same, and so there should never be any daylight between the two. But what Entous reveals is that there's a lot of daylight, and he's exposing that daylight.

Horton: Right, and so this article that you wrote, "Four Presidents Conspired to Give $100 Billion to Israel: Secret White House Letters Buttress Ongoing U.S. Arms Control Act Violations," it's about those letters and it's also about the atom.

Smith: Yeah, that's the one, and I don't think the word *conspire* is too much, because the dictionary definition is "to make secret plans jointly to commit an unlawful or harmful act."

Horton: This is the real rub here about why there's pseudo-secrecy over Israel's nukes. There are tweets of Mordechai Vanunu's talking about how they've got 200 nukes. It's not a secret, Grant. Everybody knows, but the reason that they have to pretend and be like, "Oh yeah, nobody knows if Israel has nukes or not," is that they're breaking the law. If they admit that

Israel has nukes and they're outside the NPT, then they're violating the Export Control Act.

Smith: It's the U.S. Arms Export Control Act. It's the Symington and Glenn Amendments that make it unlawful to keep giving U.S. foreign aid to any country that's outside the NPT and found to be trafficking in nuclear weapons technology, not to mention one that's building an arsenal and deploying the weapons. That would also apply to things like testing nuclear weapons technology, such as what happened with the 1979 Vela Incident [when Israel and South Africa tested an A-bomb in the Indian Ocean]. Most of the people who are honest about it within governments like Victor Galinsky, the former head of the Nuclear Regulatory Commission, flat-out say that it was an Israeli nuclear test.

Anyway, these amendments, which are now found in the U.S. Arms Export Control Act, require the president at the very least to issue public waivers to Congress saying, "I recognize that this is a nuclear weapons country outside the NPT; however, we must continue foreign aid because it's in the strategic or the national interests of the United States." I think they'd have an extremely hard time making that argument, and I think anybody who had a serious or realist foreign policy experience, like John Mearsheimer or Stephen Walt, if you put either of them up in front of that argument, they would shred it to little pieces. But the real reason that Entous does not go into why this is happening is exactly that you can't keep violating this piece of the Arms Export Control Act. I believe they'd have a very hard time adding some sort of modification to it that would somehow exclude Israel. They haven't done it. I doubt they will do it.

So what do you do if it's on the books and you have to appear to be giving aid in a lawful way? You just don't admit the obvious, the thing that everybody knows, and no president has ever come out and said, while in office, that Israel has a nuclear weapons program.

Horton: It's just like sending special operations forces to Mali after the coup. That's against the law, but that's why it's a clandestine operation, so it doesn't matter, right? I'm talking of course about the coup that was caused by the results of the war in Libya.

But now let me ask you this. What you just said is a pretty obvious answer to it, but I wonder if there's something more? There's the way the Israelis say, "Listen, yeah, we've got nukes," out of the side of their mouth, "but we will not be the first country in the region to introduce nuclear weapons into the region." That's the way that they say it. In other words, maybe what they're saying is that they don't want to threaten their enemies outright with nukes; they want to leave it a more subtle threat. I wonder if you think that that's a distinction without a difference, or if it really just comes down to this Arms Export Control Act?

Smith: Well, that didn't exist when they first came up with that formulation. That formulation is the standard response, but it doesn't really mean anything, because all they have to do to meet that extremely high bar is to simply say, "Well, the U.S. has sailed nuclear arms, aircraft carriers with bombs and missiles through the region several times, so they were the first to introduce."

Horton: And what difference does that make anyway, when again, Mordechai Vanunu came out to *The Sunday Times* in 1986? He was the nuclear weapons whistleblower. It's in *The Sunday Times*. They still have it online.

Smith: Exactly, it really doesn't mean anything. It's just something to say whenever they're asked, and I believe Netanyahu even said that recently when he was buttonholed by a reporter on CNN, no less, asking about Israel's own nuclear weapons. Then, after saying that — which means nothing — as he did, he'll just say, "And I'm not going to give you anything else on that."

Horton: Wasn't that an amazing interview, by the way? This is just what Phil Weiss is saying, that Trump has come to own the Israel issue in such a way, and he's made it such a partisan Trumpian issue, that it's really pushing liberal Americans, including liberal American Jews, away from Israel. If that's what Trump's into, that's what they're against, so CNN's party line is now so anti-Trump that it's even anti-Netanyahu, in a way. I mean, somebody was telling him in his earpiece, "Yeah, go ahead and go after him!" Which is just impossible! What an awesome time to be alive, man.

Smith: [Laughing] Well, I don't think so, because the thing that Entous doesn't go into, and which is extremely important, is why can Dermer come in, upset everybody, and still get his letter signed? Why can Netanyahu be involved in nuclear technology smuggling? It's in FBI documents, but they'll just never call him on it. There's this huge distortion, which goes back to the Israel lobby. As I mentioned in the piece, if you add up the revenues of every single U.S. nonprofit organization, not even counting campaign contributions and pro-Israel PACs, it's a gigantic industry that's always pushing a very proactive agenda, which is exactly related to what I said before, that you can never do enough for Israel. Trump can't stand up to that. Obama couldn't stand up to it either, and I spent a particular amount of time on the folding of Obama, who came in as "Mr. Counter-Proliferation" and who was going to Prague to talk about shoring up the Nuclear Non-Proliferation Treaty. There were talks of a Middle East nuclear-free zone, and then suddenly he gets handed the letter in May 2009 and immediately collapses.

I didn't have that date before, but that's Entous's report saying it was in May 2009, after the Prague speech about nuclear non-proliferation, that he yielded and signed his letter to the Israelis. Even worse, on September 6,

2012, his Department of Energy and Department of State issued a secret directive, and we've only got the title of it, called "Guidance on Release of Information" relating to the potential for an Israeli nuclear capability. Which itself is secret, but it's basically a gag order, where even quoting public domain stuff as a contractor or a federal employee, you're going to get fired, your computer's going to be searched, and people are going to look at how they can charge you criminally with leaking classified information. Even if you quote Vanunu or read from *The Sunday Times* and just quote from that, you'll still be fired.

Horton: Or even quote Colin Powell, or the current minority leader of the Senate, Charles Schumer.

Smith: Right, exactly. Sam Husseini had debriefed Schumer so thoroughly at the National Press Club, getting him to admit that yes, they have nuclear weapons. He would be subject to WNP-136, he would have been fired. If you were a high agency official, say, at the Department of Energy, and had said that, you'd have been out. He probably would have been prosecuted.

Horton: Have you done polls on how many Americans know that Israel has nukes?

Smith: We have. Those were some of the first polls we did right after we found out that we were capable of doing statistically significant polls. That was back in 2014. We asked two questions on the 26th of September. The first was: "Do you believe Israel has nuclear weapons?" And 64 percent of Americans said yes. Unfortunately, at the same time, to the same question about Iran, 58–59 percent also thought Iran had nuclear weapons in late 2014. Which, if you're the Washington Institute for Near East Policy or William J. Broad and David Sanger you'd say, "Yes! Victory! We've successfully brainwashed Americans into believing something that's not true." Public opinion polling is fascinating to see either what people know or what they absolutely don't know.

Horton: It seems like if somebody has a Manhattan Project or not, that's something that you could know. To just settle for, "My impression is that there seems to be something there because people talk about it a lot," and to settle for that and to let that serve as a casus belli, grown adults need to be more responsible than that. There's such a thing as the IAEA, and they publish a thing every couple of weeks, you know?

Smith: Right, and they do very solid work. However, there is this propaganda campaign around Iran, for instance, disparaging weapons inspectors. Aside from that, there's this whole counter-proliferation think tank universe in Washington that is about as feckless as can be. No one knows that these IAEA inspectors go ad nauseam into suspected Iranian sites, suspected explosion and implosion test sites, but whenever anything comes up about the Israeli nuclear weapons program, those think tanks aren't the ones bringing it out. They don't do any analysis. They're basically configured to be against North Korea and Iran. And by their errors of omission, they don't do any work on this at all.

Horton: Getting back to your article, what's this $100 billion in the title?

Smith: Yeah, it's a suspiciously round number, I admit. But that is the number you get if you add up all the U.S. foreign aid that has been given to Israel since Clinton signed that first letter agreeing that his administration would ignore Israel's nuclear weapons, in violation of the Arms Export Control Act. So if you add up every year and adjust for inflation, it comes out to $99.9 billion. If you go back a little further to 1976, when these amendments to the Foreign Assistance Act went into effect, banning U.S. aid to proliferators and demanding public waivers if it was to be given, then you find that the sum of U.S. aid was $222.8 billion, adjusted for inflation.

That is *a lot* of money, and it makes this a major conspiracy. This is Teapot Dome. This is Tammany Hall. This is an extremely big, bipartisan corruption scandal that's been involving U.S. administrations and federal agencies since 1976.

Horton: Let me ask you this, do you think Trump is qualitatively more committed to Zionism, or at least to Netanyahu, Sheldon Adelson and their policies, than previous presidents?

Smith: I do. If you add up all the giveaways and all the damage that's been done to U.S. commitments to international law and the United Nations — that is, if you believe that it is important to have a United Nations and if you believe that this founding principle, that there be a negotiated settlement before you did anything like open embassies or recognize Jerusalem as the capital, is important — he's basically blown up the United Nations. So yes, I do believe he is.

Horton: Now you're convincing me to really like the guy, because I hate the UN, I don't want any part of it. But yeah, it's not good enough to justify what's going on in Palestine. Not by a long shot.

Smith: If you also look at it from the side of fairness, what about at least trying to have some sort of settlement looking at the original division?

Horton: I look at it as the individual rights of the Palestinians. It's simple as that. That's what's really the question here.

Smith: Absolutely, and in terms of U.S. interests, this article says they've been bad since Clinton. They've all basically undermined the United States. Maybe you could make the argument that we shouldn't have been giving this foreign aid, and if we hadn't, the U.S. would have a lot more leverage and probably a lot more peace in the region. Instead, we continue to fund one side by violating our own laws.

Horton: Alright, well, thanks very much for coming back on the show and talking about this very important subject with us. I know it gets through to people. People tell me straight to my face, "Hey, on this Israel-Palestine stuff, I get it now from listening to you," and I know that that means listening to me interview you, so thank you.

Smith: Thank you, Scott.

Gordon Prather: Iran and North Korea's Nuclear Programs August 3, 2009

Scott Horton: Our guest today is Dr. Gordon Prather. He worked formerly in the U.S. Navy at Lawrence Livermore in Sandia National Laboratories, was an adviser to Senator Henry Bellmon and was the chief scientist of the Army. Basically, he spent a career making, testing, and knowing all about nuclear weapons. He's not a poser like you read in the media all the time, but the real thing. Welcome to the show, Doc. How are you doing?

Gordon Prather: Alright, I think. How about you?

Horton: I'm doing okay. Glad to have you here. I thought this was funny, a good friend of mine sent me this link from *The London Times*: "Iran Is Ready to Build a Nuclear Bomb, It Is Just Waiting for the Ayatollah's Next Order." I don't know if you want to comment on this, but I thought it was important that they don't cite even an anonymous source for the following assertion of fact: "The Iranian Defense Ministry has been running a covert nuclear research department for years, employing hundreds of scientists, researchers and metallurgists in a multi-billion-dollar program to develop nuclear technology alongside the civilian nuclear program." They don't even pretend to cite a source for that assertion, but there is apparently a secret military nuclear program in Iran, and we're just supposed to accept that as plain fact. What do you think about that?

Prather: Well, in the first place, if it does not involve Treaty on Non-Proliferation of Nuclear Weapons proscribed materials, if that research program does not involve, for example, uranium in any form, plutonium, or a number of other things that are associated with nuclear weapons, then it's none of the IAEA's business. However, when the Iranians agreed in 2003 to abide by the terms of an Additional Protocol, they signed it, but the Iranian Parliament has never ratified to this day. And after some of the John Bolton-inspired stuff that went on in 2006, the Iranian Parliament not only announced that it was not going to ratify this Additional Protocol, but they specifically told their nuclear agency to quit cooperating with the IAEA, as they had been doing for three years. It was a direction not from Khamenei, but essentially the same thing.

Horton: Under that Additional Protocol, when the IAEA was doing all that and had more access to the Iranian nuclear facilities than they had under the basic IAEA Safeguards Agreement, did they find any evidence that there was a massive, parallel secret program?

Prather: Well, no. But the Additional Protocol provided for it, and the Iranians essentially complied with it for more than two years. It essentially suggested to the Director General at that time, Mohamed ElBaradei, that he "attempted to form a complete picture of the Iranian nuclear program." That means, trying to figure out intent and things like that. So he might have been legitimately interested in figuring out whether or not the Iranians did have this program to develop the components that did not involve nuclear materials, and their intent. Were they attempting to make, for example, a high explosive implosion multipoint detonation system? He would have had some reason for making some discreet inquiries of the Iranians, and he did. The Iranians complied with his request in most cases, and he and his inspectors went to the sites where they thought maybe something like that was going on. That would not have been a violation of their basic Safeguards Agreement, and it wouldn't have been running in violation of the Additional Protocol, but it would have enabled ElBaradei to form a more complete picture of what their intentions were. Under the basic Safeguards Agreement, what they're doing that doesn't involve NPT-proscribed materials is literally none of his business. Anything that he finds in his investigation and in his inspections that is of a proprietary nature — for example, the design of the Iranian P-Tube gas centrifuges — that information he's required by the IAEA statutes and by the Iranian Safeguards Agreement to hold as confidential. He's not supposed to allow anybody access to that information. That's in the statutes, and that's in the Safeguards Agreement.

Horton: In the reporting of all the best reporters on this issue, or in the unclassified version of the National Intelligence Estimate released in 2007, did any of that say that there was a secret parallel program? Do you know of any good journalism anywhere? Anything that even says, "Some agents at the CIA really, really believe that there is one"? Do you know of any evidence whatsoever that a secret parallel program exists?

Prather: If there was any evidence, it would have been leaked. You had David Albright on your program a while back, and he said he considered it his goal or duty or something like that to publicize IAEA confidential reports so that everybody would know about them. And you can bet that if there was anything in there of a nuclear weapon-associated program in Iran, he would have blabbed that all over the world. But actually, as you noted, and to his credit, when the "smoking laptop" information was first floated around and the *New York Times* said that it contained information about a program to develop ICBM-delivered nuclear warheads, Albright had a fit and wrote a number of letters to the *New York Times* saying, "No. I've seen the smoking laptop information." He doesn't explain how he managed to see it.

Horton: What he said was that at best, they were plans for a re-entry vehicle, but that doesn't necessarily mean there would be a nuclear bomb inside of it.

Prather: He specifically says that the word *nuclear* in English, Farsi, or whatever language is never even mentioned anywhere on the smoking laptop.

Horton: By the way, if people just Google "'David Albright,' 'warhead,' 'New York Times'" or something like that, they can read that exchange back and forth. You can read the letters to the editor that he wrote and the original David Sanger, William J. Broad column that he's talking about.

Let's move on to the meat of this article. This part will get a little bit complicated, but it'll be worth it in the end. In that *London Times* article, they write: "Iran scientists have been trying to master a method of detonating a bomb known as the multipoint initiation system, wrapping highly-enriched uranium in high explosives and then detonating it. The sources said that the Iranian Defense Ministry had used a secret internal agency called Supply, led by Mohsen Fakhrizadeh, a physics professor and senior member of the Iranian Revolutionary Guards counsel. The system operates by creating a series of explosive grooves on a metal hemisphere covering the uranium, which links explosives-filled holes opening onto a layer of high explosives enveloping the uranium. By detonating the explosives at either pole at the same time, the method ensures simultaneous impact around the sphere to achieve critical density."

Now, let me set up my question this way. You've told me before and we've discussed how the Hiroshima bomb was a simple, gun-type nuke, where the inside of the bomb is a shaft; it's basically a giant shell, a giant bullet of uranium that gets shot into another piece of uranium, and that's what creates the detonation. As you told me before, they didn't bother to test it here in the desert of the United States because they knew it would work. That would be what most people would assume would be the easiest, quickest path to a nuclear bomb that the Iranians could take, if they were trying to take the path to a nuclear bomb. But what this *London Times* article is saying is that no, they're going for this much more sophisticated implosion method. So, Dr. Prather, you've actually made real nuclear bombs. Does this sound like a plausible story to you?

Prather: Well, I've never actually made them, but I've tested them. When I was in the Navy, I was the head of a test training group. In those days, nuclear weapons had to be assembled at a stockpile and prepared for use. So I'm intimately familiar with that kind of relatively simple nuclear weapon — first-generation, maybe even second-generation nuclear weapons. Now, having said all that, when you asked me to be on this program, I went and did a Google search for something that I knew existed. I would now call your attention to the column I wrote where I concluded by quoting from a final IAEA report on the Iraqi nuclear program — not Iranian, Iraqi.

Horton: And we're talking pre-1991 war, in the 1980s, when Saddam was backed by the United States, and he had a secret program.

Prather: That's correct, and according to the final report by the IAEA to the UN Security Council, Iraqi engineers had developed but had not tested, in about a two-year effort, a complete 32-point implosion system, including an electronic firing system, detonators and associative high explosive lenses. In other information that we got after the first Gulf War, when the IAEA got to look at all the documents for Iraq's nuclear program which had all been destroyed by then, it was discovered that in 1989, before the first Gulf War, some Iraqi scientists had attended a symposium in Portland, Oregon, where they learned for the first time about these multipoint detonation implosion systems. They didn't know anything about that until then, and that's when their work began. In 1991, when the Gulf War came along, that's when the two-year program ended, and they destroyed all this stuff.

Since 1951, there have been 14 or 15 international unclassified symposiums on high explosive detonation. Many of the articles that come out of those symposiums describe in considerable detail how you go about designing a workable, sophisticated, multipoint high explosive detonation system. It's fairly obvious that after attending that symposium in 1989 that's what the Iraqis did, but they never got to the point of testing it.

Horton: This morning, I'm sitting here drinking my coffee, and I get an email with the subject "Iranian scientists teach pigs to fly," and you call this thing a Rube Goldberg machine. The point that you're making here about the IAEA report about these symposiums is that if the Iranians were actually going to make an implosion bomb, they would use some of the blueprints that are already all over the world instead of making up this ridiculous brand-new Iranian invention of a better way to detonate an implosion bomb, is that the point you're making here?

Prather: Yeah. I mean, this is *Looney Tunes*. If the Iranians really have perfected this thing, which, of course they haven't, we ought to close down Los Alamos and start outsourcing to the Iranians.

Horton: You told me on the phone earlier that it's possible that the way that these internal implosion devices are set off now could be improved upon, but it wouldn't be the Iranians who did it. It might be us, or maybe the Russians or the Chinese.

Prather: The Italians have been really big on the use of shape charges with high explosives, and that's what this is. It's a special kind of shape charge. And there have been all these international unclassified symposiums that go into incredible detail. If you go and look at these papers that have been given

— you can find a list of them online — you say to yourself, "What's the big deal?"

So, now back again to the Iranians. ElBaradei and the Iranians were proceeding along the lines of the Additional Protocol. The Iranians went and they showed ElBaradei a lot of these things that he wanted to see, some of which were military sites, which he had no business going to, but he asked, and they said okay. They were completely transparent, and they let him in. In many cases, these were snap inspections, but some of them weren't, because they had some stuff going on there that they couldn't let him know about. But they would clean those places up and then let the inspectors come in. Even under the Additional Protocol, he still had to ask.

Regardless, if the Iranians had done anything like what was reported in *The Times*, you can bet your bottom dollar that there would be a report in the IAEA files and that David Albright would have published it.

Horton: You told me before that if they were testing any kind of implosion device, whether it be the kind that everybody uses or this newfangled design as described in *The London Times*, we're not talking about something that the American intelligence agencies would be unaware of. As you told me, they would have to test the method and the perfection of the implosion device over and over and over and over again to get it just perfect. They would have to use some depleted or natural, not enriched, uranium in the core as part of the test. There's no way any of this could be going on without the American intelligence agencies knowing about it. Isn't that right?

Prather: I wouldn't say that about the American intelligence agencies, because they've been wrong so often in the past.

Horton: Well, but that's usually deliberate.

Prather: The best source of what's going on in Iran by far is absolutely, without question, the IAEA. After all, they've had inspectors on the ground. Until the summer of 2006, they could go basically anywhere they wanted to, and they could interview anybody they wanted to.

Now it needs to be stated again that safeguarded materials in Iran include the uranium mines themselves. They mine uranium in several different mines in Iran. From the time that it's mined, all of the uranium is under IAEA safeguards, and their bookkeepers and accountants keep track of every gram of it. Every gram of uranium, throughout its entire history from the mine on, is tracked. That includes natural uranium metal that might be used in the testing of one of these implosion systems. All the uranium metal and all the depleted uranium are NPT-proscribed materials that are subject to safeguards and the inspectors continually, several times a year report to the IAEA Board of Governors that all of the NPT-proscribed materials in Iran are accounted for and that none has been diverted to a military purpose.

Horton: Let's be pretty clear about this for people to understand. The Iranians are the Branch Davidians, and the IAEA are the ATF. And just like the ATF, the IAEA are itching for an excuse. They're the cops with their hand on their gun. If they had a way to say, "Aha! Look! The Iranians diverted their uranium," they would be taking it. It is not that "Mossad says Mohamed ElBaradei is an Iranian spy," as a previous *London Times* article put it. No, actually, Mohamed ElBaradei is the head of the ATF, and he's itching to go in there. If he had an excuse, he would.

Prather: No, I don't agree with any of that. I think you mischaracterized the International Atomic Energy Agency and its Secretariat. Now, the IAEA Board of Governors is a very different animal, and there are people on that Board of Governors, including the U.S. Special Envoy, who would love to have access to some IAEA report which said things like that — and they would have access to that kind of report. Specifically, if the IAEA Secretariat, the Director General and his inspectors, were to investigate the design of an Iranian gas centrifuge that is going to be used to enrich uranium, all of that information would be proprietary. They would be required to keep it secret from even the members of the Board of Governors.

Now, that being said, Albright seems to have made himself the official biographer/leaker, so you can be certain that if there were any IAEA reports, even though they'd be held confidential from the Board of Governors, Albright would have access to a lot of those, and that stuff would end up on the front page of some newspaper, but it hasn't. They are the authorities. The U.S. intelligence agencies are not competent to make those kinds of determinations, and I doubt that the Mossad, who have got a vested interest in this sort of thing, are capable of making those kinds of determinations. You've got to know what the hell you're looking for. You've got to be an expert, and I don't think there are any people in the CIA or in the Mossad who are experts on the design or the research and development of nuclear weapons.

Horton: There have been times from Seymour Hersh's reporting; in his article "The Next Act," he says that all of his intelligence sources inside Mossad say they don't have any information about a secret Iranian nuclear program. They maybe, kind of, sort of think there is one, but they don't have any evidence. Also, it's been in *Haaretz* that the director of the Israeli Mossad has said that the nuclear program they're talking about is the declared one, the one that everyone knows about, the one that is enriching uranium to industrial grade of 3.6 percent, not weapons grade of 90 percent-plus. But they still want to bomb them anyway because industrial grade is bad enough, he said. So the official story coming out of the CIA seems to agree with what we hear from leaks to journalists, and even candid comments by the head of the Mossad seem to agree with the National Intelligence Estimate of 2007.

Prather: Well, let me make the point as well as I can. The IAEA is not a non-proliferation agency. It is not a nuclear proliferation prevention agency. It is a United Nations "Atoms for Peace" agency, and their principal responsibility to the Board of Governors of the IAEA, to the Security Council, and to the UN General Assembly is to do what they can to help facilitate the widest possible distribution of atomic energy for peaceful purposes. That is their goal. That is their mission.

Now, when they're associated with a Non-Proliferation Treaty-required Safeguards Agreement, which is the case with Iran, the IAEA is also required to make sure that none of the NPT-proscribed materials like, for example, enriched uranium, uranium ore, uranium metal and any number of other things, are never diverted to a nuclear weapon or to a weapons program of any kind. That would include the use of depleted uranium in anti-armor artillery.

Horton: Doc, what you're saying is that the purpose of the IAEA was supposed to be this dream of spreading peaceful nuclear technology to the poor people of the world and all this wonderful stuff?

Prather: And that still is their purpose. That's why there are so many countries in the world that are members of the IAEA General Conference and have signed the Non-Proliferation Treaty.

Horton: Right. But it's also casting a shadow, because in effect it has proven the negative. It has shown that there's been no diversion of Iran's nuclear materials. So I see you're saying, the IAEA is like a park ranger who's been turned into a cop, but he's done a pretty good job over there.

Prather: Yeah, I guess so.

Horton: Well, this is important because John Bolton is on Fox News saying, "Yes, see? *The Times* of London says I'm right, Dr. Prather."

Prather: Who cares about what *The Times* of London says?

Horton: John Bolton does, and he repeats it to millions of Americans on TV. That's the punchline.

Prather: And the IAEA says he's wrong, everybody says he's wrong. Once again, as far as finding out whether or not NPT-proscribed materials have been diverted to a military purpose is concerned, the CIA and Mossad are rank amateurs. They don't have anything like the resources, the authority, or, more importantly, the cooperation of the Iranian government to be able to track every aspect of Iran's nuclear program like the IAEA does.

Horton: Let's go to one of the other major headlines of the paper today, the North Koreans. Of course, there's a whole backstory. But today, Bill Clinton is over there, according to the headlines, negotiating about those two journalists who have been convicted of spying. One of the stories said that he may try to negotiate with them or make a little breakthrough on the nuclear issue. This is such an important story for people to know, about how it is that the North Koreans ended up armed with nuclear weapons and what kind of position they're in as compared to the United States.

Prather: Well, in 1993, when he came in as president, there was already a dust-up between the International Atomic Energy Agency and the North Koreans, who, at the time, were in the process of subjecting all of their nuclear programs to a Safeguards Agreement as required by the Treaty of Non-Proliferation of Nuclear Weapons. The IAEA, headed at that time by Hans Blix, had some questions about the previous operation of that Soviet-supplied reactor. They wanted to examine the operating records of that reactor that had already been operating for several years without being under safeguards, to try to figure out how much plutonium was in the spent fuel elements that they had in a pile somewhere. They thought maybe that the way the reactor had been operated, there was more plutonium there than the North Koreans were claiming was there.

So, they made a complaint to the IAEA Board, and the IAEA Board went to the UN Security Council and said, "Make these guys comply with Blix's request," and the UN Security Council declined to do that. Of course, that's what the UN Security Council should have done back in 2006 when Bolton got the IAEA Board to make all these entreaties and appeals to the Security Council about the Iranian nuclear program. But back in 1992, the UN Security Council declined to take any action on these complaints by Hans Blix.

Well, in any case, that angered the old man, Kim Il-Sung. So he says, "You guys are still conducting military exercises twice a year in South Korea, and now you're trying to get the UN Security Council to impose sanctions on us, all because we're trying to negotiate a Safeguards Agreement with the IAEA, as required by the Non-Proliferation Treaty. So the hell with you. We're withdrawing from the Non-Proliferation Treaty, which is our right." About that time, in comes Clinton, and he had this goal — and it's also the stated goal of Barack Obama — that everybody, including Israel, India and Pakistan, should become a signatory to the Non-Proliferation Treaty. This would have meant they had to place all of their nuclear programs, including their nuclear weapons programs, subject to IAEA Safeguards Agreements.

That was never going to work then, and it's not going to work now for Obama. President Clinton had his people negotiate the Agreed Framework with the North Koreans, where they would agree not to withdraw from the Non-Proliferation Treaty, put everything nuclear in the whole damn country under IAEA lock and seal, and freeze their program right where it was. In return, Clinton promised that he would normalize relations with North Korea and finally sign a peace treaty with the North Koreans. There's been an armistice in place since 1953, and here it was in 1993, 40 years later, and we were still operating under this armistice. The North Koreans wanted then, as they want today, a peace treaty. They wanted the war to be over.

At any rate, they negotiated this Agreed Framework, and what the North Koreans got out of it was a promise that all this other stuff was going to be resolved and they'd have normal relations with the rest of the world. Well, the thing is, when President Clinton negotiated that agreement, he never expected to have to live up to the terms of it.

Horton: Because he and his administration thought that the North Korean government would just fall apart. Kim Il-Sung died and left it to Kim Jong-Il. And people always get all frustrated and say, "Aha! See? The Americans made a deal," (and it was, parenthetically, Donald Rumsfeld's company that got the contract to build a light-water reactor in North Korea). The point of that was, "Here, we'll give you this light-water reactor." Which, like you say, they never lived up to. That reactor would have been able to produce nuclear energy, but the North Koreans wouldn't be able to turn any of the waste into weapons material. Or, at least, it wouldn't come right out as weapons material like their old Soviet reactor, which produced weapons-grade plutonium as a byproduct, right?

Prather: The terms of the agreement were irrelevant as far as President Clinton was concerned, because they never expected to have to honor them. Anyway, in 1999 or maybe 2000, they finally concluded that this regime was going to last. So then Clinton started making nice with the North Korean leadership. There were some agreements, and there were actual meetings that took place in 2001 between U.S. officials and North Korean officials. There had been no official relationship whatsoever in all those years.

But then, immediately upon taking office, Bush and Cheney and Bonkers Bolton promptly informed both the South Koreans and the North Koreans that they had no intentions of honoring any of the treaties or the agreement that the Clinton administration had with North Korea. Not only that, but they started claiming that the North Koreans were not in compliance with the Agreed Framework. They started saying that North Korea has this secret uranium enrichment program somewhere over there, and that they've been fooling everybody. Finally, in September 2002, Bush unilaterally abrogated that Agreed Framework and said, "We're never going to establish normal relations with that outfit until their nuclear weapons program is destroyed and no longer exists."

Horton: Let's emphasize this part. Bill Clinton is in office. It's 1994. The old man dies right as this Agreed Framework is coming together. The Agreed Framework promised fuel oil and welfare payments and all this stuff.

Prather: The main thing it promised the North Koreans was normalization of relations between the United States and North Korea.

Horton: And even though the Clinton administration never lived up to that, the North Koreans basically stayed within the Non-Proliferation Treaty. They still had a Safeguards Agreement. Then, when Bush took office, it was Bush that unilaterally abrogated the Agreed Framework that Bill Clinton and Warren Christopher had signed in 1994, which was the deal that kept the North Koreans inside the Non-Proliferation Treaty. Then, as you say, they're a sovereign country, so they have the right, as stated in the Non-Proliferation Treaty, to be able to withdraw from it. They're supposed to give, I think, six months' notice or 60 days' notice. They did that and withdrew from the treaty. But it was George Bush and Bonkers Bolton, as you call him, that picked the fight.

Prather: It was worse than that. Almost immediately upon taking office in 2001, they refused to honor the commitments that President Clinton had made when he finally realized that the North Korean regime was not going away. So Clinton had proceeded to start making nice with their leadership and getting the North and the South Koreans to talk.

There was a new president, as I recall, for South Korea, and he came to the United States, and Bush flat told him, "To hell with all the stuff that Clinton did. There's no way in the world we're ever going to normalize relationships with that bunch."

Horton: When they finally did actually abrogate the agreement, it was in the name of this so-called admission by the North Koreans that they had a secret uranium enrichment program. Whatever became of that?

Prather: It was worse than that. The Bush administration claimed it was a secret nuclear weapons program that involved uranium. The program that they had was not a nuclear weapons program; they had never even separated out the plutonium that they produced in that Russian-supplied reactor with the intent of making a nuclear weapon with it. In January 2003, when they announced that they had completed their withdrawal from the Non-Proliferation Treaty, and they were restarting that plutonium reactor — they'd taken the seals off and the padlocks and all that sort of thing — and they were going to make more plutonium, they said, "We're going to separate this plutonium out, and we may have a different intention now than we did back in the old days."

Horton: When people read on the front page of the newspaper that North Korea has nuclear weapons, they might conclude that George Bush and John Bolton were right when they accused them of having that secret nuclear weapons program. But there's a difference in the recipe. One would have been made out of uranium, but apparently it doesn't exist. What they did detonate came from bombs made out of plutonium, which came from, as you say, this Russian-built reactor. There's still no evidence that they had this secret uranium program at all, is there?

Prather: Right. And not only that, but in 1994, when the IAEA went in to implement that freeze, there were allegations back then that they had a secret high explosive test range wherein they were attempting to develop an implosion system for nuclear weapons. The IAEA went there and found no such program and duly reported it to the IAEA Board of Governors. You can go find that.

Horton: It seems like no matter where in the world the accusations are being hurled, they're always based on a bunch of ridiculous false premises. And when you're just a regular person, "nuclear" this and "nuclear" that has got to be an expert issue left up to somebody else. Nobody figures that they can really do this on their own. It's complicated stuff, and a lot of people just figure, if it's nuclear and the government says it's dangerous, then there you have it.

They accused Iraq, who had no nuclear program whatsoever. They accused Iran, who put their hands up and opened up all their books and everything else to inspection. That's never good enough. They made false accusations against North Korea, forced them out of the Treaty and ended up pushing them to go ahead and get nuclear weapons. That's a pretty good tackling of the axis of evil there, wouldn't you say?

Prather: Except for the thing where you said Iraq had no nuclear weapons program. They did have a nuclear weapons program from about 1989 to '91.

Horton: Oh, no. I meant at the time of the invasion, but that's a whole other story about how the Israelis bombed Iraq's safeguarded nuclear program in 1981, drove it underground, and that's when it became a nuclear weapons program.

Prather: But not really a significant one until the very late 1980s. In a way, Clinton has some credibility when it comes to the North Koreans because he did attempt, near the end of his presidency, to try to honor that Agreed Framework. Maybe the Obama administration will try to honor our end of that agreement or something like it, but the genie's out of the bottle. The North Koreans have built and tested nuclear weapons as a direct result of Bush, Cheney and Bolton putting the kibosh on that Agreed Framework and essentially forcing the North Koreans to withdraw from the Non-Proliferation Treaty and kick the IAEA inspectors out.

My impression is that that's what they're attempting to do to Iran. They're attempting to make the pressure on Iran so intense that the Iranians finally say, "The hell with this. We're withdrawing from the Non-Proliferation Treaty," but they don't do that. Every time the Iranians make a statement, they say, "We want to reinforce the Non-Proliferation Treaty. We want everybody to sign on. We want Israel to sign on. We have no intention of ever having a nuclear weapons program." Not only that, but their Supreme Leader Khomeini says, "It's against our religion to build such a weapon."

Horton: Well, Dr. Prather, to wrap up this interview, I have here a short clip of John Bolton on a conference call that was recorded. If you Google "'John Bolton,' 'Iran,' 'conference call,'" it'll come right up. It's John Bolton explaining to members of the American Israel Public Affairs Committee that their purpose was to make the pressure on Iran so intense that they do what the North Koreans did, throw up their hands and withdraw from the treaty; to get Iran to say, "Fine. You know, if surrendering isn't good enough for you, we just won't even participate in this charade any longer." That's the purpose, to drive them out of the treaty. That way, it makes war a more viable option. Although Bolton denies that, he says he wants to support a revolution from within. But I'll be playing that right at the end of the interview.

Before I do that, I wanted to ask you about one more issue here. We talked about this before, and that is the idea that the Iranians or the North Koreans could somehow get a nuclear missile to our coast and then launch it into the sky over the United States. It would create an electromagnetic pulse that would turn out all the lights, destroy our civilization, millions of people would die, we'd all turn into cannibals, and I guess the Sun wouldn't rise anymore. So I just wonder, how awesome of a nuclear weapon do you have to have to actually create an electromagnetic pulse that would not just turn out the lights for a day, but would destroy the entire electrical infrastructure of a country like the United States?

Prather: As far as I know, Congress was specifically asked during the Clinton administration by the Department of Energy for permission to develop a device that might do that. That says, first of all, that we didn't have one then. We didn't have a nuclear weapon that was specifically designed to be detonated outside the atmosphere that would generate as its primary kill mechanism an electromagnetic pulse. This electromagnetic, multi-dipole pulse essentially would only affect secondary transformer loops on the ground. So if you've got a transmission line that's not shielded very well, that runs around in a circle or something like that, then you will induce in that loop a secondary pulse.

It's the same way that the old auto-ignition systems used to work. There was a primary coil and a secondary coil, and you send a pulse through the primary and a secondary pulse gets induced. That's what we're talking about. We're talking about producing an electromagnetic signal, a dipole or multipole pulse that then induces in a secondary loop a pulse.

Now let's go back again to when Congress was specifically asked for permission by the DOE weapons labs to do some studies and to develop such a nuclear weapon. That says two things. One, we did not have at that time such a weapon. Two, Congress expressly prohibited them from doing that.

Horton: But, Dr. Prather, we've all seen on the Discovery Channel that when they did a hydrogen bomb test over the Pacific, it turned the lights out. And they said, "Oh, wow! Maybe we could use that as a weapon."

Prather: Well, I was semi-involved in all of that.

Horton: Okay, so then, isn't it still true? Newt Gingrich says that if the Iranians launched one of their nuclear bombs over our country, it would destroy all the electrical circuits.

Prather: No. First of all, that was a multi-megaton nuclear weapon that was not intended for that purpose. Its intended purpose was to irradiate an incoming Soviet ICBM warhead and disable it while it was outside the atmosphere. That was the purpose of those nuclear weapons.

Horton: When you say multi-megaton, that means a thermonuclear hydrogen bomb?

Prather: Yes, and it was intended to bring down multiple independently targeted warheads.

Horton: You wrote at one point that if an atom bomb launched by the North Koreans or the Iranians — 15 years from now or something — at the United States is close enough to turn your electricity off, it's close enough to blow you apart. You wrote that your electricity's the last thing you need to worry about.

Prather: Well, even worse than that, if al Qaeda or somebody gets ahold of a nuclear weapon…

Horton: They'd have to be able to deliver it to outer space, right?

Prather: Yeah, and that's really absurd. But once again, the prohibition on their development by the DOE's weapons labs was lifted on bunker busters, but the ones that were supposedly designed to produce an electromagnetic pulse, that prohibition was never lifted. No funds have ever been requested for making such a device. They can't. They can't do something that Congress especially prohibits them from doing.

Horton: Well, obviously, there are a lot of exceptions to that rule.

Prather: I don't know of any exception in this case. Congress has said that no funds shall be appropriated under this or any other act to be used for this or that.

Horton: We have to cut it there because I want to play this clip. This is John Bolton on the phone with the American Israel Public Affairs Committee, verifying what Doctor Prather just said, that their purpose in lying and misusing the Non-Proliferation Treaty and the IAEA Safeguards Agreement with respect to Iran is to drive them out of it:

Let me now turn to the question of Iran and what I think the situation is there. The Security Council just passed a resolution at the end of last month, imposing certain limited sanctions on Iran — obviously the product of a long effort based on Iran's refusal to comply with the earlier Security Council resolution that gave them until August 31st to cease their uranium enrichment activities. I have to say, because I'm a private citizen and therefore a free man again, that these are my personal views:

This sanctions resolution is very disappointing. It is not as tough as I would have liked to have seen. In many respects, the Russians did an outstanding job from their point of view in protecting Iran, in narrowing the scope of the sanctions, and limiting the effectiveness of many of the things that we wanted to try and do to prevent the Iranians from continuing to make progress on their nuclear ballistic missile program. I think the Iranian reaction to the sanctions resolution has been very telling in that respect. Although they've passed a resolution in Parliament to re-evaluate their relationship with the International Atomic Energy Agency, they have not rejected the sanctions resolution. They have not done anything more dramatic, such as withdrawing from the Non-Proliferation Treaty or throwing out inspectors of the International Atomic Energy Agency, which I actually hoped they would do. That kind of reaction would produce a counter-reaction that would be more beneficial to us.

Lt. General Robert G. Gard, Jr.: Loose Nukes and Iran's Program April 15, 2010

Scott Horton: Our guest on the show today is retired Lieutenant General Robert G. Gard, Jr. He's the Chairman of the Center for Arms Control and Non-Proliferation. We talked to him a couple of weeks ago about possible consequences of war with Iran, and now it's time to catch up on nuclear policy. We've got a New START Treaty with Russia, we have Iran news, and we also have the Summit on Nuclear Security where Barack Obama sat at the head of the table the other day. Welcome back to the show, General Gard. How are you doing this morning?

Robert G. Gard, Jr.: Pleased to be here.

Horton: Let's start with this Summit on Nuclear Security. Almost every leader and every country was invited to this thing. What did they accomplish?

Gard: Well, I think what was done was to raise the realization among many of the leaders worldwide of the dangers of an attack by terrorists with a nuclear explosive device. This has not seemed real to many people, and we've tended to treat this in a routine way instead of giving it the priority it deserves.

Horton: How much priority do you think that does deserve? The idea that a stateless group could get together an actual fissionable nuclear weapon and set it off in the United States, do you think that is a pretty big danger?

Gard: I would say it is a relatively low priority, but it is not only possible — over time it could become probable, unless we secure the fissile materials that they could use to fashion a nuclear explosive device.

Horton: There's a guy named Richard Mayberry who's a financial prognosticator with a newsletter, and he's been pretty good. He predicted the fall of the Soviet Union back in 1985 or '86, and his prediction is that at some point, somebody is going to nuke Washington. If it's not a nation state like Russia or China, it would be a terrorist group, because you can't just go around the world making millions and millions and millions of people want to kill you without some of them actually doing it at some point. He wasn't necessarily saying that Ayman al-Zawahiri is going to do it or something like that, but rather as a matter of probability. Here we have a world with thousands of nuclear weapons, and here we have a world with millions of people who have reasons to take revenge on the United States.

Gard: Yes, and the problem is that there is enough loose plutonium and highly enriched uranium, some of which is not secured very well at all. Should terrorists get their hands on it, it would be extremely difficult to intercept it if they tried to move it into one of our ports or across our borders and explode it in this country. So what we must do, and what the summit was all about, was to secure that material at its source and make it much more difficult for terrorists to buy or steal any of it.

Horton: Since the end of the Cold War, Senator Dick Lugar has spearheaded this effort. There was originally the Nunn-Lugar-Domenici Bill right after the end of the Cold War in which they said, "Let's buy up all the extra fissionable material we can find on the planet, plutonium, highly enriched uranium. All the former Soviet Republics that had nuclear weapons stationed there? The Russians need the money, and we don't want a bunch of loose nuclear material around."

Yet, as far as I can tell, Bill Clinton and George Bush, Jr. did basically nothing to further that policy. My understanding is that Barack Obama was kind of taken under Dick Lugar's wing in the Senate. I think Lugar even brought him on a tour of Russia and really impressed on this young man that something's got to be done about all this extra plutonium in the world, didn't he?

Gard: Yes, Barack Obama, who was a senator at the time, did accompany Senator Lugar on a tour of Russia. The Nunn-Lugar program that you refer to was essentially to secure Russian weapons and to get those weapons out of 30 of the states from the former Soviet Union and back to Russia, but there's material elsewhere. Nunn-Lugar has made a good deal of progress in securing Russian weapons and, to some extent, Russian fissile materials, but there's more out there than just in Russia and the United States. Unfortunately, Nunn-Lugar's writ did not go beyond Russia. Consequently, there is now, in places like South Africa, Brazil and other countries, highly enriched uranium that is not as well-secured as it needs to be. What the summit was all about was to gather world leaders together, try to point out dangers, and to try to get them to commit to taking positive action to nail down these materials before terrorists can get their hands on them.

Horton: I know that highly enriched uranium could mean 20 percent medical-grade, but to be weapons-grade, it has to be roughly 94 percent or better uranium-235. So when you're talking about highly enriched uranium in Brazil and in South Africa, are you talking about weapons-grade? I thought at least the South Africans had given up all of their weapons-grade material along with their weapons.

Gard: Well, they've given up all of their weapons, but there was, just a year or two ago, a raid in Pelindaba in South Africa, which is a facility east of Johannesburg, that had enough highly enriched uranium in it for 25 weapons. An armed group of four men got through the outer fence, got through the electronics, got into the control room, shot an armed guard and spent 45 minutes there. So by no means have we policed up all the highly enriched uranium from all these countries, even in the case of South Africa, which has given up its weapons.

Horton: The agreement that they reached yesterday, it's not a new treaty, and it doesn't add anything to the Non-Proliferation Treaty or anything like that; it's basically just a general agreement by all these states that they're going to give up the rest of their weapons-grade uranium or plutonium. Is that right?

Gard: Some states did make a commitment. For example, Vietnam and a couple of others that I don't recall right now are saying that they would repatriate their highly enriched uranium. The problem is that in the 1950s and '60s, we had something called an "Atoms for Peace" program, where we provided highly enriched uranium to fuel research reactors in some 26 countries. Not to be outdone, the then-Soviet Union responded by doing the same thing in some 17 countries. So you're quite right that you can produce medical isotopes with uranium enriched only up to 20 percent, but unfortunately, the highly enriched uranium that we exported was quite concentrated and definitely weapons-grade.

Horton: Yeah, I guess some of the older reactors — we have the light-water reactors and so forth now — really did run on the very highly enriched stuff, right?

Gard: Yes, and there's still a lot of it out there. There is around the world right now some 3.5 million pounds of highly enriched uranium. Now, that does count the U.S. and Russia, but we're not immune. Our security has not been as stringent in some of our facilities as it should be, and the Russians have highly enriched uranium spread in some 50 locations. Plus, there's always the risk of an inside job of selling it to terrorists. So a lot of work remains to be done.

Horton: Now, I was surprised to hear you say that Vietnam had a nuclear program. I wasn't aware of that at all.

Gard: Well, I believe that that's one of the countries which the Soviet Union provided research reactor materials to. Vietnam did agree to repatriate its highly enriched uranium that was used in a research reactor back to Russia.

Horton: One thing that was really surprising was that during the news conference, Barack Obama called on Israel to join the Non-Proliferation Treaty as a nuclear weapons state. What do you think is going to happen there?

Gard: Well, I don't think that Israel is likely to join the Non-Proliferation Treaty. You mean as a non-nuclear weapons state or as a nuclear weapons state?

Horton: As a nuclear weapons state. I mean, they'd have to admit that they have nukes if they're going to join the NPT.

Gard: Well, that's right. Technically, in the NPT, there are only five so-called nuclear weapons states. The others that have developed a nuclear capability since then are not recognized by the United States or by the adherents to the treaty as nuclear weapons states, even though they have them.

Horton: Yeah, and predictably, the first news stories said that the Israeli response was, "Yeah, right." That's basically the end of that. They're not going to join the NPT.

Gard: No, and I don't think it's likely that India and Pakistan will either.

Horton: I was watching Wolf Blitzer's show yesterday, and they did a whole segment on the Iranian nuclear program that made no mention at all of — it didn't even offhandedly imply the existence of — the Non-Proliferation Treaty, the International Atomic Energy Agency, the Safeguards Agreements, or any of the inspections going on. They basically just said, "Well, Iran now has enough uranium to make a bomb" — even though it's safeguarded, and it's almost entirely at 3.6 percent. Very small amounts of it are at 20 percent, and none of it is at 94 percent weapons-grade. None of it could be diverted to be made into weapons-grade without the IAEA knowing about it. Yet they talked about this for 10 minutes yesterday with no mention whatsoever of any safeguarding of the Iranian nuclear program. Basically, they ended with, "It's a big mystery, and for all we know, they'll have a nuclear bomb put together by the day after tomorrow and kill us all."

Gard: Well, I think you're correct that the Natanz facility, which is the location that the Iranians used to enrich uranium, is under IAEA scrutiny at the present time. What we don't know is whether they have other abilities that have not been exposed. You will recall that several months ago, there was an announcement that we knew they were constructing a second facility, and sheepishly, the Iranians admitted it, even though it is not yet operational.

Horton: Well, I'm not so sure about that. I mean, isn't it the case that they declared it to the IAEA at the end of October? Then it was only after that when Obama pretended to bust them for something they had already declared to the IAEA?

Gard: No, they hadn't declared it until we publicly announced it and exposed it.

Horton: I'm not so sure. I remember it all being hinged upon the assertion in the *New York Times* that was made by anonymous officials, that the only reason they came forward is because they knew we were about to bust them. But they never even proved that.

Gard: Maybe they announced it a few hours before we did, but that would have only been because we'd been tracking it for over a year.

Horton: Well, I knew that part, but I like parsing this because, according to Mark Hosenball of *Newsweek*, the CIA had just in September 2009 put papers on the president's desk saying that they stood by the NIE that said they believed that the Iranians had not made a decision to embark on a nuclear weapons program. That was well after, as you just said, the CIA knew good and well about the construction of this facility at Fordo (or Qom).

I would go back and check the timeline on who leaked what, because it seemed to me like they made a big deal out of saying we busted them when in fact they had already reported what they were doing to the IAEA. It was only then that the administration said, "Well, they only admitted it because they knew we were about to bust him," but they never proved that.

Gard: Well, we have been tracking that. Unless officials in our government have lied publicly, we'd been tracking this for about a year, and were about to announce it. I think you're correct, now that I think back, that Iran quickly admitted that they had it. I think it was the same day or the day before we exposed it. The worrisome thing is that they were working on that thing for quite a good while, but they failed to fulfill their obligations to notify the IAEA when the protocol under which they're operating required them to do so.

Now, I'm not an alarmist. I recognize that Iran does not yet have enough highly enriched uranium to produce a weapon. What worries me is that if they can enrich it from about four percent up to 20 percent, then it gets easier to take it up to 90 percent. You can produce a weapon, an explosive device, with less than 90 percent enriched uranium. In fact, the bomb we dropped on Hiroshima was about 80 percent. Most of the weapons that advanced countries have now, you're quite right, have 90 percent-plus highly enriched uranium. The concern is that right now, they have enough low enriched uranium at around four to five percent that if they took the material already enriched and proceeded to enrich it further to get it up to 80–90 percent, they would have enough to make a warhead. The concern is that they'll do what the Koreans did, namely withdraw from the treaty when they enrich the uranium, and then put the highly enriched uranium together in such a way that they have an explosive device.

Does that mean they have the missile to stick it on and launch it against the United States? No, that's not going to happen for a number of years. But the concern is that they will go right up to what Mohamed ElBaradei calls a "breakout point," that is, when they have enough highly enriched uranium to make a weapon, they're proceeding to develop a missile, and then at time X they will decide to withdraw from the treaty, claiming that their supreme national interests require it. There are stipulations in the treaty that permit withdrawal.

Horton: This is the crux of the matter here, that they could — assuming that they have all their centrifuges working well enough and they obtain this breakout capability, as you said — withdraw from the Non-Proliferation Treaty. But the point is, all of their uranium is accounted for, and they cannot enrich that uranium to weapons-grade levels without withdrawing from the treaty, kicking the IAEA out of the country like the North Koreans did, and announcing to the world, "Now we're making bombs." This is the problem with the narrative of Iran and their nuclear program in general, that they cannot divert that uranium to military purposes without announcing to the world that that's what they're doing. One could just as easily say that they have enough uranium in the ground to make a bomb. Well, first they have to transfer it to uranium hexafluoride gas, then they have to enrich it up to 90-something percent or at least 87 percent, as you say, to get something workable at all for a gun-type nuke.

What I'm getting at is I agree with your details, I'm just saying these are the details that they always leave out. We're talking about a push in the United Nations for brand-new sanctions. Congress's version of those sanctions would amount to a blockade, which is an act of war, as you well know, General. We're talking about a serious policy based on a giant half-truth, right?

Gard: Right now, they're enriching uranium. We have offered to swap fuel rods for their medical facility for their low enriched uranium. They initially agreed and then backed off. I don't care whether we do it on their soil or elsewhere; I think we should try to follow that up so that they won't have enough low enriched uranium that if they continue to enrich it, they can produce enough for a weapon. I'm not so much concerned that they'll suddenly attack somebody with a single nuclear weapon — that'd be suicide — as I am over the impact that it will have in the region and cause others to feel that they need to develop weapons so they can deter Iran. The problem is that they can continue to enrich uranium right now, even with the IAEA watching them do it, and get it up at higher levels of enrichment. In other words, no one has stopped them yet from continuing to enrich uranium, despite the UN Security Council resolutions urging them to cease and desist.

Horton: I'm like you. I'm against anybody having nuclear bombs. I'm also against America enforcing that rule on other countries, but it seems to me that the UN resolutions demanding that the Iranians stop enriching uranium are illegal. The Non-Proliferation Treaty, which we are signatories to and are bound to respect, and that they are signatories to, protects their unalienable right to pursue nuclear technology for peaceful purposes. If the United

Nations Security Council said the United States can't enrich uranium, I'd grab a rifle. It's none of their damn business.

Gard: You're absolutely right. Article 4 of the NPT gives all states an unalienable right to pursue atomic power for peaceful purposes. The problem in the case of Iran was that even though they were signatory to the NPT, they had a program that was concealed for 18 years before a defector blew the whistle on them. So you've got a little bit of a different situation in the case of Iran, who violated their obligations under the NPT. That's what triggered the UN Security Council resolutions.

Horton: Except — I'm sorry to keep arguing with you, sir — it could very well be that I don't quite have my facts straight on this because I am just a layman radio show host here, but I always thought that basically what happened there was that Bill Clinton violated the NPT repeatedly by refusing to let China and Russia sell turn-key facilities to the Iranians. They were forced to the black market, and then all they were doing was buying uranium enrichment equipment from the heavily CIA-infiltrated A.Q. Khan Network, that the CIA knew had been around for generations, which had a booth at all the arms shows where they handed out their brochures. There was nothing even that black market or secret about the A.Q. Khan Network, and we forced them to do it anyway. The Chinese were ready to sell them a reactor back in 1996.

Gard: We gave them a reactor when the Shah was there. The fact is, when they developed a uranium enrichment facility, as a signatory to the NPT, they had an obligation to reveal it to the IAEA as a non-nuclear power under the NPT, and they didn't do it.

Horton: I thought that the rules say that it's six months before you introduce nuclear material into the equipment. It's not necessarily required that a country disclose that they have equipment or that they plan to use it at some point.

Gard: But they were enriching uranium.

Horton: No, Natanz was empty. In 2005, they took the BBC on a tour of it. They didn't start enriching uranium until at least late 2005 at Natanz.

Gard: Well, you may be more knowledgeable than I.

Horton: I'll send you a link. They got a picture of it. It's a giant, empty, underground, Walmart-sized warehouse where the Israelis did bust them, and ISIS (Institute for Science and International Security) did bust them for having the Natanz facility there, but it was still just under construction. They were still months and months away from that facility having the ability to do any enrichment. So what I'm getting at is that, yes, there is a nuclear program, but there's so much hype about it that even for General Gard and myself, despite the fact that I deal with this stuff every day, it's hard to parse it all. There's so much deliberate mis- and disinformation on the issue that it just seems like the Iranian government must be guilty of something, but nobody could figure out what other than, "Well, maybe they have some secret stuff that we don't know about," which is the Rumsfeld standard for starting a war, you know?

Gard: It could be that we phonied up all of the information that was taken to the UN Security Council that caused the Security Council to pass the resolutions, but I doubt it. There have been three of those resolutions, and when you get Russia and China to go along, there's probably some validity to the information that caused the resolutions to pass without being vetoed.

Horton: I really appreciate your time on the show. Everybody, that is Lieutenant General Robert G. Gard, Jr. He is the chairman of the Center for Arms Control and Non-Proliferation. It's ArmsControlCenter.org.

Seymour Hersh: Iran and the IAEA November 22, 2011

Scott Horton: Seymour Hersh is the author of the book *Chain of Command* about the torture regime in the Bush years, and he's written a great many articles about Iran, their nuclear program and American war plans for *The New Yorker* magazine, including "The Next Act," "Preparing the Battlefield," and, this year, "Iran and the Bomb." He's got a follow-up to that last one called "Iran and the IAEA." Welcome back to the show, Sy. How are you doing?

Seymour Hersh: Okay.

Horton: Alright, well, thanks very much for joining us today. So to get right to the meat of this thing, as far as your particular reporting that we can't find other places, you have sources who have told you the results of CIA and JSOC boots-on-the-ground investigations of Iran's nuclear program inside of Iran, right?

Hersh: Yeah. I've been writing stories for *The New Yorker* about the Joint Special Operations Command operations inside Iran that started sometime in late 2004. Vice President Dick Cheney, during the Bush-Cheney years, was convinced that Iran was cheating, even though it was a member of the Non-Proliferation Treaty community and all of its declared nuclear enrichment facilities were declared to the International Atomic Energy Agency, which is the sort of the UN watchdog group for monitoring nuclear development.

Anyway, Cheney was convinced they were secretly building the bomb somewhere. He was convinced that they were doing more than just enriching, that they were actually manufacturing weapons. So he had the Joint Special Operations Command send teams in, and most of the work was done with the help of locals. They worked with a series of dissidents, people who were against the Mullahs, the religious regime in Tehran. They worked with Iranian Kurds, they worked with Jundallah, which is a hardline Sunni fundamentalist group that was also very much in opposition to the government. The idea was to inspect any place they thought there might be secret weapons manufacturing going on and also to get tough with certain people if they had to. I don't know all of exactly what was done, but I do know that they targeted certain people. Whether this was why certain people got assassinated or not — as you know, three or four nuclear scientists have been assassinated. It's not clear who was responsible, but the fault seems to suggest Israel did it. But who knows? I certainly don't know; I just know they were targeted. I don't know whether it accomplished something or not, but that happened, and that was going on until the end of the Bush-Cheney years.

What I've done now is not really an article for the magazine; I just did a blog piece in response to the report a few weeks ago by the International Atomic Energy Agency.

Horton: I'm sorry, let me hold you right there for a second. Because the part I wanted to get to about JSOC on the ground there wasn't just about their possible involvement in the assassinations, but what you reported in "Iran and the Bomb," namely that they had put sensors everywhere, fake street signs and sensors in the roads, they had done this long-term investigation, and they still were standing by their conclusion that they really couldn't find any evidence of a secret parallel nuclear weapons program inside Iran. Is that correct?

Hersh: Yes. That is not only correct, but also in 2007, the sixteen agencies of the American intelligence community got together and produced what they call the National Intelligence Estimate, a highly classified paper, elements of which are always sort of spread out and leaked a little bit. They did a study that said that as far as they could tell, Iran dropped all interest in weaponization, even the consideration of looking into weaponization, in 2003, when we attacked Iraq. The NIE was updated in 2011, which confirmed what they said in the 2007 and the 2001 estimates, that there was nothing there.

But here's the critical point. The American estimate was that Iran may well have looked hard at the possibility of building a bomb after its eight-year war with Iraq. Iran and Iraq fought a terrible eight-year war from 1980 to 1988 in which we supported Saddam Hussein's Iraqi government against them. In the years after that war, there was a lot of talk about Iraq getting a bomb. As you know, there was that whole WMD issue. So the Iranians looked at it, but once we hit Iraq in March and April 2003 and took down Baghdad, the American estimate was that the Iranians said, "Okay, no sense worrying about Iraq anymore." And there was no way, with Iran's small arsenal. The most they could have made was one or two bombs in the next five years, whatever the estimate was. There was no way that Iran's weapons system, if there was one or if there were to be one, could neutralize or balance the 1,000 or more nuclear weapons we have or the hundreds of nuclear warheads the Israelis have.

So, the only reason they even looked at the prospect of nuclear weapons, according to the American intelligence community, the only reason Iran even studied the possibility of making a weapon up to 2003, was as a deterrent against Iraq. When that didn't happen, they were out of the business. That's still the position of the American intelligence community. You wouldn't know it from what you read in the newspapers, but that's still what the American intelligence community says.

Horton: Well, and there's nothing in this recent IAEA report that the CIA didn't know about, is there?

Hersh: There's nothing in the report that the IAEA didn't know about for many years. My problem with the report is that it's not a scientific document. This is what I wrote in *The New Yorker* on the blog page, that the report is not a scientific document; it's a political document. The most important thing that happened at the International Atomic Energy Agency in the last few years was the retirement of Mohamed ElBaradei. He was the Egyptian who ran the agency for a dozen years. He won a Nobel Peace Prize.

Anyway, when ElBaradei was there, he initially was very angry at Iran for its early work, but then worked out a lot of problems. There was always stuff put in papers, always stuff published in the London press and in America, and allegations about Iranian cheating. The IAEA would investigate under ElBaradei, and they would write reports every quarter, and they would say, "Hey, two things are important. We looked at this allegation, and we can't verify the substance of the claim. We don't believe that the documents presented are real. They're fabricated. And we're going on." Also, what the IAEA said during the last 10 years, and even in the new report, the one that includes sort of a catch-all, what the IAEA said is, "Everything that we know in Iran is under camera inspection and also individual inspection. We've never detected any evidence of a diversion of uranium." In other words, if the Iranians are going to build a bomb, they have to build it with enriched uranium, and none of the uranium that's now being enriched — to low levels, by the way, nowhere near weapons-grade level — has been diverted, so they're going to have to get the uranium from someplace else. Weapons-grade is 90–95 percent enriched uranium. Right now, the most the Iranians enrich anything is a small amount of 20 percent, for medical purposes. Most of the stuff they're enriching is around 3.8 percent, to run a power plant.

Meanwhile, the critical issue is that in 2009 ElBaradei retires. The new Director General is a Japanese politician named Amano. I'm sure he's an honorable guy, but the first thing he did — there are WikiLeaks documents to support it — is that he told the American embassy in Vienna, "Thank you for helping me get elected." There were six rounds of votes before he got in with our help, and he said, "I want you to know I share your strategic values, and I share your views on Iran." He's much more of a player in our sphere. He's one of our guys. So you have a change in leadership that results in a much tougher tone on Iran, and since he's been in, they've been taking a lot of old allegations, recycling them, being very careful to caveat what he says. A report was published of possible military uses, and the language of the report is littered with "maybe," "perhaps," "we believe," and all these caveats.

I talked to a number of people who worked at the IAEA: one who couldn't be named, a senior guy who's still there, one who just retired, and an American named Kelly, who's a very competent guy, who for 30 years worked in the American nuclear warfare business before going to the IAEA. These guys are quoted in this blog item as saying, "There's nothing new. We don't understand what's going on here." Privately, they're saying — which they don't like to say in public because they don't want to diminish the IAEA — that they have a lot of grave doubts about Amano and his integrity and about his sense that he must do the American bidding, which he probably wants to do. I'm not suggesting that he's corrupting his opinion; he's just got a different opinion. He's less objective.

Horton: He's taking all the things that ElBaradei just refused to include in the report, including the "alleged studies" documents from the so-called "smoking laptop." When you talk with your sources about that laptop, do they agree with Gareth Porter's reporting that it was manufactured by the Israelis and funneled into the stream by the Mujahideen-e-Khalq? Or do they still just shrug and say they don't know where it came from? Do they believe it or not?

Hersh: I did do some reporting on that topic, and I think I wrote a piece a few years back for *The New Yorker*. It was delivered to an allied intelligence agency. A laptop suddenly appeared. They passed it off to the U.S. The allied intelligence agency was very skeptical. It included about a thousand pages or many hundreds of pages of crude drawings, basically fantasy drawings, drawings about making a warhead and then putting it on a certain kind of missile, etc. Most of the paperwork that was done was crude, it was amateurish, and it didn't make much sense to the experts. It certainly wasn't something done in a sophisticated weapons laboratory.

That doesn't mean somebody at the University of Tehran didn't draw something, but basically the conclusion of the IAEA was that they weren't very credible. Therefore, when Mr. ElBaradei reported about it over the years, what you're saying is correct. He would list a series of new allegations that appeared either in the press or were relayed to the IAEA by Western powers, most of which were from America, the main funnel for this kind of information, and he would dismiss them. What you're getting in the new report was most of these allegations recycled without any notion that they weren't credible, unless you happened to read the early reports from ElBaradei's time in which they had been discredited. So this new report leans heavily on the laptop, which was very much discredited by most people in the community. It was a laptop from where? There was no providence.

Horton: Here's the confusing part to me, though. For example, Hillary Clinton's understanding of the situation. She may be in disagreement about her suspicions about where they mean to go from here, but the reality is the reality. If it's the Americans and the Israelis who are always trying to push the IAEA to come up with this nonsense, then they know it's nonsense too. So I wonder, you said in your last article, "Iran and the Bomb," that you have at least some sources in Israel. We see at least two, maybe three, former heads of Mossad saying, "Don't attack Iran," and I just wonder whether it's just blatant that this is merely a pretext, that all sides in Israel and America understand that there's not really a nuclear weapons program in Iran, and that they would have to withdraw from the Non-Proliferation treaty, kick the inspectors out, and announce they're making a bomb just to get started on one? So what is the point of this? Are they really afraid of a nuclear electricity program and a light-water reactor at Bushehr?

Hersh: I can't get into the minds of Israel. There's a great division in Israel which the former head of the Mossad, Meir Dagan, who's as tough as they come, was the guy that was involved in the attempted assassination in Dubai a year or so ago, a really tough character. He's been saying since he retired four months ago that this is crazy, and that they're not near a bomb. He doesn't dispute the fact they may be wanting a bomb and they may be planning to make one, but he says they're not near a bomb and that bombing or attacking Iran would be insane.

On the other hand, you have both Ehud Barak and his one-time bitter enemy, Bibi Netanyahu, the two leaders of Israel, in agreement on this. You have a Defense Minister and a Prime Minister both agreeing that this is an existential threat. I have no idea what they really think. I have no idea if they're just talking through their hat, but my guess is they are. I'm pretty sure that this administration, despite its tough talk about Iran, does not want to see Israel attack Iran. What's troubling to me is that if Israel does do it unilaterally, the Middle East would always think the U.S. were involved, no matter what we said.

Horton: Well, I'm sorry, because I really phrased the question too broadly. What I should have asked you was: What do the intel guys that you talked to over there think? They basically agree with the CIA about this, no?

Hersh: It's complicated for me because I did talk to senior people there, and the ground rules are so prohibitive that I couldn't really write what they said. I can tell you in general that there are smart people in Israel, in intelligence and in the military, who do know two things. They know that they do believe that the Americans are probably right when they said Iran stopped building a bomb. Iran's ambition about building a bomb dwindled in 2003 when we took down Baghdad. Clearly, whatever efforts they had were aimed at Iraq and not at deterring the U.S. or Israel, because how could you, with one or two bombs? They also agree that Iran is not near a bomb. They do think they still might make one. But the most important thing is — and I've had people say this to me, in essence — that it's also understood by many in the military and the intelligence community that Iran, with 2,000 years of Persian society, understands that if they got a bomb and they threw it at Tel Aviv, they would be incinerated; that the Israeli response would be that of 300 warheads hitting Iran. So there are people in Israel who understand that.

The Iranians have said — Ahmadinejad, Larijani and others have said — among other things that there's a fatwa against the bomb, which we always dismiss. Ayatollah Khamenei issued a fatwa about 15 years ago, saying that it's against the Quran to build such a weapon. I'm prepared to believe that there are many people in Iran who take that seriously. I know in America that doesn't sell, but that's what Iran says, they do take it seriously. The other issue is they're also totally aware that the bomb doesn't buy them any security. I mean, did our huge arsenal help us in Iraq? Did it help us in Vietnam? Did it help us in Afghanistan?

But I'm a finger in the dike because the mainstream press has pretty much gone along with the sort of exorbitant language in the IAEA report and all the Doomsday talk. They've gone along pretty totally. So it's a little depressing to me. But, you know, there is another side to everything, and maybe common sense will prevail. Or maybe some diplomat, some official, or somebody on the inside will stand up and say, "This is crazy," but so far, it's not happening.

Horton: Well, there was one guy that you said you talked to, Kelly, who also spoke to the *Christian Science Monitor*. He seems willing to maybe get on TV.

Hersh: Well, I'll tell you something about Bob Kelly. It's interesting to me. He's a former inspector, he's retired now. I knew him as a tough guy inside. He was very skeptical of Iran, and he still is skeptical of some things. But this new report is a non-starter, and I quote more than him. I quote others in the American non-proliferation movement. Most of the people who know anything about proliferation basically see the problems with the report, that it simply was way overblown. In any case, I'm off to lunch. Good to talk to you.

Horton: Yes, thank you very much for your time, I really appreciate it.

Gareth Porter: The Ayatollahs' Fatwas Against WMD October 17, 2014

Scott Horton: Our guest today is the great Gareth Porter. He's an independent historian and journalist, as well as the author of the book *Manufactured Crisis: The Untold Story of the Iran Nuclear Scare.* He's a writer for Inter Press Service as well as Truthout.org, where he's won awards for his work on the war in Afghanistan. Here he's got a brand-new one at ForeignPolicy.com titled, "When the Ayatollah Said No to Nukes." Welcome back to the show, Gareth. How are you?

Gareth Porter: I'm good, Scott. Thanks for having me again.

Horton: Very happy to have you here. So wow, this isn't even about the current Ayatollah. This is about the mean old former Ayatollah Khomeini, who's twice as bad as Khamenei, right?

Porter: That's right. Khomeini was the one who really played an extremely central and key role in establishing the fundamental position of the Islamic Republic of Iran on weapons of mass destruction, and that's really what this story for *Foreign Policy* is all about.

Horton: You went to Iran, and now they're publishing your book in Farsi over there, right?

Porter: Exactly, yes. That was one of the things that I did while I was in Tehran last month and into this month for about a week or so. I had a launch for the Farsi edition of my book. I did some interviews, and I had several very interesting interviews. I also participated, at least for a few minutes, in this rather strange conference that you may or may not have read about.

Horton: Yeah, well, I saw you smeared by Rosie Gray over at *Buzzfeed* for participating in a thing that would include truthers and such, and they went on and on with their guilt by association, which I thought was really hilarious. I defended you on Twitter where it really matters, Gareth. [Laughing] Anyway, let's talk about this awesome journalism that you've done. While you were in Iran, you interviewed at length in his office the former head of the Iranian Revolutionary Guard Corps, and you learned some things. Is that right?

Porter: Well, he was not head of the Revolutionary Guard Corps; he was the Minister of Sepah, which is the minister of Revolutionary Guards, and that's a slightly different position. But he was a very key person with regards to the supply of weapons during the Iran-Iraq War.

Horton: Okay, so rewind everybody back to the Reagan years, when George W. Bush's father was the vice president, and Saddam Hussein worked for us against Iran because they'd overthrown America's sock puppet dictator in 1979.

Porter: Yeah, so this was a very interesting interview, as I knew it would be, by somebody who had actually met with Khomeini on two occasions. He may have met with him on other occasions as well, but there were two occasions which were crucial to understanding the wartime policy of Iran with regards to the question of chemical weapons. First of all, this issue arose because the Iraqi forces began to use chemical weapons to attack Iranian forces as early as 1982 or 1983, and then a little later in the war actually began to use them against civilian targets inside Iran.

So, as a result of the Iraqi use of chemical weapons against Iranians, the head of the Ministry of Sepah, whose name is Mohsen Rafighdoost — who is now retired from the government and whom I met — recalled how he had prepared a report for Ayatollah Khomeini, then-Supreme Leader of Iran. He prepared a report in which he had organized groups of specialists on different kinds of military needs that Iran had for its war against Iraq. One of the groups that he had organized were young specialists with various technical expertise in the areas of biological, chemical and nuclear. So he went to see Khomeini to ask his reaction to this list of special groups. Particularly, he was interested in that particular group of young specialists because he wanted to do something to respond to the Iraqi use of chemical weapons.

What happened in that meeting, as he related to me, was that Khomeini said, "What's this?" when he saw this mention of the chemical-biological-nuclear group. So Rafighdoost explained his plan to have this group work on various kinds of weapons that might be needed against Iraq, and the first thing Khomeini said was, "Well, forget about nuclear weapons. We're not going to do nuclear weapons. That's forbidden by Islam." As for the chemical weapons, he said, "No, you can work on defensive means to defend against the use of these weapons by Iraq" — meaning to have gas masks and other things that would detect the use of the weapons and give some protection against the weapons, but not create or use chemical weapons. That was a key decision by Khomeini, and he based it on Islamic doctrine and Islamic jurisprudence, saying that it was not consistent with Islam.

Time passed, and in 1987, the Iraqis used chemical weapons against the civilian targets in a place called Sardasht, a city of some 10,000 people. At that point, Rafighdoost went into action again, and this time he went further. He actually got the precursor chemicals for mustard gas and set up a facility with which to manufacture the weapons. At that point, he went back to meet with Ayatollah Khomeini to ask him whether he would approve the manufacture of chemical weapons. In that second meeting, Khomeini again said, "This is Haram. This is forbidden by Islam. You cannot do it," and that was the end of the work on chemical weapons. They were never weaponized. All this was recorded in a document which was made public in a WikiLeaks release of a diplomatic cable that reproduced the entire document that Iran had passed on to the United States government in 2004.

Horton: As far as I know, the most in-depth and, I think, the best study of America's support for Saddam Hussein's use of chemical weapons in the war against Iran is also at ForeignPolicy.com. It was written, I think a year ago, by Matthew Aid and Shane Harris, and it's called "CIA Files Prove America Helped Saddam as He Gassed Iran." For people who are unfamiliar, these guys have real fact-checkers, and they don't publish articles that aren't fact-checked.

Now, Gareth's story is about the Iranian side of things. He just got back from Iran, where he interviewed one of the leaders of the Iranian Revolutionary Guard Corps who, when America was helping Saddam target Iran with those chemical weapons, went to the Ayatollah twice and said, "Let me please use chemical weapons and/or nuclear weapons," and was told no. And this was by Ayatollah Khomeini, not Khamenei. This was Khomeini, the one who was involved in the coup back in 1979, who people think of as the Great Satan himself for thwarting American will and embarrassing America so badly with the hostage crisis back then.

Anyway, I want to make sure I understand this right. I believe what you said, Gareth, was that the first time he brought up nukes also, and the Ayatollah said, "Forget it," but the second time he went back to him, he was really only asking about chemical weapons. He had already gotten some of the precursors together and was really hoping for a green light, but he didn't even really try nukes at that point. Or did he bring up nukes again at that point and the Ayatollah told him no again?

Porter: My recollection is that the nuclear issue was not on the table in the second conversation. The primary purpose of it was clearly to try to get Khomeini's okay to go ahead and produce mustard gas in order to respond to the Iraqi attacks on Iranian civilians.

Horton: You say here that he said no to chemical and nuclear weapons for religious reasons; that he said, "This is Haram, you can't do it," but he also cited the relationship with the United States, saying, "Are we going to have a bad relationship with America forever, for 1,000 years?" He says something like that here, and he knew that if they embarked on a chemical weapons program, that that would just be making matters worse as far as any possible rapprochement with us. He really said that?

Porter: That's right. This was not necessarily something that was said in that same meeting. In fact, I don't think it was. It would have been soon after the revolution, when Rafighdoost was looking for a place for the headquarters of the Revolutionary Guard, and proposed the former U.S. embassy site. That's when Khomeini said to him: "Why go there? Do we expect to have a hostile relationship with the United States for a thousand years? Just don't go there." So that was an indication of Khomeini's pragmatism, even at the very beginning of the Islamic Republic of Iran.

Horton: So the conservative argument here is going to be: "Yeah, right. Politicians say lots of things, and theocrats even more, but that doesn't mean that it's true. Why should we believe that some fatwa is actually legitimate rather than just the cover story?"

Porter: The reason that we should believe this fatwa was real and effective is precisely the fact that Iran did not go ahead and weaponize the precursor chemicals for mustard gas, which they could have easily done, to retaliate against the Iraqi chemical weapons or even to say, "Okay, we now have the chemical weapons, Iraq. You cease and desist, or we will strike back." They didn't do even one of those things. In fact, until the very end of the war, Iran was unable to pose even a threat of using chemical weapons against Iraq. That is a situation which cannot be explained in any other way, and has never been explained in any other way, than the effectiveness of the fatwa by Khomeini. To me, that is a very convincing argument that it was indeed Islamic jurisprudence that determined Iran's policy against weapons of mass destruction during that war.

Horton: In the article, you talked about the difference between a religious fatwa about how you should pursue your courtship or whatever, as opposed to the Supreme Leader of the Islamic Republic issuing an actual religious edict over the political government. Talk about that really fast.

Porter: Exactly, that's an extremely important point. There's a lot of nonsense that has been published about the question of fatwas, where people argue: "Oh look, all these Ayatollahs have put out many fatwas where they changed their mind later on, or they just put out obvious nonsense, and therefore fatwas are not to be taken seriously." Well, I'm sure that's true. There are plenty of fatwas that are not worth anything and which are only in fact binding, even in the theory of Islamic legal practice, on those people who choose to follow that Ayatollah.

However, in the case of the Supreme Leader of the Islamic Republic, the Supreme Leader's fatwas on matters of government policy are binding on the entire government. They have a status which is higher than legislation in the Islamic Republic. So that's a very different situation, and all those arguments that have been put forward for paying no attention to fatwas simply don't apply in that situation.

Gareth Porter: The Iran Nuclear Agreement August 18, 2015

Scott Horton: Alright, guys, it's our friend Gareth Porter, live from Lebanon. Hey, Gareth, how are you doing?

Gareth Porter: Hi, Scott. I'm fine, thanks for having me again.

Horton: Very happy to have you here. You wrote a thing, and that's why I wanted you to come on the show, so you could talk about the thing that you wrote: "Don't Expect Much Change in Post-Vienna U.S. Middle East Policy." You go through quite a few different recent statements by the president, beginning with his big speech at American University in favor of the deal, analyzing what it is that Obama says and doesn't say about the future of America's relationship with Iran. So go ahead.

Porter: Right. A lot of people in Washington who are strong supporters of this agreement would like to think that Obama is going to begin — once this is safely approved by Congress and he doesn't have to follow the script that is written essentially by the Israeli lobby and by the Saudis — to talk like a rational head of state and talk about what needs to be done to reduce these conflicts. So my piece is really looking at the evidence and saying, "Well, no. I'm sorry, the evidence really doesn't show any sign that this is about to happen in a couple of months if this agreement does in fact go through as we all hope it will."

On the contrary, his words, starting with the speech at American University, seem to suggest the absence of any clear-cut conception of any role the United States might play after the Vienna agreement. The American University speech, I thought, was particularly telling, because if you're ever going to say anything that's going to suggest an intention of playing a role as peacemaker, of changing the direction of U.S. policy in the Middle East, then certainly the 52nd or 53rd anniversary of the American University speech by JFK would be the time to do it. But as I read that speech, I find absolutely nothing suggesting that the thought of acting in the spirit of JFK's American University speech and its general central notion — that it's time to move from confrontation to negotiation with the main foe — is going to apply to the Middle East. On the contrary, it seemed to me that all he did was to say that we're going to be better off with this agreement than we would be without it. It was entirely devoted to defending the agreement per se, and not a word of it really went beyond that.

Horton: Then again, maybe it's just that in this current climate, trying to mix any other issues with the nuclear deal is a deal-killer. So let's just deal with this, and then we'll worry about the fact that we're fighting for Iran in Iraq this whole time.

Porter: Well, I think that's correct. The key point that you made was to emphasize that it's the political climate that determines what Obama is able and willing to say about regional politics and policy in the Middle East. And that political climate is so much more decisively right-wing today than it was in 1963 that, I would argue, it does in fact impose very tight constraints on what we can expect reasonably from President Obama.

Horton: Now, I wasn't around, but really? As far as…

Porter: I'm not suggesting that it was a liberal climate in 1963 by any means. And let's bear in mind that…

Horton: Maybe just a little bit more grounded in actual facts rather than pure propaganda like the War Party now?

Porter: Well, one can say that there are a lot of parallels between 1963 and today in the sense that Kennedy felt it necessary to subvert his own administration's policy toward Vietnam, for example, as I document in my book about how the United States went to war in Vietnam.

Horton: Right, right.

Porter: So that definitely is a parallel, but I think that the president's constraints go even further than that.

Horton: My thing is, Gareth, I don't really want the U.S. to get along better with Iran more than just want the U.S. to stop threatening them with aggressive warfare all the time. Stay within the NPT, and we're cool. In fact, even then they've got the right to withdraw from the NPT, and what the hell do I care? I want America to butt out completely, and I'm afraid that if the U.S. government does start getting along better with their government, that will mean fighting for the Dawa Party, the Supreme Islamic Council and the Badr Brigade. Just like before, again, more. In fact, probably in a way that'll just make the problem with the Islamic State worse, if everything we do just plays directly into their narrative that it's the Americans, the Shiites and the Jewish Israelis lined up against them.

If all our government is doing is proving that they're right about us in the way they frame it in their rhetoric, then doesn't that just make them more powerful? If Iran wants to fight the Islamic state, shouldn't we just butt the hell out rather than work with them on such a project?

Porter: Well, frankly, I don't differ with you very much with regard to the issue within Iraq, of the U.S. cooperating militarily to defeat ISIS. I think that it is a dangerous notion and a dangerous situation. I would just point out that the Middle East also involves Syria, Yemen and Lebanon, and that, in fact, what we're talking about here is a region-wide situation where the Saudis are on a rampage. They have a Sunni coalition of states that is in one fashion or another going along with, if not actively militarily participating in, a war in Yemen that is wrecking that country, bringing about a humanitarian disaster of massive proportions. So we're talking here about a situation in the Middle East which demands diplomacy; that does not have to do with military cooperation at all, but rather as the opposite. It has to do with the United States playing the role of saying, "Look, Iran and Saudi Arabia, it's time to cut out the bullshit" — pardon the expression — "and sit down and get this done." The United States must have an understanding that is going to end these conflicts. That means leaning on the Saudis very heavily.

It's necessary to stop the demonization of Iran. That's the minimum requirement for this kind of policy to emerge. What I'm suggesting in my article, although I don't go into it in detail, is that the rhetoric that we're getting from Obama is really still moving in the opposite direction. It's still moving in the direction of the demonization of Iran. That's really the problem that we face. Some people might read some of the extremely ambiguous or impossible-to-penetrate rhetoric in a couple of the interviews where he says, "Well, if everybody in the region were to shape up and realize that their main problem is chaos and ISIL, that would be great and we should welcome that." I'm not exactly sure what that means. I don't think it means anything. One might optimistically read that as meaning that he's really intending to play that role diplomatically once he's free and clear of the danger of losing the vote in Congress. I doubt that, because of the weight that he gives to continuing the demonization theme. There's simply no letup on that. His main argument in defending the agreement is that we're better off in a region where the Iranians are continuing their misbehavior without nuclear weapons, and that has become the primary argument now for him to combat the attack on the agreement from AIPAC and its friends.

Horton: That's certainly a very dangerous game to play there, buying into the War Party's narrative 90 percent but then with a different conclusion. "That's why we ought to do it my way." "It's this or war," rather than, "It's this, or hopefully they'll still remain within their Safeguards Agreement in the NPT. Then maybe we could work out a better agreement later someday or something."

Porter: The other angle that I want to mention is that without a shift in the diplomatic posture of the United States toward the region, particularly toward Iran and Saudi Arabia, the United States is still going to be defending whatever the Saudis do and whatever the Saudis' clients do in order to make sure that we can hold on to our military bases. In other words, the tight linkage here between the fealty to the Saudi line in the Middle East and the interests of the national security state. Making sure that we're going to hold on to those bases throughout the region — all in the Sunni Arab states which in one way or another are associated with the Saudi strategy — that is going to continue.

Horton: By the way, the last good thing that I read about America's role in the Yemen war was in the *Wall Street Journal* like six weeks ago, saying more or less that yeah, America's running the thing. They don't say that it's our planes and ships, but they do say that it's our spies picking the targets and our officers who are the command and control. They're the air traffic control for the Saudi coalition, which does include a lot of different countries who have sent bombers to go and help them in their war over there. But can you tell me to what degree Obama, the U.S. and the DoD own that war over there?

Porter: Absolutely, there's no question about it. This has become not an American war per se, but a war that the United States is up to its neck involved in and supports. The U.S. has said nothing, not one word, to indicate that it frowns upon, is worried about or intends to pull the plug on this terrible war that the Saudis are waging in Yemen. I think that that is part of the price that Obama is prepared to pay to maintain the close alliance with the Saudis. That's really where the rubber meets the road. Is he going to be willing to put real distance between himself and the Saudis in order to have an independent diplomatic posture in the region? So far, no sign of it.

Horton: Here it is in the *Washington Post* today: "In Parts of Yemen, Rebels Have Lost Control. No One Else Has It Yet." Yuh-huh, al Qaeda.

Porter: Right, exactly. There's no question that al Qaeda's taking advantage. Nobody's really doubting that. Nobody's expressing the slightest doubt.

Horton: Yeah, you can read it right here. Gareth Porter, everybody. *Manufactured Crisis* is the book. Thanks. Appreciate it.

Porter: Thank you, Scott.

Jim Lobe: Pro-Israel Supporters Working Against the JCPOA September 4, 2015

Scott Horton: I've got Jim Lobe on the line. He is from LobeLog.com, and he has got a couple of very important pieces for us to look at here. In fact, that entire blog is very important. Nothing gets published there that is not very important, so please go and look at it. These two are "AIPAC's Plan B" and "Cardin, the Iran Deal and the Future of AIPAC's Plan B." Welcome back to the show. How are you doing Jim?

Jim Lobe: Hi, Scott. How are you?

Horton: I'm doing really good. I appreciate you joining us on the show here. We have enough Democratic senators now not necessarily to filibuster, but definitely enough to prevent the overriding of the president's veto of the presumably soon-to-be-passed Republican resolution against the Iran deal. Therefore, its passage is secure now. It will be implemented. It's virtually certain and locked up now. So this is the reason for the term "Plan B." That didn't work, so now what are they going to do, Jim?

Lobe: Actually, there is kind of an update to my last post on this, because Senator Cardin said that he will vote against the Iran deal. He said this in an op-ed that was published on the *Washington Post* website just about an hour ago. In it, he says that he is going to introduce legislation that I think is Plan B, although it's not entirely clear because he doesn't provide much in the way of detail. This Plan B is essentially designed to "strengthen" the JCPOA — that is, the nuclear deal between the P5+1 and Iran — by adding some provisions which he would like to see enacted into law. I think that in this respect he is acting as a cat's-paw for AIPAC. He laid out four provisions of the legislation he intends to introduce, but they're not very specific. He also said that these provisions would be consistent with the administration's interpretation of the JCPOA.

So until we actually see the details of that bill, we can't really know if in fact such provisions will be consistent. But this op-ed does tend to confirm that Cardin wants to introduce something that will probably be backed by — and probably drafted in part by — AIPAC which could still derail the deal, particularly by strengthening hardliners in Iran who argue that the United States cannot be trusted and will not act in good faith.

Horton: In the previous piece, you have some draft legislation that somebody furnished you with. Do you think that what Cardin is going to be introducing is different? It sounds like it's probably along the same lines, even down to the number of different hurdles they're going to introduce, no?

Lobe: I think it will definitely overlap. My suspicion is that this legislation is still being negotiated because…

Horton: It's an important taste of what direction they're going, trying to figure out ways to sabotage the deal…

Lobe: Right. Essentially, what they want is legislation that they say will strengthen the JCPOA but will actually probably include a number of poison pills — that is, provisions that would be unacceptable to Iran or to Washington's negotiating partners. Again, he lists four general points, but we don't know the specific provisions. We do know that some of the specific provisions that appeared in the draft I published a few days ago would be poison pills. For example, providing Israel with the latest bunker busters and the means to deliver them, or immediately extending the Iran Sanctions Act, Iran has already said that those would be unacceptable to them.

So we have some idea of what would in fact be a poison pill if it were to be included in this legislation, but Cardin doesn't provide much in the way of detail. There are certainly hints that it may include provisions that would be poison pills, and undoubtedly AIPAC would be pushing him and whoever backs this bill — Republicans in particular — to include poison pills. The issue then will be whether they will gather enough Democrats who are uncomfortable with the JCPOA even though they will end up voting for it. Will they attract enough of those Democrats to create a veto-proof majority?

Horton: In other words, as far as those Democrats go, even if politically they don't want to oppose the president on this, they're sort of on the line. Well, maybe we can get them to vote for these amendments. That way they can say that they supported the deal, but they are actually supporting things that will ruin the deal and will obviously force Iran out of the deal.

Lobe: Well, it wouldn't be in the form of an amendment, because what the Congress will be voting on is a resolution. What they're suggesting is that there would be, immediately after or even at the same time, additional legislation designed, in their words, to strengthen the JCPOA, but it actually may be designed to sabotage it.

Horton: Politically that's smart, right? That sounds like exactly the kind of thing that they will be able to pressure some Democrats on the fence into making their compromise.

Lobe: Right. For example, yesterday, both Senator Mark Warner of Virginia and Senator Cory Booker of New Jersey announced that they will vote against the resolution that would reject the JCPOA, but then in their statements they announced that they would favor legislation that would, for example, deliver the bunker busters, which Booker wanted, or extend the Iran Sanctions Act, which Warner included in his statement.

So these Democrats might be attracted to this legislation that Cardin is talking about, and it's significant that it's Cardin, because Cardin is the ranking Democrat on the Senate Foreign Relations Committee, so he has influence. The question is: What will be the details of this provision? Will there be poison pills? How will the White House react? Can the White House keep the Democrats in line? So the war over the JCPOA is not over. Remember also that the Supreme Leader of Iran, Khamenei, announced yesterday that he wanted this to be debated by the Iranian parliament — the *Majlis*. Hardliners, who are overrepresented in the *Majlis*, are going to use any effort by Congress now to pass additional legislation that could be interpreted as sanctions — or as threatening military action by Israel or whatever — in the debate in Iran. What Cardin has done will essentially strengthen hardliners who will be arguing against ratification of the JCPOA by the Iranian parliament. So the JCPOA is not a done deal, at least as far as this country is concerned, and possibly as far as Iran is concerned.

Horton: To try and look on the bright side of it, with Booker and Warner — and it says here Heitkamp — now that they have come out in favor, we're actually getting pretty close. How many votes away are we from the ability of the Democrats to possibly filibuster the negative resolution?

Lobe: It wouldn't be a filibuster, because there is an agreement that this resolution would pass under regular order. That means that it has to be negotiated between the majority and the minority leadership, and the minority leadership will probably say that since it is being done under regular order, the resolution needs to pass by 60 votes. So technically, it wouldn't be a filibuster.

I think there are now four outstanding Democrats who have not committed one way or the other. I don't know. I think it's very difficult to predict. I think that Senator Wyden from Oregon is the most likely to line up with the Republicans in opposition. The side that wants the JCPOA needs three more votes out of the four who remain uncommitted. As I say, I think Wyden will oppose. I think a lot depends now on Senator Blumenthal from Connecticut. I think he is probably the key swing vote. But I haven't done the math this morning, because Cardin was still up in the air, and, as I said, he just decided to vote no.

Horton: That's really too bad about Wyden. He seems like a decent and honest enough guy that he could be intellectually honest enough about this issue to come down on the right side of it, but I guess not.

Lobe: My understanding is that he is very tight with AIPAC and the lobby, but I personally have no great inside knowledge about it.

Andrew Cockburn: U.S. Support for Pakistan's Nuclear Weapons Program
December 15, 2009

Scott Horton: I'm happy to welcome Andrew Cockburn back to the show. It's been a little while. He is the author of *Rumsfeld: His Rise, Fall, and Catastrophic Legacy*. And I'm looking at a great article that he wrote last June at CounterPunch.org, called "How the U.S. Has Secretly Backed Pakistan's Nuclear Program from Day One." Welcome back to the show, Andrew. How are you?

Andrew Cockburn: Good to be with you.

Horton: I appreciate you joining us on the show today. Let's talk about, well, when was day one of Pakistan's nuclear program, and how did American support for it start?

Cockburn: It was a long time ago. It's a not-unthinkable story. We pay lip service to the great cause of non-proliferation and how it's more important than anything else to stop nuclear weapons spreading around the world, and yet at every turn, when an ally of ours or someone of use to us is developing a nuclear weapon, we always looked the other way. Israel is one notable example, of course. But Pakistan as well, I think people would be more surprised to hear.

If you go back to the time of mid-'70s when Mr. Khan, the father of the Pakistani bomb, was working in Holland as a nuclear enrichment intern, he was actually filching designs for nuclear uranium enrichment and sending them home. Dutch security people got onto this and said, "Hey, this guy is stealing nuclear secrets that could help his country make a bomb and is sending them home." They checked with the CIA, and the CIA said, "No, no, leave him alone. We want to find out more about the network." So he continued his work unmolested, the blueprints went home, and Pakistan went on working on its bomb.

We get to 1979–1980, when Pakistan and the U.S. together start the great jihad against the Soviets in Afghanistan. The Pakistanis, by this time, were working hard on their bomb, and the U.S. knew all about it, but it was more important to fight the godless communists in Afghanistan than to stop Pakistan from building a bomb. When Ronald Reagan was asked, "What about the Pakistanis? Should we do anything to stop the Pakistanis from building a bomb?" his answer was, "I think it's not our business."

Horton: Well, let me stop you there. Was that necessary? I mean, was it the case that the Pakistanis wouldn't cooperate with American support for the jihad against the Russians in Afghanistan unless we turned a blind eye to their nuclear program? Do you think that the Reagan administration thought that that was the deal, or did they have another reason for turning a blind eye to this?

Cockburn: I think if you asked them, "Is it very important for us for Pakistan to have a bomb?" Well, there are a number of things. Pakistan, among other things, sits on the Iranian border, and this was a time when Iran was just emerging as a big enemy. Also, for an ally of ours — like Israel or all the European countries — to have nuclear weapons is an instrument of control for the United States. A very important part of U.S. control of NATO, in particular the European nuclear nations, is that the U.S. helps them a lot, sells them components and things like that. By that measure, they are rather dependent on the U.S. That was certainly true of the French for a long time when they were allegedly completely separate, but in fact, they depended on the U.S. to help them get their nuclear force up and running. So there may have been that thought, too.

Horton: I'm trying to nail down the timing here. This was after Nixon went and shook hands with Mao and the Sino-Soviet split. Is it correct that Pakistan and China are sort of allies in the containment of India from their point of view?

Cockburn: Absolutely.

Horton: And India was more in the Soviet camp, so China and Pakistan were considered to be in the American camp as versus the Soviets back in the 1980s, is that right?

Cockburn: That's right. In fact, a huge proportion of the arms we bought for the Afghan Mujahideen, who are now the people we're fighting, were bought from China, AK-47s and so forth. We're talking about hundreds of millions, billions of dollars' worth of weapons.

Horton: [Laughing] They were deniable because they were AKs, not M-16s, right?

Cockburn: Exactly. They did the same thing with Egypt. This is a usual practice for the American government.

Horton: I'm sorry because I was diverting you. You were kind of going through the chronology here, and I stopped you when you were talking about the motivation for the blind eye that Reagan turned.

Cockburn: Well, that's okay. Things carried on. Two themes were interesting. On the one hand, we were turning a blind eye. On the other hand, we continually said that non-proliferation is a sacred cause, and we will leave no stone unturned to stop nuclear weapons from spreading around the world.

As often happens in these cases, there were some people in the bureaucracy who didn't get the real message. For instance, there were people in the CIA and in the Defense Department who thought that they were in the Non-Proliferation Department, who thought they should do their job. Michael Barlow, for example, who was studying the part of Pakistani illicit components network — there were six men buying components from around the world — was trying to close it down. At one point, he and Customs had organized a sting for a major Pakistani illicit nuclear components-buying operation, but it didn't work because the State Department tipped off the Pakistanis that this was coming. That's how schizophrenic the policy was. One of the lead people for making the world safe for the Pakistani bomb at that time, in the late 1980s and early '90s, was Secretary of Defense, Richard B. Cheney.

Horton: In your article, you have the American War Party talking about: "Oh, my God! Pakistan has kind of fallen apart, and there's a danger that these crazy people could take control of the nuclear weapons! So maybe we need to take control of them." Yet, as you say in here, American money is going to finance the advancement of Pakistani nuclear weapons to this day. Not just in the Bush years, but since January 2009.

Cockburn: That's exactly right, and their logic is that we need to make Pakistani weapons more secure. So we have all sorts of fancy technology to make sure they don't go off if you drop them, which is our way of getting into the Pakistani program and getting them to cooperate with us so we can learn more about it. But in the process, we're making the Pakistani weapon more operationally useful because the more secure it is, the more they can afford to have it all in one place. At the moment, Pakistanis keep it in bits, partly to stop people stealing it, but also because it's inherently unsafe. The phrase in the industry is "one point safe." When people describe a weapon as one point safe, that means that there's only one chance in a million that it will go off accidentally. Well, Pakistani weapons have not been one point safe, but we're helping to make them so, which means the Pakistanis will feel more confident about keeping the warheads on the missiles. And the U.S. is financing that.

Horton: Whereas right now they keep them in pieces, we're giving them better technology so they can go ahead and keep them ready to go.

Cockburn: Right. This obviously makes the Indians more nervous, so the Indians are moving to be more operationally ready, and the world is moving closer to the hair-trigger.

Horton: Well, no problem, we'll just help the Indians with their nuclear weapons program as well, right?

Cockburn: We already are. Despite them not being signed onto the Non-Proliferation Treaty, we've sold them billions of dollars' worth of nuclear technology. So the U.S. is back to their old game of arming both sides.

Horton: It's interesting to me that at the same time that we're beating the Iranians over the head with the Non-Proliferation Treaty as though they're not within it, here we are helping the nuclear weapons technology of India and Pakistan, who are both armed to the teeth with nuclear weapons, who have gone to war four times since their division in 1949, who have a longstanding feud that continues in Kashmir, and we're helping both of them with their nuclear weapons. I mean, tell me I'm confused. This can't be right. America? The land of the free and brave and peace and all that stuff?

Cockburn: Well, we've done it before, in 1986, during the Iran-Iraq War. In January 1986, the Iraqis launched an offensive against the Iranians in the al-Faw Peninsula. Both sides were operating with satellite intelligence maps supplied by us.

Horton: They're like Israel. It's just India, Pakistan, Israel and North Korea that have nuclear weapons but are not members of the Non-Proliferation Treaty, right? Here we are, beating Iran over the head with it, while we completely violate it.

Cockburn: Well, I'm not thrilled with Iran joining that club, but I agree there is an element of hypocrisy.

Horton: Thanks very much for your time today, Andrew. That's Andrew Cockburn. He is the author of *Rumsfeld: His Rise, Fall, and Catastrophic Legacy*, and he writes at CounterPunch.org.

Doug Bandow: North Korea's Nuclear Weapons December 13, 2017

Scott Horton: Introducing the great Doug Bandow from the Cato Institute. He's been all around the world a hundred times, and he's good on just about everything, as far as I can tell. Welcome back to the show, Doug. How are you?

Doug Bandow: Happy to be on.

Horton: Happy to have you. I should mention that you write for *The National Interest* and for *Forbes* as well as Cato.org.

So, North Korea. Do you think there's really going to be a war, or we're going to get close? It seems like Donald Trump has drawn, like Barack Obama, a red line that's unenforceable without a major catastrophe.

Bandow: Well, that's the scary thing. If you listen to what he said, it sounds like we're going to have one. Basically, he says he's not going to let the North build nuclear weapons and missiles, but the North says they want them for defense. You put those two statements together, it looks pretty bad. My hope is that he's bluffing and that he's trying to intimidate them, but we don't know. I mean, nobody really is inside the head of Donald Trump, so exactly what he plans on doing, we just don't know.

Horton: The parallel to Iraq War II is not there, right? Where they're hell-bent on lying us into that regime change policy one way or the other? It doesn't seem like that. Eric Margolis was on the show, and he said, "I'm really afraid they're going to blunder into this thing, that Trump thinks he knows what he's doing, but he doesn't know what he's doing, and he's going to end up making a series of mistakes that causes this thing to ignite."

Bandow: Yeah, the greatest danger here probably is mistaken misjudgment. It's not like they're going out and selling this to the American people, making a case for why they need to do this. The Bush folks did that very strongly. They went out of their way to mislead and to lie about everything to get us into that war. My sense is that Trump seems to assume that if he talks tough, everything will work out. I worry that what he's going to do is convince the North Koreans that war is coming. If they think it's about to happen, then they have an incentive to strike first. The worst thing they can do is to wait for the U.S. to attack. So he could very well trigger a war that nobody wants simply because of his kind of blundering and blustering.

Horton: Good cop/bad cop makes sense, right? Nixon telling Kissinger, "Tell them I'm drunk and I'm going to do something crazy." I don't approve of that, but I understand it. But if Trump is playing bad cop, and he sends Tillerson over there to play good cop, then what sense does it make to tweet, "Don't bother, Tillerson"? Which is another way of saying, "North Korea, don't bother listening to him. He doesn't have my confidence, so he can't really make any promises on my behalf."

Bandow: That's obviously the problem. A good cop/bad cop works only if you actually believe the good cop has some authority. If you believe that the good cop is irrelevant, then it doesn't matter. So that suggests that if that is his strategy, he's blowing it. I suspect it's probably not his strategy. He just doesn't think talking will actually help, even though during the campaign, he said he'd be happy to sit down with Kim Jong-Un. So this is somebody who I think gets up one morning and has one feeling, and maybe he gets up another morning and feels something very different. Again, if you're trying to figure this out — and frankly, if you're South Koreans or North Koreans trying to figure this out — you have no idea. It's kind of scary because you just don't know what might happen.

Horton: Trump and Tillerson, and you also have Mattis and McMaster up there, they must have some kind of policy that they've agreed on? Not that they can necessarily get the president to stay within it, but they have had at least back-channel direct talks with the North Koreans, right?

Bandow: It's been very limited. Apparently, they've been primarily through the human rights ambassador who's a holdover from Obama, so he doesn't have much authority. There's been some informal stuff with private Americans talking, but it's not at all clear what comes of that. Unless you have the decision-makers sit down, it's not clear, you can't make much progress. If you send in somebody from the previous administration who doesn't meet with the current president, what can he achieve in those talks? Then of course the whole thing with the death of Otto Warmbier, the college student, really kind of set things off there as well, so I don't know if we're getting very much talking going on. That's scary to me. We talked with the Soviets during the Cold War, nasty people, but it would have been dumb not to. But as far as I can tell, we don't have anything regular going on with North Korea.

Horton: I don't understand. Why not, Doug? It seems like Donald Trump is not a Clinton or a Bush; he ran on being divorced from their stupid policies and wants to do things different. As you said, he said he would talk to Kim Jong-Un; he said he was a "smart cookie," which was kind of a weird thing to say, but it was a compliment. He was trying to thaw the ice a little bit there, but then what happened?

Bandow: I think that Trump is somebody who has feelings but not policies. He's somebody who during the campaign rightly talked about the importance of diplomacy, but then also clearly has this urge to go mano-a-mano, be the tough guy and perhaps feels that that's the only way he can sell his presidency, by appearing to be the tough guy. But you look at what he said about Europe and other things. He's complained about the Europeans living off the United States on defense, then Tillerson and Mattis kind of do all the same stuff. He recently gave a speech, and he again complained about the Europeans not doing enough.

You wonder why, it's a year after you've been elected, you're president, why don't you do something? It's very hard to understand. So there's nothing coherent going on. Certain things bug him, he gets up one day, wants to emphasize that, he talks about it, but there's no follow-through. There's no strategizing. He's very good at playing to the public in certain ways and recognizing what bugs people. A willingness to say what he thinks and not feel the need to filter himself gains a certain amount of public appeal, but I don't think he has a strategy. I don't think he's ever sat down and thought, "I want to achieve X," whatever X is, "Now, how do I do it?" Instead, he just kind of does this stuff, people pay attention, they applaud, and he moves on to the next one.

Horton: It just seems like all he ever does is ratchet tensions up by one degree or two and then move on to the next thing. [Laughing] He could be ratcheting them down and then get distracted, but anyway…

Bandow: That's right, I agree. Unfortunately, it seems to all go in the wrong direction.

Horton: Talk to me about North Korea's new missiles. It used to be that, "Well, maybe they can detonate a nuke if they dig a hole real deep in their own country, but they can't deliver it." But now all that's changed. Now they have a true deterrent. Supposedly they can even hit Austin or Washington.

Bandow: Probably not. I mean, the missile they sent up didn't have a regular warhead on it, which would have been a lot heavier, and that would have had an impact on its range. The other thing is that it's not at all clear that they have missiles that can hold together well enough and won't fall apart after re-entry. We also don't know how they're targeting. At least until recently, we didn't think they had very good targeting. So they're still not at the point where they can drop one down on the White House into the Oval Office, but that's coming. The point here is the North has shown that it's done a lot better than people predicted, so it's serious and it's going to continue. I don't think they're going to give that up by negotiation. I just cannot believe that they're going to say, "Oh, after all that effort, we're going to give it away," especially because what's the guarantee for them? They've watched that movie in Iraq and Afghanistan and Libya. They don't like what happens to the guy in the country that the U.S. attacks, so they really do have an incentive to continue with the program.

Horton: Jonathan Schwarz keeps pointing out that the *Washington Post*, the politicians, and the media, too, keep mis-paraphrasing this one particular statement, where they say, "We will not negotiate our nuclear program with the Americans," but the rest of the sentence is "as long as they continue to make threats like this." In fact, I think that's even first: "As long as the Americans continue to threaten us, we will not put nukes on the table for discussion." Not to be too utopian about it, but doesn't it sound like they're willing to negotiate their possession of nuclear weapons, or at least we'll talk about talking about it?

Bandow: When I visited in June, their answer to that was basically, if everybody else is willing to give up their nuclear weapons, they'd be happy to negotiate. Now, obviously, the Supreme Leader, as Kim Jong-Un is known, can change that. It's tough. It's in the Constitution. They declared themselves to be a nuclear weapons state. He has invested a lot in this, all the imagery of him with missiles and nukes. But the point is, unless you sit down and talk to them, you don't know. So if we demand that they concede everything before we start talking, that's a non-starter. Tillerson at least has said, "We'd be happy to talk without preconditions." It's kind of "talks about talks," you know, but at least sitting down would be a start, because if you don't talk to the other side, you don't know what they're willing to discuss and where they might be willing to go. And that really is stupid because you really cut everything off.

Horton: How many nukes do you think they have by now? Is there some kind of reasonable estimate that you can rely on?

Bandow: A common estimate is 20 to 30. That seems reasonable, based on not having seen the weapons; it's based on how much plutonium we think they probably have and how many weapons that might make. So it's a small arsenal by the standards of most nuclear powers, but it's real, and if there is a war, assuming that they could deliver them, they could do a huge amount of damage, which is why we certainly don't want to start a war. Imagine if they could drop one of those on Seoul, one of those on Tokyo, one of those on Guam, one of those on Okinawa. It would still be pretty ugly even if they couldn't reach the United States.

Horton: Let's say, to spin it in a positive way, how about the theory that they can't deliver any of their nukes yet because they can't make them small enough? They don't have a trebuchet big enough, so all they can do is sit on their nukes, but if war breaks out, what kind of damage do you think the North could do to the South before the Americans were done carpet-bombing them off the face of the Earth?

Bandow: The major damage they could do would be to Seoul. Seoul is about 35 miles from the border. The North Koreans have a lot of Scuds, they have a lot of artillery, they have chemical weapons, biological weapons, and they could use all of that. So they could do mass destruction of Seoul. They have a lot of tanks. They're older tanks, but if they struck first, it's possible they might be able to take Seoul. That would be a kind of a catastrophic mess. The Seoul metropolitan area has about half of their population, something like 25 million or more people there. So you can imagine the chaos and the devastation that would come from that. They would lose any war, but even most hawks admit that this would be a protracted fight. If we wanted to conquer the North, you look at the terrain, you look at the mountainous territory, this would be a real mess. It'd take a real effort. There'll probably be an insurgency. No one wants to do this. This would be crazy to start it out for no reason.

Horton: So your problem is that you think that the North Koreans possess it, right? But the American narrative, and we've seen this before — whether it's David Koresh, Saddam Hussein or whoever they want to target — they go: "This guy is crazy. He can't be dealt with. The only way to deal with him is by fighting him. Because if you talk to him, he's going to say something crazy, and if you make an agreement, he's not going to keep it. He's an irrational actor." Jeez, he might start a nuclear war with the global superpower, Doug.

Bandow: The irony is that our South Korean friends are probably more worried about Donald Trump than they are Kim Jong-Un. They've heard North Korean rhetoric for years about turning Seoul into a lake of fire, and it never happens because they're not going to start a war that they're going to lose. Even the U.S. intelligence agencies think that Kim is a rational actor. Everybody is worried about what kind of judgment he has, and of course who wants to tell him bad news, that he can't do what he wants to do? But there's no evidence that he or anybody else there is suicidal. I tell everybody, "Look at the behavior of Kim, his father and his grandfather. They all wanted their virgins in this world. They live very good lives compared to their compatriots. There is no evidence they want to go out on a funeral pyre lit by American nuclear weapons. That's simply not on their agenda. Their agenda is survival." In that way, nuclear weapons make a lot of sense, and that in many ways shows their rationality. They want a deterrent. They want it for rational reasons.

Horton: Obviously, a hot war would be a complete catastrophe, but even this cold war and this state of tensions, and the high likelihood of an accidental war breaking out at this point, with all these red lines being drawn and crossed everywhere and all this. How about I'm President Horton and you're my Secretary of State, and I give you orders. Fix it, Doug. I want to solve this problem. What's your program?

Bandow: The first is, you talk to the North Koreans. At the very least that helps reduce the tensions a bit. The fact that you isolate them and refuse to talk to them itself seems threatening to them. The second thing I would do is that I would certainly sit down and talk to the Chinese and say, "Look, how do we work together on this? We know you have interests. We're not going to tell you what we want. We're going to sit down and figure out what you are willing to do and how that works with us." If you come up with an attractive set of carrots, backed by the Chinese, to offer the North Koreans as part of negotiations, you can try to use sanctions as part of that, but don't have any illusions that that's enough.

Basically, you step back from the military option. You make clear that we will defend ourselves, but we're not going to start a war. That's something where you're going to make it easier to work with the South Koreans who don't want a war started. This is going to be a long-term thing, but I think the starting point, one of the things I would advocate, is what the Chinese have been pushing, the "freeze-for-freeze": North Korea stops missile and nuclear testing, we stop military exercises. Again, you've toned down the tensions. If you stop the testing, you've suddenly stopped the urgency. So we have a moment to breathe and to think about how we solve this, as opposed to this seeming relentless rush towards confrontation.

Horton: Do you see a possible Agreed Framework in the future for really normalizing relations and making peace, or is it just too much in the American empire's interest — and too much in the interest of the regime in North Korea — to keep the conflict going?

Bandow: I would love to see that, but I'm not going to hold my breath. I do think one of the possibilities here is that Kim Jong-Un has actually talked about a parallel policy of economic development and nuclear development. He wants economic development, so I do think that's an interest of his. And if you get China on board where China to some degree can credibly offer some security guarantees, you might very well be able to put together something that works. Again, I'm not going to hold my breath because there's a lot of distrust, and there are a lot of interests. There are reasons why the North would like to have nuclear weapons, and there are reasons why a lot of American hawks would never want to deal with the North Koreans. Nevertheless, I think it's worth a try. I don't think it's hopeless, but I do think we have to move quickly; otherwise, we might get this war through mistakes or misjudgment, and that really would be unfortunate.

Horton: Thank you for coming back on the show, Doug. I sure appreciate it.

Tim Shorrock: The Prospects for Peace with North Korea December 18, 2020

Scott Horton: On the line I've got the great Tim Shorrock. He wrote the book *Spies for Hire*, and he's an expert on Korea because he used to live there. I learned that he's been writing about it since I was in the second grade or third grade in 1983, the year *Return of the Jedi* came out. You started covering…

Tim Shorrock: Actually, it will have to be when you were in nursery school because I actually began writing on Korea in the late 1970s.

Horton: Oh, it was for *The Nation*. You started writing for *The Nation* in 1983.

Shorrock: That's correct. My first article in *The Nation* was about an incident in 1983, when North Korea set off a bomb in Burma at the place where the South Korean dictator and his cabinet were meeting, killing most of the South Korean cabinet. So I've been covering conflict between North and South for *The Nation* since 1983, and Korea in general since the late 1970s.

Horton: Well, I already knew how lucky I was to have you as a regular guest on this show to cover these issues for us, but now I think so even more. I'd like to go back and read your old stuff now, knowing it goes back that far. Do you have books about Korea too?

Shorrock: Funny you should ask. I'm actually in the middle of writing a proposal for a new book on Korea that would be a kind of hidden history of the U.S. role in Korea, North and South.

There is a book out on Kwangju that was published quite a long time ago and has been republished in several editions. It's called *Kwangju Diary*, and it's about the 1980 uprising in Kwangju, South Korea. The text is written by a Korean participant in that uprising. I had an epilogue that sort of talked about the U.S. role in South Korea at the time. That's the only book on Korea I have out, but, like I said, I'm working on one.

Horton: You know what? I bet that if you could just wrangle up a good editor and put together a compilation of your magazine articles and so forth that you've written over the years, that could be a book on its own — highly informational.

Shorrock: I've actually thought about doing that because there's a lot of really interesting pieces that I've written about areas that most people don't know much about.

Horton: I'll say.

Let's talk about this thing in Responsible Statecraft. You had a great one in *The Nation* recently too, but this one in Responsible Statecraft, the Quincy Institute — called "Old Obama Hands on Korea Policy Could Pose New Problems for Peace" — reminds us that the status quo was a nightmare. Then when the nightmare came, he was actually a lot better on this issue than the establishment that he replaced.

Now that they're back — in many cases, the very same people who ran this policy in the Barack Obama government. It seems like we have lots of good reasons to anticipate which way they're going to go with this policy. From what I can tell here, it doesn't look good.

Shorrock: From what they've done in the past, the future doesn't look good, but you never really know until they actually start implementing policy. So who knows? I sure hope that some of the progressive voices that have become active on Korean issues can. I know they've been meeting with the new Biden team, and I know that there's been some discussion by peace groups with the Biden team about what to do in Korea. So at least they're open to it. You would never get that kind of audience with the Trump administration. That's for sure.

Horton: Yeah, that's true.

Shorrock: Although that's not quite true, because Women Cross DMZ, which has been very active and one of the most prominent of the peace groups, did meet with Stephen Biegun, who is Trump's negotiator with North Korea. He has held several meetings with peace groups and tried to get their views. So I think there are people in government who are very serious about actually trying to come to a negotiated peace.

Horton: After all, Trump — he didn't do it right, but he did make some real progress, shaking some hands and doing his photo-op at the border. They destroyed some bases near the DMZ and performed some other symbolic actions. He dialed back the exercises. In other words, to go back to the status quo would be to ratchet tensions back up quite a bit. Or another way to put it would be that Trump, in a way, has set Biden up to say, "Jeez, I kind of have no choice but to follow through on really seeking peace here, because I'm in a much closer position to peace now than I was four years ago."

Shorrock: I think Biden has actually acknowledged that. He was very critical during the campaign, as you saw in the debates. He was always the one to call Kim Jong-Un a thug and to criticize Trump for meeting with him, putting him on the world stage and giving him so-called legitimacy. But, as everyone knows who knows about Korea, those meetings between Trump and Kim actually changed the dynamic and got some real negotiations going.

Unfortunately, the Trump hardline of maximum sanctions and maximum military pressure didn't work, and the talks fell apart in March 2019, almost two years ago. Since then, there's not been very much discussion between North Korea and the U.S., and relations have really gone downhill between South Korea and North Korea. In fact, it's at a real low point now. So it's really important, I think, that the Biden administration starts picking up the ball and trying to come back to a negotiated situation, rather than continuing this hostility and military confrontation, which has never gotten us anywhere.

Horton: Can you talk a little bit more about the deterioration of the relationship between North and South in the last year since Trump quit trying to deal with them?

Shorrock: If you recall, the Trump-Kim meeting was actually set up in large part by South Korea, by President Moon Jae-in after his peace and diplomacy initiative during the Winter Olympics of 2018. It was almost three years ago that he invited the senior leadership from North to South to observe the games and to start some negotiations. Actually, he held his first meeting with Kim Jong-Un in March 2018, in Panmunjom, at the border, and they declared that they would try to work out peace between North and South on their own. After that, Moon gave the word to Trump and the Trump administration that Kim Jong-Un was willing to meet with President Trump, and that's how that got started.

In September 2018, Moon Jae-in went to Pyongyang, and they had a major summit meeting with the North Korean side. They came to some very important agreements, and that's where that agreement to de-escalate the situation on the DMZ came about. They had a military-to-military agreement. As I've told you before, they shut down a lot of border posts, they no longer carry some of the weapons that people used to carry, and there was a real softening of the situation on the border. The real progress was that they made plans for economic projects — cross-border projects like the South Koreans working with North Korea to rebuild the railroad system and link North and South. They started on that.

That's where the Trump hardline started making things difficult. There's the U.S.-controlled United Nations Command in Korea, which is responsible for administering the armistice that ended the Korean War, and it has complete control over anybody who crosses the border. Anyone who wants to cross the border North to South, or South to North, has to get the permission of the United States military through the UN Command. They blocked certain projects, certain plans, that the North and South had made. One of the projects that they stopped was this cross-border railway project.

Moon Jae-in kept trying to persuade the Trump administration and Secretary of State Pompeo to kind of pull back on the sanctions and maybe lift some sanctions to make it easier for the North and South to work together as they proceeded with negotiations on the nuclear issue. That didn't happen, and the talks between Kim and Trump fell apart. After that, North Korea started saying to South Korea, "Look, you're not helping this process because you're too close to the U.S. Everything you want to do is vetoed by the United States. So why should we trust you?" That's how things really went downhill.

Then it was earlier this year when Kim Jong-Un's sister, Kim Yo-jong — who's become quite a powerful figure there — apparently through her control of the military arranged a liaison in North Korea where North and South negotiators would meet and talk about their plans. They actually blew up that building. That was a real sign that "We've given up on you," but then a few days later Kim Jong-Un said, "That was a mistake," and by that he's sort of saying, "Well, there is room for negotiation."

So we're on thin ice in terms of the North-South situation. But if there is some outreach by the Biden administration, it's possible that the situation could start moving forward again.

Horton: You've got to appreciate the antics, at least. Blowing up the meeting house and then coming out the next day going, "Nah, we might have left the meeting house alone. That would've been okay to keep that."

Shorrock: [Laughing] Yeah, I don't know what it is, if it's a bad cop, good cop routine or whatever. Right now, as anyone who reads my tweets or follows the Korean press would know, there's a huge dispute between all these U.S. Democrats, neocons in Washington, and all these so-called human rights groups that focus on North Korea. South Korea just passed a law that actually makes it very hard for these defector groups to send these balloons across the border into North Korea. That's becoming a real serious problem in terms of negotiations between North and South. It is also a problem for South Korean people who live on the border because it sort of escalates the situation in the border region.

South Korea passed a law basically banning these kind of cross-border balloon tricks and these other propaganda operations that were mounting there. Now you have all these people in Congress — including actually some Democrats in the House — who are going to have a hearing to criticize South Korea for this law. A lot of South Koreans are just like, "Look, this is a question of our sovereignty as a country. If we want to make a law to protect our borders, who are you to tell us not to do this?" So there's increasing tension now between South Korea and the United States. It's not a simple situation at all now.

Horton: It is simple if they wanted to solve it, right?

Shorrock: Yeah.

Horton: This is something we've talked about for a long time. You brought up Stephen Biegun. He was the guy that worked for Trump, and as we've discussed, he gave this one good speech where he said, "Listen, we know that what we have to do first is to end the war with a real peace treaty. Then we'll drop the sanctions, we'll open up more diplomatic relations, and we'll make friends with them. Then we'll figure out how to denuclearize from there" — because we know that they're not going to give up their nukes as long as we're enemies. Only after they're so reassured of our new status that they feel comfortable doing so would it even be possible to consider it, so we're just going to do the right thing.

So my question for you is: Was that his position? Was that what he wanted to do? Do you know whether he was telling Trump that this is the way we have to do it? Or, at some point — at some principals or deputies' meeting or something — did they say it was okay for him to give that speech but then decided that, no, it's definitely not? But it's like a light switch, right? It's either on or off. Either you want to make peace with Korea, and you do it the way he described in that speech, or you don't, and you stick with, "Give up your nukes first," which everybody knows is a poison pill and is going nowhere.

Shorrock: Biegun made that speech a few weeks before the Hanoi Summit that fell apart. A lot of people, including myself, read that speech, heard that speech and thought, "Wow, at this next meeting, they may have some real solid proposals that maybe the North could accept," because he was talking about being flexible on the issue of sanctions. That would mean allowing South and North to proceed with different projects that are now banned by these sanctions and that the UN Command there blocks.

It seemed like that's where they were moving, but John Bolton was Trump's national security adviser. When the meeting in Hanoi happened, one of the first things I noticed was that before it collapsed, when they showed the table, on one side is Trump, and on the other side is Kim Jong-Un and his team. On the U.S. side, it was Pompeo and Bolton. Biegun was sitting in the back. He was on the side. So the main negotiator had been pushed aside.

Then when it came to a deal, Kim Jong-Un offered to close this one major facility that's been around for a long time. It wasn't all their facilities, but he was proposing to do that. He wanted to have some of the more recent sanctions that had been imposed in 2016 and 2017 by the United Nations — at the urging of the U.S. — which had basically completely stifled his economy by banning coal, as well as imports and exports of basic goods. He wanted those sanctions lifted.

As Bolton wrote in his book, he strongly opposed this and demanded that Trump not agree. Trump acceded to Bolton. When you read Bolton's book, you will see the utter contempt he held for South Korea and its own vision for unification. I mean, he just insults Moon Jae-in up and down: "He's just this far-left..." etc., etc. He just did not take the Korean interest in peace seriously at all. His position was: "You can't trust North Korea." So the talks fell apart, and that was largely Bolton's doing. I guess Biegun just got cut out of the decision-making and could not make any of these proposals. They've never really changed from that hardline of, "No, we're not going to give up any kind of sanctions until you definitely show you're getting rid of all your nuclear weapons," and like you say, that's a non-starter.

There's been a lot of talk recently about how, if it's not going to be denuclearization, it's at least going to be about arms control. Starting there, right? Let's maybe recognize that North Korea is a nuclear power and then try to have an arms control agreement. At least start there. There has been some talk from the Biden side — including from Tony Blinken, who's going to be his Secretary of State — along those lines, but if you read what Blinken has said, it's really not much different from what he was saying during the Obama administration. They just thought that North Korea was going to collapse, as I wrote in this Responsible Statecraft article. They just sort of thought it was going to collapse and go away, and the U.S. role would be that if there was a war — I talk in this article about Avril Haines, who's Biden's nominee to run U.S. intelligence, and a speech she made about military intervention with South Korea, Japan and China — it would seek to replace the regime in North Korea and to clear the way for a united Korea.

Horton: What?!

Shorrock: Yeah, read the article. I mean, I've got this amazing speech in it.

Horton: Wait, so she's saying that she thinks that China's going to take our side with Japan and South Korea to invade North Korea together and install a new government?

Shorrock: If there was a war, yes.

Horton: Has she run this by Beijing to see whether they said they were interested in that?

Shorrock: When she made that speech, it was early 2018, and relations with China hadn't deteriorated like they have now, but yeah. What actually stunned me more was that she actually thought it was a good idea to have Japanese military involved. I can't believe a senior American intelligence official would actually take that seriously. Anyone who knows anything about Korea knows that having the Japanese military involved in the Korean Peninsula is just outrageous. Ridiculous. South Koreans would go after the Japanese soldiers. It's just outrageous and stupid, so who knows what they're planning. When they're in power, the situation is very different.

I don't know what to think about it, really, until we hear what they have to say in the Senate confirmation hearings. Then we might hear clearly what they think and what their plans might be. I'm going on some of the things Biden has said. He did send a message to President Moon before the November election. He reached out and wrote this op-ed in a Korean public wire service, Yonhap — which is owned by the Korean government — which was very conciliatory, and said he will work for negotiations for principled diplomacy to resolve this situation.

So maybe there is some hope there, but if he comes in and starts making the kind of demands and preconditions for talks that Obama made, or if North Korea decides to do something like test an intercontinental ballistic missile, then the situation could really deteriorate. That's why peace groups like Women Cross DMZ are urging Biden to make an initial reach-out to the North Korean side. They can do that easily through the UN embassy in New York. It would be easy for the incoming Biden administration to make contact and get the ball rolling on negotiations. If that's his intention, I think that some kind of statement would really help the situation.

Horton: I sure would like to see that too.

Let me ask you this. John Bolton, first of all — it's all his fault that North Korea has nuclear weapons in the first place. Do you have a specific article about that? Because I have a couple, and I could recommend Gordon Prather.

Shorrock: In my view, the party responsible for the nuclear crisis is the United States. After all, we're the ones who introduced nuclear weapons into the Korean Peninsula in the 1950s…

Horton: Well, in 2002, we had a deal. Bill Clinton had a deal, and John Bolton personally ruined it. Deliberately, right?

Shorrock: I wouldn't say it was just Bolton. This was the Bush administration…

Horton: Well, I didn't mean to say that he didn't have Bush's permission. I just mean that he was the hatchet man, and the proud one in charge of ruining it, right?

Shorrock: He was the UN ambassador, right? Under Bush? But I think it was…

Horton: He was in the State Department, the Under Secretary for Arms Control or whatever in 2002, when he ruined the Agreed Framework deal and accused them…

Shorrock: Well, yeah, they went over there for a meeting to follow up on this Agreed Framework that had been agreed to in 1994, after which North Korea stopped its nuclear production and its production of plutonium altogether for 12 years. So that agreement actually held for quite a long time.

Then, as you say, what happened was that at the end of the Clinton administration, in 2000, before the election and before Bush came in, they were very, very close to an agreement with North Korea. This agreement picked up on the earlier agreement to stop their nuclear production. They almost reached an agreement on ending all missile tests and ending their missile program. They came within a hair of that agreement. If you recall, the Secretary of State under Clinton, Madeleine Albright, went to Pyongyang and met with Kim Jong-Il at the time. That was the meeting before the final agreement. Then after the election happened, Clinton was thinking about going to North Korea and finalizing it, but he never made it for various reasons. So that agreement was never made.

Bush came in and said, "We're not going to negotiate with North Korea. We don't trust them." It began to fall apart from there. Then, like you said earlier, he sent this lower-level State Department guy over there, and he accused them of building a uranium bomb, when actually, all they were doing at the time was thinking about it. They were importing parts for a uranium facility, but they didn't have a uranium production facility at the time. Even if you read South Korean intelligence that have talked about this period, they'll say that they believe that was a mistake by the U.S. The North Korean side did not at the time have a uranium facility. They do now.

Horton: And none of their nukes have been made out of uranium. They've all been plutonium nukes. Then, what was funny was that it was aluminum tubes, right? Gordon Prather called it "Aluminum Tubes: The Sequel." Only this was some stuff that they had bought from the Pakistanis from A.Q. Khan, but none of it was in operation.

Shorrock: Right. That's true. Then they just tore up the agreement after that. It's important to note that it was shortly after that, during the Bush administration, that they exploded their first nuclear bomb.

Horton: Well, I just like filling this in. First, the Americans, in the name of the non-existent uranium program you mentioned there, abrogated our side of the deal. So the Agreed Framework deal was off. Then they added new sanctions. Then they created the Proliferation Security Initiative, which said that we have the right to seize all of their boats on the high seas if we think that they might be selling missiles to Iran or whatever. Then they released the Nuclear Posture Review that said that we have North Korea on the short list for who gets a nuclear first strike. It was only then that Kim withdrew from the Non-Proliferation Treaty — or announced that they were withdrawing in six months, as per the deal — and kicked the inspectors out. Only then did they start making nukes out of plutonium.

Four big provocations by Bush and Bolton — to do what? What did they think they were going to do? They're going to go to Baghdad, and then they're going to be in Pyongyang by August? So we're going to kick them out of the treaty, but don't worry, they never will get a nuke? Did they even have a plan? All they did was push them into the possession of A-bombs!

Shorrock: A lot of it was Cheney and Rumsfeld. In the Clinton administration, in the late 1990s, there was this big study group on missile defense. Rumsfeld was the chairman of that study group, and the whole thing was that the U.S. has to build missile defense, etc. So they came into office thinking, "Well, we've really got to confront North Korea." That was Cheney's position, too: "We're not going to negotiate. We're just going to go with the missile defense, confront them, try to contain them, and not look for a negotiated end to this."

At the end of the Bush administration, after Cheney had lost clout because of Iraq and after Condoleezza Rice had come in as Secretary of State, she got negotiations going. As you recall, it was the Six-Party Talks.

Horton: Right.

Shorrock: Bush, at that time, agreed to participate in those. He actually agreed to those talks only a couple weeks after North Korea had exploded its first bomb. So his approach really damaged and totally ripped apart that previous agreement.

Horton: He had a bit of a deal there in 2008 before he ruined it and put them on the terrorist list, which ruined his own diplomacy.

Shorrock: Right, he took them off and then put them back. Then Obama came in saying, "I don't talk to dictators." Early on in his administration, the North Koreans tested a rocket. They tried to launch a satellite, and that was after they had signed an agreement as part of the Six-Party Talks. Then they said this satellite did not violate any of the sanctions or any of the previous agreements on missiles. They said it was just a civilian satellite, and that blew it up for Obama. After that, they just said, "We can never trust them." Then they actually successfully launched a satellite not long after that. That was seen as a test of a missile that they were just pretending was a civilian satellite, but they actually did just launch a satellite.

But that became the reason to move away from any kind of negotiation. Then after that, when it looked like talks were just completely off the table, they tested two or three more nuclear devices under Obama. There were many overtures made by the North Korean side, and then some delegations came from the U.S. — former officials — to talk with the North. At that time, under Obama, they were reaching out saying, "We'd like to sign a peace agreement, a peace treaty," and so on. But by that time, the Obama administration — with these people who are now going to come in and work for Biden — were convinced that that would not work, that North Korea was on the verge of collapse, and that there was no point in talking to them.

So by the time Trump came in, the situation was very, very dangerous. Actually, I wrote an article for *The Nation* about a month or two before the 2016 election, and it was called "Hillary's Hawks." I had talked to all these people and was quoting people who were advising Hillary Clinton and would have become part of her government. They were saying, "Well, maybe we have to consider preemptive strikes," and this kind of thing. So the U.S. and North Korea were actually quite close to war at that time, which is something that Trump has said. Of course, the liberal press said, "Oh, that's not true. We weren't anywhere near war, blah, blah, blah..." But actually, we were.

Horton: Yeah, the national security adviser at the time, H.R. McMaster, was pushing what he called "bloody nose strikes," where we'd just hit their missile program real hard but then we'd stop, and they would know better than to respond. That was where he was going to place his bet. Trump, of course, at the time, was threatening nuclear war: "fire and fury like the world has never seen before."

By the way, this is a really important point. On pages 71 and 72 of Bob Woodward's latest book, *Rage*, he quotes James Mattis — who was Secretary of Defense at the time, a four-star Marine Corps general, and former head of Central Command — who was talking about the situation where the North Koreans' missiles were getting better, and they were still testing nukes. He said that he had to consider killing "a couple million people." He says, "No one has the right, but that's what I have to confront in my job as Secretary of Defense. Not that I have the right to, but I might anyway kill a couple million people."

What's funny about that to me, honestly, is that I really don't understand whether he is implying that all of their entire nuclear weapons facilities are right in downtown Pyongyang, and there's nothing that we could possibly do to destroy them other than to kill every man, woman and child in the capital city. What is he even talking about? Two million people have to die for America to be able to take out North Korea? Why would we need nukes at all, with all of our bunker busters, all of our B-1s, and everything else we have there to bring to bear?

Shorrock: Well, another thing in Woodward's book, which I noted in this Responsible Statecraft article, was that he said they had revived this plan called OPLAN 5027, which had been in a planning stage for a long time. This plan called for strikes on the North's "core" military facilities and the removal of its top leaders. The plan had actually been originally drafted years ago, in the 1970s. They were studying it, and Woodward said he was told that the U.S. response might have included the use of 80 — eight-zero — nuclear weapons. So, under this plan, it would've been totally destroying North Korea. Like Trump was talking about at that UN speech, it was a really, really dangerous situation.

You mentioned McMaster. At that time, he was talking about these so-called "bloody nose strikes," but it would've been a unilateral U.S. military strike. I reported around that time, and of course NBC — which is always getting all these intelligence leaks on everything — ran this story saying that the U.S. was planning unilateral American strikes from B-1B bombers. These bombers would be off the Korean coast, of course. They wouldn't be flying over Korea, right? They can launch cruise missiles from offshore. The plan was to launch weapons from these B-2s and have a unilateral strike on some of their nuclear facilities.

When South Korea president Moon Jae-in got wind of that, he went public and made this very strong speech, saying that there cannot be a war on the Korean Peninsula and that there cannot be any kind of attack without consulting South Korea. That was a real turning point. I think it was from that point on that the Trump administration saw that South Korea was not going to put up with this, so we're maybe going to have to take advantage of this opening and start talking to Kim Jong-Un. That's what we did.

Horton: Back where we started, with the failed talks. One last question. John Bolton to everyone represents some kind of winger, right? He's not really a neocon, but he sure does love neocons because he's just as bad as them on every single thing. So it seems like his position here in sabotaging Trump's diplomacy with North Korea actually represented the centrist point of view in D.C. that they would rather not have peace. They would rather keep a divided Korea.

They have Bill Clinton's same policy from 1993, which is that North Korea will fall pretty soon, even though that was almost 30 years ago and they haven't fallen yet. It seems like a pretty stupid bet to base a policy on. I just wonder whether you think his sabotage of Trump there in Hanoi was kind of a right-wing hawk sabotage? Or was it really just the entire national security state's consensus, and he was coming through for them?

Shorrock: I think it's the latter, and I think it's a position held by lots of liberals and Democrats in Washington. The distrust of South Korean progressives and left is sky-high in Washington. The contempt that Biden held toward Moon Jae-in is held by all these Korea experts at CSIS and all these other think tanks in Washington. I've heard Moon Jae-in, and he is on the left. He was a progressive activist during the period of authoritarian rule. He was arrested by the military junta in South Korea in the 1970s. He was a labor lawyer and a human rights lawyer. He was the chief of staff for Roh Moo-hyun, who was the second progressive president, who had a summit with North Korea. So Moon was very involved in that summit in 2007. The people around Roh Moo-hyun and Moon Jae-in at the time had all come from the Korean left of the 1980s, organizing against the dictatorship. They wanted to have a different approach to North Korea.

I have heard people at CSIS, these big experts that are always quoted in the *New York Times* and the *Washington Post*, make incredibly insulting remarks about these people. One time I heard this guy, Michael Green from CSIS, laughingly call the people around President Roh Moo-hyun the "Taliban at the Blue House." The Blue House is the Korean equivalent of the White House. He associated them with terrorists and the Taliban. That's how he saw the Korean progressive movement.

So yeah, you're right. It's a commonly held view. Basically, they all think that South Korea is our colony, it's always been our colony, and they should do whatever we say. Things have changed.

Horton: One thing that we do have going for us is that Joe Biden is no Barack Obama when it comes to flash and a big smile and making liberals and progressives swoon and stuff like that. He is known to the left as being one hair to the left of the Republican Party, which otherwise he would've belonged to for the last 50 years. He's like George W. Bush if he'd been a senator his whole life.

Everybody knows that. Nobody really likes the guy, and everybody doesn't believe in his smile at all. So I know that there are a lot of leftist groups that, for example, are loaded for bear right now for the afternoon he's sworn in, to hold him to his promise to get the hell out of Yemen right away and to call off all support for the Saudi-UAE war there. For that matter, Korea and the rest of these things, he's no Obama. He's more like Lyndon Johnson.

So, everybody on the left, put up your dukes. And for all libertarians and good antiwar right-wingers, too, we can get these guys in a real pincer strategy where this is the thing that we agree on: We want to quit messing around in the Middle East and Asia and Europe and everywhere. Enough already, right?

Shorrock: Well, Biden keeps talking. One of the things he's talked about is supporting allies, right? Let's reinvigorate our alliances. Well, South Korea…

Horton: Let's start with South Korea. What do they want?

Shorrock: This is supposed to be your most close, close ally. So let's listen to our South Korean ally, and let's move these negotiations forward. One of the differences I see between Biden and Obama is that, yeah, Obama had this way of really making liberals swoon by making all these kinds of promises and talking in a very progressive way, but behind that was always this hardline on everything. I think there's less of that with Biden. Yeah, he does have the smile, but he makes it really clear that he doesn't support the Green New Deal and that kind of policy. Whereas Obama would probably have said, "Yeah, I support it, and we're going to move that way," but then he wouldn't do it at all. At least Biden's honest about his opposition on certain things like that. I think that's the difference. I don't think he fools people as much as Obama did. I think it's clear that he has a more conservative position, and that's true for a lot of issues.

On this one, we'll have to see. Obviously, they're going to have a policy review of Korea like they will do on everything. Hopefully they're going to listen to some new voices, and there are some new voices in Congress. The first Republican House member recently signed on to a resolution to end the Korean War and have a treaty that the U.S. would support.

Horton: I didn't know that.

Shorrock: I forget his name, but you can find it on Women Cross DMZ, because they've been lobbying for this. There's now something like 25 co-signers — or maybe it's even up to 40 people — who have signed on to this bill. That's a huge step. Recently, Christine Ahn of Women Cross DMZ was telling me that a few years ago, they could get no more than two or three Democrats to sign on.

So I think times are starting to change, and there have been some good people, good Democrats, elected to the House, who have much more progressive ideas. But some of the people who are supposedly super-left on the Democratic side of the House are not very good at all on foreign policy. They've been silent on Korea, so we shall see.

Horton: Well, listen, man, I can't tell you how much I appreciate your time on the show. Thanks as always, Tim.

Shorrock: Thanks a lot.

Gareth Porter: Israeli Fabrication Almost Led to War with Iran May 1, 2020

Scott Horton: Introducing my very favorite reporter, the great Gareth Porter, this time writing for The Grayzone at thegrayzone.com. It is also reprinted at Antiwar.com: "With Apparently Fabricated Nuclear Documents, Netanyahu Pushed the U.S. Toward War with Iran." Say it ain't so, Gareth Porter! Welcome back to the show, sir.

Gareth Porter: I'm glad to be back. Thanks again, Scott.

Horton: Netanyahu telling lies to increase tensions between the U.S. and Iran, who could have imagined? But you reminded me that Donald Trump took America out of the Iran nuclear deal in May 2018, just a couple of weeks after Benjamin Netanyahu's big press conference — or, well, publicity stunt; I don't know if he took any questions — where Netanyahu claimed to have revealed all of Iran's nuclear documents. That's what you're talking about here, these apparently fabricated nuclear documents. Those documents certainly played a role then in changing the narrative in the time leading right up to when Trump repudiated the deal, so we can talk about the consequences of that. But first of all, tell us why you're so sure that these documents were not in fact liberated by the Mossad from a top-secret facility in Iran.

Porter: Right. You know, this is a story that I love in part because I'm able to show that there are multiple levels on which the Netanyahu tale — of the Mossad going in in dark of night and stealing half a ton, supposedly, of the most highly classified, top-secret nuclear documents out from under the noses of the Iranians — is totally false. Of course, the first level, which I think is really crucial here. If you accept the idea that he was really fabricating this story about the Mossad going in and stealing these documents, then obviously, the fabric of this entire story is highly questionable itself. All the documents themselves become highly questionable, but we can talk more about that.

The point that I make in starting out the story is that there is really no good reason to believe the Netanyahu tale of Mossad's stealing of the documents, mainly because they make such an extravagant claim as to make it impossible to believe. The claim was that they were able to find these documents because they had such a sensitive source within the Iranian government. This source was among only a handful of such people, according to both Netanyahu himself and a Mossad official who briefed

Ronen Bergman who was at that point writing for an Israeli newspaper but who then joined with *New York Times* staff to write a much longer account later on. The explanation was that they had this source who was so sensitive that he was among only a handful of people who knew the warehouse in which these documents supposedly resided. Not only that, but he could steer them precisely to the two or three safes that were in that warehouse that held the most important documents from the point of view of Israelis — the ones that the Israelis would find most politically important and most lucrative, shall we say — to get and exploit once they were able to translate them and everything. So this is the story that one has to believe in order to credit the entire fabric of the yarn about the documents that Netanyahu talks about in his on-camera briefing.

I am able to quote two former senior CIA officials, both of whom were the top CIA analysts on the Middle East at different times over the past few decades. Paul Pillar was the top Middle East analyst, and he was the National Intelligence Officer on the Middle East during the period from around 2001 to 2005. Graham Fuller had the same position, National Intelligence Officer for the Middle East, much earlier, back in the mid-1980s. So they are very far apart in terms of timing, but both of them agreed in emails to me in response to my queries that there was something really not quite right and not believable about the story, specifically the notion that the Israelis had this very sensitive source, whom they then "burned," as the intelligence people call it. They burned their source publicly by bragging about him to the press and to the public. That simply wasn't credible. If they had had such a source, they would never have burned him, since he was so valuable. He would be able to give them presumably the most highly classified documents having to do with Iran's nuclear program or perhaps other aspects of Iran's defense policies as well.

Of course, according to the story, they chose to burn the source in order to prove just how important these documents were and how sensitive they were. Both Graham Fuller and Paul Pillar agreed that it simply wasn't credible. I quote Paul Pillar as saying, "This seems somewhat fishy." Graham Fuller says, "The story seems somewhat fabricated." So I think that's a quite extraordinary set of parallel responses by people who are in a very good position to judge this kind of problem. This really discredits the story quite definitively.

Horton: Now what did they say about the location of this warehouse on the outskirts of town there? That's either a perfect hiding place or a completely ridiculous one.

Porter: I don't talk about that in my story, but the fact is that there were plenty of places where documents could have been stored that would be highly secure, obviously — unlike this warehouse, which was out in the middle of nowhere and supposedly, according to Netanyahu's story, didn't even have any security at night. There were no guards at nighttime? That just doesn't hold water. It's not believable in the least. To my mind, it's just totally incredible that nobody in the U.S. major news media who covered this story stopped to think, "How credible is this story?" It was simply given a pass. That's the way they operate, covering these kinds of stories so routinely nobody even gave it a second thought.

But definitely what adds to the incredibility of the story was that the documents were supposedly stored in this warehouse in the middle of nowhere in a part of town that was not used by the government very much at all, and which basically lacked the security typical for these kinds of sensitive documents. They could have had them in the Atomic Energy Organization of Iran (AEOI) — which would have been the logical place for anything that had to do with a nuclear program — or they could have had them in the Defense Ministry. If indeed they had nuclear weapons work that they wanted to hide, they could have hidden it in the Defense Ministry. That would have been the logical thing to do. But no, it had to be somewhere that they could sort of unfurl this yarn about sending a Mossad team in, breaking the lock on the door, then presumably using blowtorches to open up these specific safes, and finding precisely the folders that they want to take back home to make public.

Horton: What did those documents supposedly say that was so damning about the Iranians that it helped lead to Trump's withdrawal from the Obama 2015 nuclear deal?

Porter: Well, there were two big finds, supposedly, that were given a lot of publicity and were covered in the U.S. and global media. One was the story that somewhere around the late 1990s into 2000, there was a plan that was written up that called for Iran to have five nuclear weapons designed, fabricated and tested by the year 2003. That was obviously a spectacular claim that would show they had these designs on having nuclear weapons way back when before they had ever even begun to spin the centrifuges. Not a single centrifuge had started spinning. Indeed, that's such a totally incredible tale because Iran was nowhere near having the capability to think that far ahead, to having nuclear weapons. They would have had to be much farther along in terms of their plans for actually enriching uranium, unless they had access to enriched uranium, which they didn't. Nobody has claimed that they did have such an access to highly enriched uranium. At that point, it would have been completely out of nowhere. It makes no sense whatsoever.

The other document, which was given even more publicity, claimed that there was a decision by the defense minister in the spring of 2003, which said, "Okay, now we're going to hide that part of our nuclear weapons program that would cause us some problems potentially with the West.

We're going to keep them covert, and we'll only have an overt program that has to do with the part of it that is legal, above board and under IAEA supervision." Of course, that makes no sense either because, in fact, they had nothing already that was not known about by the West. There was nothing to hide. So both of these documents, highly lacking in credibility, were the ones that they were pushing successfully in the media. They got quite a bit of coverage of those things.

Horton: But you and your CIA experts all say that you don't even believe in the documents anyway.

Porter: I have to take primary, if not exclusive responsibility, for actually calling the documents fabrications, because nobody else thus far has been able to speak up and express this. I can tell you that there are a couple of people I have talked to who in the past have expressed a lot of reservations about the documents' authenticity, but at this point, nobody's willing to go public.

What I have done is essentially show two things that I think are really important. One is that there is no evidence of authenticity that has been provided. Normally, a document is shown to be authentic by having people have access to the original. The forensic analysis would look at the paper, the ink, the typewriter used and so forth; the absence or presence of evidence of official government sponsorship of the document; and whether that sponsorship is provided and how credible it is. All those things would make up a forensic analysis of the authenticity. In this case, we know from Netanyahu himself, as well as from visitors to Tel Aviv and Jerusalem, that nobody has actually been given access to the originals. Nobody's even been given a binder and told, "Here, you can sit here and look at these and examine." None of the people who visited Tel Aviv, the people from the Harvard Belfer Center or David Albright's ISIS — what he calls the good ISIS, the Institute for Science and International Security — have been granted access.

Horton: They're not much better than the other ISIS, really, but go ahead.

Porter: No, not really. But anyway, none of them have the ability to do any real forensic analysis, nor do they have the desire to do it. Nevertheless, none of those people were given access to the originals. We know that the U.S. government and the IAEA have been provided copies, but no one from IAEA or the U.S. government has been given access to the originals. So again, there is reason for suspicion on that level because of the Israeli government's unwillingness to give anybody access in a way that would allow them to do that sort of analysis.

The other thing is that I have been able to analyze one of the key documents which shows up in the Netanyahu video: the show-and-tell slideshow. One of the slides depicts a technical drawing of a Shahab-3 missile, and it shows the actual drawing that is in the documents themselves. It shows a Shahab-3 with the design of the re-entry vehicle, which is shaped like a dunce cap — just a straight cone coming to a point at the end — and we know from the studies that have been done on the Iranian missile program, and from other evidence, that the Iranians had already discarded that design of the Shahab-3 in redesigning the missile from 2000 to 2004. The first thing that you redesign is the re-entry vehicle. The re-entry vehicle that was shown in this drawing is dated 2003, according to the IAEA, in an unpublished paper that David Albright published on his site. By that time, it's clear that the Iranians had moved on and that they had a new re-entry vehicle design which had a baby-bottle shape and which bore no resemblance whatsoever to the shape shown in this drawing that is now public. So I mean, this is really strong evidence that it wasn't the Iranian secret team of missile designers and nuclear weapons designers who came out with this drawing in 2003; it was a foreign intelligence agency that didn't know the truth about what the Iranians were doing.

Part of the storyline that I tell in this article is that the Iranians deceived the outside world — the Americans and the Israelis, in particular — by announcing that they were producing the Shahab-3 in 2002 and 2003. Instead, they had no intention of really making that their main weapon. They had, as I've already said, abandoned it in favor of a new design, which they finally tested for the first time in 2004. No one had ever laid eyes on the new design, so they didn't know that it had a completely different re-entry vehicle shape. This is the evidence that I put forward here. No one has ever refuted it. I've published this story before, and no one has ever refuted it, although people have certainly done their best to ignore it.

Horton: I'm kind of sad that the Israelis would go ahead and use the same lie again, after you've completely debunked this in your book *Manufactured Crisis* and in previous reporting that you've done. Once we know…

Porter: I'm shocked that they would do so.

Horton: You know, I'd like to give them a little bit more credit, that they'd at least forge some new documents or come up with a new lie that hasn't already been debunked. You've already shown where the IAEA admitted that they got the documents from the Mujahideen-e-Khalq, and that means from Israeli Mossad. Case closed already, right there. So that's another strong indication that none of this whole thing two years ago, this publicity stunt that Netanyahu did, was legitimate at all. These papers weren't stolen from Iran. In fact, I remember that there were people out front on Twitter and on YouTube the next day, at the place where this supposedly all went down, laughing and mocking the idea that this warehouse was a top-secret government facility of any kind, full of any kind of documents or anything like that.

So we've got to talk about the importance of all of that because they got us out of the nuclear deal so that they could institute a policy of maximum pressure in order to bring the Ayatollah to his knees and force him to sign a whole new deal that would include limits on their missiles, no sunset provisions, and a suspension of all support for Hezbollah. How's that working out?

Porter: Yeah, this is a key point. I'm glad you've come back to the larger picture, because it is very important for people to understand just how the Israelis and their friends in the United States were using this supposed revelation of the secret Iranian nuclear weapons planning in order to advance a strategy to maneuver the Trump administration into military confrontation with Iran. That's what they were after. I wrote about this in my book at some length, that Netanyahu tried a number of ways to maneuver the Obama administration into a kind of confrontation militarily with Iran and failed to do that, but that had been the intention of the Netanyahu government for many, many years.

They found in the Trump administration a much better opportunity to do it. In fact, in 2018, when Netanyahu was carrying out this plan in the spring, they were also getting Mike Pompeo as Secretary of State. So from then on, Pompeo was helping the Israelis to advance a strategy of trying to maneuver Trump into a military confrontation to use force against Iran if at all possible. We know that he was successful in doing that on a couple of occasions — in conjunction with Netanyahu in one case, and without him in the other case — to persuade Trump to respond to the situation with Iran by threatening or actually using force. In one case, Trump changed his mind and decided not to do it. In the other case he did, with regard to the Soleimani assassination. So this little plan that they had cooked up with regard to the Iran nuclear documents was part of a much larger design that was put into effect at various levels and in various ways over the next year and a half. It's very important to understand the full impact of that, and it's not over yet.

Horton: Well, it was in early March when there were some strikes against American forces. There were some rockets launched toward American forces in Iraq, and those were blamed on Iranian-backed militias. Some reports had it that Pompeo and Defense Secretary Esper were both pushing for strikes and that Trump refused just because he said it would look too bad from a public relations point of view to hit Iran when they were in the midst of such a bad Coronavirus crisis. Not that he'll lift the sanctions or anything like that, but that he turned down their push for war at that point.

Porter: And that was the second time around for this kind of ploy by Pompeo. He had done the same thing back in December 2019 and had succeeded in maneuvering Trump into a position where he was then able to push the idea of the assassination option. And we know that the Iranians responded to the assassination in their own very clever way. It was nuanced. In attacking this Iraq base where the Americans were present, they showed the capability of killing Americans, and at the same time they made it clear in their response that they were not intending to do so.

So basically, this is part of a much broader fabric of Israeli strategy in which Pompeo plays a key role. But Pompeo is not the only one. They also had somebody who had been at the Foundation for the Defense of Democracies (FDD), who was moved into the White House at around the same time, and who was the one who was designing the all-out pressure campaign that was clearly a part of the Israeli strategy of putting the maximum pressure on the Iranian economy in the hope that this would be much more likely to bring about a military confrontation between the United States and Iran. Of course, that's exactly what we have seen. We saw it in the spring of 2019.

Horton: There's this piece by the hated David Sanger in the *New York Times* from a few days ago about Pompeo's new scheme to get America back into the Iran deal in order to accuse Iran of breaking it now. Is that going to work?

Porter: That's the craziest idea that I've heard so far, I must say. I can't believe that anybody, even in the *New York Times*, would find that even minimally credible. How do you stay in the agreement and outside the agreement at the same time? You can't be. I mean, it's such a stretch that I don't see that anybody would take it seriously. Certainly the Iranians wouldn't take it seriously. I don't think the Europeans would take it seriously for a moment, either. I just think it's dead in the water from the very beginning.

Horton: Well, it's complete nonsense. No wonder David Sanger believes in it. That makes perfect sense.

Porter: I suppose you're right, yeah.

Horton: Thanks again, Gareth.

Porter: Thanks, Scott, as always. My pleasure.

Part 4

Hiroshima and Nagasaki

"The Japanese were ready to
surrender, and it wasn't necessary
to hit them with that awful thing."
— Dwight D. Eisenhower, 1963

Daniel Ellsberg: Hiroshima and the Danger of 100 Holocausts
August 5, 2011

Scott Horton: Tonight's guest on the show is Daniel Ellsberg. He famously liberated the Pentagon Papers from the RAND Corporation. He's the star of the documentary *The Most Dangerous Man in America.* He is the author of the book *Secrets: A Memoir of Vietnam and the Pentagon Papers.* He keeps a website at Ellsberg.net, and he's written these two great pieces for TruthDig two years ago: "Hiroshima Day, America Has Been Asleep at the Wheel for 64 Years" — it'll be 66 now — and "A Hundred Holocausts: An Insider's Window Into U.S. Nuclear Policy." Welcome to *Antiwar Radio*, Dan.

Daniel Ellsberg: Good to be on, thank you.

Horton: Well, I'm very happy you could join us this afternoon. Tomorrow is Hiroshima Day, and you had a career inside the government and are familiar with nuclear weapons planning strategy from that point of view. But you also had a window into nuclear weapons policy that most Americans never had a chance to before they were ever used on Japan, isn't that right?

Ellsberg: Yes, that's an odd story. In that it's in a way kind of coincidental, because ironically, despite the abhorrence of nuclear weapons that I started out with, I found myself working on nuclear war plans years later at the RAND Corporation, which is a consultant to the Defense Department. What made that so ironic was that my introduction to the subject happened before that of most Americans. It so happened in a sociology class during my sophomore year at my high school near Detroit, when my teacher raised the subject of what was then known as "cultural lag," the notion that our politics, our morality and our ability to work together as a species and in societies lagged behind our technical ability of destruction and technology in general.

As a possible illustration of that, he raised the possibility that — this is now during the war, in 1944 — that there might come into existence a uranium bomb, a U-235 bomb, as he did describe it. This is something that was underway at the Manhattan Project at that time, but that was super-secret, and it really didn't leak out at all. What happened was that from prewar scientific papers, some people followed up during the war a couple of times with articles, including one in the *Saturday Evening Post* that described the possibility of a bomb which was otherwise a very great secret, and nobody was talking or writing about it. But he had come across one of these articles, and he said that this would be a bomb that would be 1,000 times more powerful than the blockbusters that at that time had been used against Britain and were being used against Germany. Some of them were 10 tons of high explosives, and that was about the limit of what our bombs were. They were called "blockbusters" because they were said to destroy a city block of buildings, not just a house. This would be 1,000 times more powerful, and the question would be: What would be the implications for humanity, for human society, for civilization if there should be such a bomb? The emphasis wasn't on who had it. It wasn't on the notion of whether the Germans or whoever got this first. It was just on how humanity would relate to a bomb like that.

And I think all of us in the class, high school sophomores at that time, were asked to write a paper on it. We spent some days, and I think we all came to pretty much the same conclusion: that this would be bad news for humanity, this would be a bad thing. We weren't really able to deal with that kind of destructive power in a way that would preserve cities and civilizations.

When Hiroshima Day came along, August 6th, about nine months later, I looked at the headlines about a city having been destroyed by a single bomb in a manner unfamiliar to everyone not involved in the Manhattan Project. Namely, I knew what that was — that's the bomb we studied — and it had come into existence. We got it, and we used it on a city. I didn't have a sense at that time what the alternatives might have been, or whether we needed an invasion or didn't need an invasion. I was really thinking of the long-run consequences, and I had quite a chill, thinking that this was an ominous development.

Horton: As you write in the article, "Hiroshima Day, America Has Been Asleep at the Wheel For 64 Years" — it's two years old, at TruthDig.com — it really sounds like this could have been written by Joseph Heller and included somewhere in *Catch-22*, where somewhere is a high school sophomore who has thought very deeply about the profound consequences of this new technology, and here it is deployed by Harry Truman, who could just as easily be a character in *Catch-22* saying, "Hurray, everybody! Look what we did!" And as you say in your article, you heard from his first announcement, you could tell in Truman's voice, that he hadn't thought about this nearly as hard; he didn't have the same wisdom as a single one of you guys in your sophomore high school class.

Ellsberg: I think the point of my piece, and the way I really think about it, was that it did not take any moral giant to see this. When you thought about it in the context that most Americans had then and have ever since — of the assertion that it had ended a war, that it had saved American lives from dying in an invasion, maybe a million lives — and I revered Franklin Roosevelt, who developed this bomb, essentially. Harry Truman dropped it. All of those things gave it an aura of legitimacy, especially the efficacy. It was a savior. I should be grateful to it, as well as recognize it as an amazing American technical achievement.

Whereas, thinking about it beforehand nine months earlier, which, as you say, almost no one had done that, certainly not Truman. When I was reading about it, Truman did not know as vice president that this weapon existed. He wasn't told until he inherited the presidency from FDR. He hadn't spent a minute thinking about it.

And the scientists, amazingly enough, didn't really give strong thought to the post-war implications of this and what it would mean for humanity if we tested it and then used it on humans. The bomb was dropped in August. Around April and May, scientists, especially in Chicago, who were no longer at the forefront of the technical work, really sat down and began discussing the long-run implications. Their conclusion was — and these are people who all worked on the bomb, including Szilard, who had first proposed to Einstein and actually drafted a letter from Einstein to FDR calling for a crash program on it — and James Frank and others on the committee said, "Even if it would save American lives" — and they didn't know the strategic situation that by May, June and July it would no longer look as though an invasion would be necessary at all; that's another story, but they didn't know that — they said, "Even if it would save American lives, we should not drop this." They made two forms of this point, and another partition was that we should at least demonstrate it and make proposals, specifically that the emperor could stay — a proposal that we eventually did agree to — but we should make that in advance and try to get a surrender without using this, because it will make an arms race inevitable, with very grave implications for the human species. That was the consideration that, as far as we know from the records, was hardly in the minds or the discussion of the people in Washington who were dealing with this.

So, all I'm saying is that the scientists who knew where this was heading and knew that ultimately and before long it would lead to a hydrogen bomb, an H-bomb, a thermonuclear fusion bomb which required an atom bomb, a Nagasaki-type bomb, as its trigger. They knew what they were building was the trigger of another bomb that would be a million times more powerful than the blockbusters of World War II. They could see that, whereas the people in Washington were not thinking in those terms. But basically, they came to the same conclusion that a bunch of high school sophomores could easily come to nine months earlier. Unfortunately, their letter to the president, their report, did not get into his hands; it was bottled up by General Groves and others. It didn't even get to the Secretary of War Stimson until after the bomb had been dropped.

Horton: Well, on the other hand, it looks like MacArthur and Eisenhower and a great many others in a position to know better told Truman so beforehand. It was really just down to Truman and Stimson, right?

Ellsberg: Eisenhower in particular said that he told first Stimson, then Truman himself, that he said he saw no reason to drop "this terrible thing." We should not be the first to do it. He said it was not needed. The Japanese were near surrender. He knew what the rest of us in the country did not know, that we were intercepting the Japanese communications and were aware that they were looking for surrender terms. They were not by any means — even the military wasn't — determined to fight to the death as they had done on Okinawa. That's not what they were thinking at that point, and therefore, no invasion would be necessary. So Eisenhower said it wasn't necessary and we shouldn't do it. He felt very sick at the thought, really. He reported that later, in 1960.

Horton: I kind of figured this out today, talking with Greg Mitchell, author of *Hiroshima in America*, and his new book is called *Atomic Cover-Up*, about the footage suppressed. But I grew up in the 1980s when Reagan was perceived as a hawk. It was the end of *détente*, and the start again of brinksmanship. I was in elementary school when the movie *The Day After* came out, and so my conception of what a nuclear war might look like was an absolute nightmare, the end of mankind as it's portrayed in that movie. Yet, as Greg Mitchell explained on the show today, for the first many decades — the first 40 years, I guess, after the nuking of Hiroshima and Nagasaki — a great amount of truth was covered up regarding those attacks and their effects. Basically, Americans could only see a few different kinds of pictures. There were a few different black-and-whites of the mushroom cloud from the air, and the one with the dome on top that we all see that's now a peace monument there. Then maybe a wide shot of Hiroshima after the fact, where a lot of it has been flattened, but that's about it.

For decades, the American people were deceived. There was a massive lie by omission about the destructive power of these weapons and the deaths and suffering of the Japanese people who survived it. All the deaths, but also the people that survived it and what they went through.

Ellsberg: Let me go one step beyond that. I would say that to this day, the American public and people of the world on the whole have not really absorbed the implications of the next nuclear revolution that came along, some nine years later. That was the H-bomb. The first feasible design of it was conceived, I think, in 1952 or '51, but it was first tested in 1952 with a device. Then a real bomb was tested in 1954. It didn't come into our arsenal until the mid-1950s.

That was another factor of a thousand. The Hiroshima bomb had the explosive power of 13–14,000 tons of TNT — in other words, more than a thousand times the 10 tons that were in the World War II blockbuster. What I don't think many Americans realize to this day is the difference between an A- and an H-bomb, the difference in scale. The first real test of a big bomb in the mid-'50s had the explosive power of 15 million tons of TNT. In other words, a thousand Hiroshimas.

And I don't think the American public has ever realized that all of the strategic weapons, the hydrogen bomb, the long-range weapons — missiles, warheads and bombers — are H-bombs. They're thermonuclear weapons. Each one of them has for its trigger, its detonator, a Nagasaki-type bomb. So when you look at those pictures that Greg Mitchell is talking about that have now, I think, become available in color, of that flat landscape, I don't think many people would get it in their minds that they were looking at the effect of dropping the trigger to a modern large nuclear weapon. The detonator, the detonator cap, that's what you're looking at. An H-bomb is an entirely different order, and we have thousands on alert, with more thousands in reserve and in stockpile.

Horton: You're saying these are all tens of megatons in yield strong?

Ellsberg: A megaton is a million tons of TNT explosive equivalent, and these things are half a megaton. The individual warheads are smaller than they were in the late 1950s and the early 1960s because those were carried by heavy bombers. Now we carry a number of them on a single missile, so they're smaller than those early ones. They might be a third of a megaton, half a megaton, 0.6 megatons, that sort of thing. There are still some very large weapons in stockpiles, but the point is that with the numbers that are involved — thousands on both the U.S. and Russian sides — we're talking about a capability that was not ever possible in the world before, that was not even conceived in 1945, '55, '65, or '75.

That is the ability to cause nuclear winter, which was first imagined and then calculated and analyzed very much starting in 1982, with more recent research on that just in this last decade confirming the more ominous implications of the earlier research. That is, with these fusion weapons, the largest part of the energy goes off in the way of heat, so their immense heat that they let out will result in the burning of cities. When you burn cities, smoke — an aspect that was not even calculated until the 1980s — smoke rising simultaneously from all these burning cities will coalesce into great clouds that will blot out the sunlight. And if it's in the summer, it will lower the temperature by 15 to 20 degrees centigrade, freeze lakes, freeze rivers, kill all the crops and cause famine. Plus, there's the potential of that moving even into the Southern Hemisphere, covering the Earth and ending most complex life on Earth. That's a real possibility. Not a small possibility, not a small probability.

Just the thought that we have on alert systems that, if set off by a false alarm or an escalation from a conventional war, become a limited nuclear war! And then if they threaten to escalate and each side starts preempting the other, what you have are two Doomsday Machines on hair-trigger alert, each one interested in preempting the other's strike, and each with a high potential of ending life on Earth. It's inexcusable, it's outrageous, and it could not be more ominous that this possibility is allowed to exist. But it is, and that's true under President Obama as much as under any of his predecessors. Either side should make it impossible to cause nuclear winter. Neither has even come close to moving towards that direction.

Horton: I wanted to ask you about something that I read in Andrew Cockburn's book *Rumsfeld*. It was about the war games, the "continuity of government" program and so forth; how in the 1990s, during the Bill Clinton years, they would run all these drills about: "What if we did get into a nuclear war with Russia?" Donald Rumsfeld would always play the emergency "continuity of government" president, and every time that they did the drill, he would always nuke everyone to death and exterminate all of mankind. Even if they built into the war game a few chances to try to negotiate a settlement, to try to call some kind of ceasefire, he would always end up escalating it.

And I just wonder, with your inside knowledge of the war plans, at least from a certain date back in the past, is there even a possibility of such a thing as a limited nuclear war between the major powers? Never mind India and Pakistan for the moment, but say, America and Russia?

Ellsberg: Well, it's never been very likely that a nuclear attack on the homeland or even the forces of either one of the superpowers by the other one could stay limited. Certainly, Eisenhower didn't believe that was possible, McNamara did not — I know from talking to him — did not believe it was possible that it would stay limited. And under Eisenhower, the plan that was operational when I came into the picture in 1959, '60, '61 did not allow for the possibility of limited war between the U.S. and the Soviet Union, either conventional or nuclear. He assumed that any such war would go nuclear. Therefore to save money on divisions, and to refuse the Army any basis for asking for divisions to match the Soviets, he directed the Joint Chiefs that there should be no plans whatever for war, beyond more than a couple of brigades being engaged, that did not involve all-out attack on the Soviet Union.

Now, that has been changed to some extent, but not with any high probability of escaping an all-out exchange. But, you know, the idea of using nuclear weapons first against Russia could hardly arise, except over a dispute over Ukraine and/or Georgia. I mention this because leading people like Senator McCain, other Republicans and a number of Democrats have called for Ukraine and Georgia to be admitted into NATO. If that were the case, that would be a commitment to the first use of nuclear weapons against Russia in the case of defending Ukraine or Georgia. That's the nature of the NATO commitment, it's the heart of it. And although the Germans and others have tried to propose a no-first-use commitment and unhook NATO from the Doomsday Machine, from this threat to initiate nuclear war, the U.S. has always thought that. Partly because it's our nuclear weapons, it gives us the rationale for a critical role in Europe because we are the guardians of NATO's nuclear weapons, the French excluded, and so we don't want to give up that role. Therefore, the idea of extending a NATO guarantee to Ukraine and Georgia means that if there were a conflict, as did arise over Georgia, in McCain's accounting, "Well, we would have known what to do there. We could use nuclear weapons against the Russian troops." A really good idea. In other words, "Let's use the weapons that we have and we've spent so much for." So we're still keeping tactical nuclear weapons in Europe, hundreds of them, with no clear purpose other than that they would be the detonators to the Doomsday Machines on both sides. So he's talking about ending the world.

Horton: George W. Bush was the thin line of civilization in this circumstance, holding Dick Cheney back, that was it. If he had been able to convince Junior…

Ellsberg: That's why we have to change our view of the relationship. I think it is true that Cheney was not calling the shots by the end of Bush's term on that issue. Because by every account, Cheney wanted to attack Iran with nuclear weapons on their underground site right through the term, and it was Bush, of all people, that we relied on. Well, I think that shows the situation that humanity is in, when our survival depends on the prudence and judgment and wisdom of a George W. Bush, wherein the species — and this is not rhetorical — the species is in very deep trouble. There's no *if* there; that's where we are.

And I have to say, with President Obama, all his talk of abolition I think is as meaningful as his promise to close Guantanamo, and I don't say that rhetorically. I mean that there is essentially nothing in it. I haven't seen anything there at all. And to talk of lowering the number of weapons that each side has operational to some 1,500 — which is far more than needed to cause nuclear winter in the world — that's just laughable if it weren't so tragic. Whether you can abolish nuclear weapons, at any particular time period, is a controversial question. You could unhook and destroy that Doomsday Machine and make it impossible to cause nuclear winter. That leaves you with nuclear weapons, but it should not be measured in thousands; we should get down to dozens. Or something like what the Chinese have. They've had the least outrageous and demonic nuclear weapons of any of the major nuclear powers. Of course, that is what should be done, and it should be done right now. We should dismantle all our land-based missiles. That could be done before the end of the year. There hasn't been a real rationale for the land-based missiles in a world where we have

Polaris and Poseidon Trident submarines. We haven't had a strategic use for those for a good 50 years, and I say that as somebody who did work on nuclear plans in that period. Yet they remain in large part in order to — it would sound too absurd, I almost hate to say this — but I would say the major political reason for keeping those land-based missiles — Minutemen, which are targets for Russian missiles in a Russian preemptive strike and have no practical usefulness in reducing damage to the United States under the circumstances — the major reason is that senators and congressmen from the states in which those missiles reside are concerned about the jobs of the barbershops and the support groups in the little towns that service the servicemen and those missiles. They don't want to lose it. That's their jobs program. But that kind of sums up the solemnity and wisdom of our nuclear posture for the last 60 years.

Horton: Alright, well, we're all out of time, Dan. I'm sorry we'll have to leave it there, but I want to thank you very much for your time.

Ellsberg: Always good to talk to you. And remember on Hiroshima Day the slogan written on a monument near Hiroshima, in Japanese, to the people who died there: "Rest in peace. The error shall not be repeated." That translation is in English for the benefit of the Americans. The Japanese tell me that a better translation would be: "The *crime* shall not be repeated." Unfortunately, that promise has not been kept. It hasn't been dropped yet on other people, but the threat has been there continuously. I don't think we can keep our commitment to those spirits, but that's what we should work to do.

Peter Van Buren: The Hiroshima Myth August 23, 2019

Scott Horton: I've got Peter Van Buren, former State Department whistleblower, author of *We Meant Well* and also *Hooper's War: A Novel of World War II Japan*. How are you doing?

Peter Van Buren: I'm doing well, Scott. Thanks for having me back. It's always a pleasure to be here.

Horton: How would you describe your book *Hooper's War*?

Van Buren: It's a book about moral injury and about what happens to people in war. It's set in World War II as a so-called neutral setting that will hopefully allow us to have a conversation about these topics without bringing in the external politics of more modern wars. Because, unfortunately, the things that wars do to human beings, whether they're the targets or the trigger-pullers, are universal and haven't changed a lot since the early Greeks; and whether it's Afghanistan or Okinawa or a fictional battle as I created in my book, the stories are very much the same.

Horton: Everybody knows it, and Sherman said it: "War is hell." And everybody knows that guys come home from these wars and then stand at intersections, asking people to help them get by. I grew up with Vietnam veterans on the side of the road constantly. But at the same time, we also know that war is what makes us great, war is what makes us. Without that, what are we, except just a bunch of disconnected, individual, community-less beings? Our army, that's the thing that makes us the U.S.A., kicking butt and stopping bad guys.

Van Buren: It comes as close to a national religion as anything else, certainly a national obsession. The idea that the few things that pull us together are when the United States is "under threat" from abroad, and whether that threat is somewhere on the real scale or somewhere on the completely made-up scale is largely irrelevant.

If you want to look back — I'm not a big fan of the Single Theory of History, where one event or one theme controls everything. The *New York Times* this week has released their 1619 Project, which posits that everything in America is based on slavery, and that type of analysis, whether it's slavery or it's war, is simplistic, but at some point, there's validity buried in there. The idea that America must exist in a state of conflict, that an external enemy is always necessary, has a lot of validity.

Our country was founded through fighting the British external enemy, and there's been relatively few periods in our history where we haven't done that. About the only significant time where we haven't summoned an external enemy is when we used an internal enemy during our Civil War. The problem with substituting the internal enemy for an external one is that the casualties are usually multiplied by double, and making out the bad guys when they speak the same language gets a little tricky. So we've luckily corrected that mistake and haven't repeated it since 1865, so some good news is mixed in as well.

Horton: [Laughing] Yeah, a little bit, I suppose. But it's always seemed to me that history in American education is so neglected that really people don't know anything about George Washington or even Abraham Lincoln or anything when it comes to America's founders. Really, what we have is FDR, who led America to war against Hitler and saved the economy by doing so, they say. He brought the unemployment rate down by conscripting 16 million people, and they stopped one of history's greatest tyrants. So that's really where we've been stuck ever since is World War II, right?

Van Buren: When you talk about American history, it's very interesting, because I would be amused to find someone to point out a high school in America that accurately and thoroughly even gets into the history of America post-World War II. It's been a couple of years since I've been in high school, but I've watched my kids go through it, I've seen friends' and neighbors' kids go through it, and American history is incomplete. And as poorly taught as it's taught from the Founding up through World War II, it sort of tapers off into a gray, fuzzy zone: "And then World War II and Korea, Nixon was bad, and that was kind of it."

There have been some wonderful attempts. I'll put in a quick pitch for an Oliver Stone project. It's called *The Untold History of the United States*, and it piggybacks on Howard Zinn's work and some other fine scholars. It presents in a fairly lively documentary form the history of the last 80 years of the United States. In a little bit more objective point of view, it talks about the underlying themes that we've just touched on very briefly and sarcastically here. This idea of dominance, of empire, of the need for an external enemy, and it even looks into the value of the Cold War if you're a politician running for office. Now, it is an Oliver Stone project, so there's places where it can't help itself but to go over-the-top, but there's a lot of very good history in there, and if someone is saying to himself, "Well, alright, Scott, you've convinced me, I need to educate myself," he could start in worse places than watching a couple hours of that.

Horton: One of the benefits of being a libertarian is that you can indulge all you want in leftist historical revisionism, or for that matter, right-wing historical revisionism, on whatever issues, and you don't have to abandon your identity in order to look into something like that. Whereas a conservative might say, "Oh, my God! Oliver Stone and his commie historian co-author. I can't bear to peek through my fingers at what they might say! I don't want to know, and I don't want to think that I have anything in common with anybody like that!" But when you're a libertarian reject like me, it's no problem. There certainly is great value in that series. In fact, once they started getting into the terror wars, I turned it off because I didn't want to copy them in any way. I was in the middle of writing *Fool's Errand* at the time, and there's too much good stuff in there. I wanted to prevent myself from sounding too much like what they were doing.

Van Buren: I'd like to touch on the word *revisionism*, because I understand exactly how you mean it — and it's a valid word — but be cautious about that word. Oftentimes, it's flung about when you think someone is sort of revising things, telling you that what you knew was wrong. The word *incomplete-ism* is, I think, a better word to say the same thing.

Horton: I like the confrontation in that. It's like, "Guess what, the official history of everything as written is all false, so if you want to be straight about anything, you're going to have to go back and look at critics." Look at the way they lie to you every day on the news. That's next year's history. You could read an American history book right now in probably any college in America, and it'll tell you that the Branch Davidians all committed suicide. Well, you know what you need? Some historical revisionism, because that's not what happened.

Van Buren: I just came back from Germany, and for part of that trip I made a visit to Dachau and spent a little time poking into the roots of Nazism. I think if someone is out-and-out putting out falsehoods, if you're willing to engage with the material, you can kind of poke those down fairly quickly. What is more dangerous is incomplete-ism, which is the idea that we'll tell you enough true stuff, but not all of it. We'll give you a set of facts that, on their own, are more or less accurate but lead you to a false conclusion.

For example, those who claim that America now is Germany in 1933 know nothing about the Germany of 1933. You can pull up little points of connection, and you can try to argue that silly elementary school tweets are the equivalent of the Nuremberg speech, but in fact, those are false conclusions. They give you bits of facts that lead you to something, but they are so incomplete that they lead you to a false end. That becomes very dangerous, because if I come right out and say something completely false and silly, then you can pretty easily shoot that down — even a quick drive through Wikipedia might be enough.

But if I give you some real information with the idea of walking you toward a particular conclusion, then it's more difficult, because you go back and you say, "Well, gosh, it looks like the Japanese weren't really ready to surrender, so maybe blowing up Hiroshima and Nagasaki was necessary." If someone simply presents the fact that there were problems with arranging the Japanese surrender, which was true, but present only that bit of information and nothing more, you might be led to the conclusion that the annihilation of two civilian cities was perfectly justified. That's where things get nasty and a little bit difficult.

Horton: One thing that I learned very recently in the annual discussion of the atomic bombings is an anecdote that right after the war, Stimson, the Secretary of War, said that we ought to go ahead and give the atomic bomb plans to Stalin because they're going to get them anyway. He argued that by hoarding the secret, we're just going to make the USSR distrust us more, we're going to get the whole post-war era off on the wrong foot, and we could end up in an arms race against them. He was overruled, of course, and the Russians got nukes two or three years later anyway.

In fact, there's plenty of good reason to believe that Americans did transfer nuclear technology to them regardless, but it just goes to show that that was the Secretary of War's advice to the president at the time. One doesn't have to agree with that at all, but it sure is important to know that that's there because at the very least, it shows you that the way things turned out, it didn't necessarily have to be that way. It could have been some other way. Whereas, the way we're presented, especially when we're kids in school, is that essentially everything that happened had to happen. As in, "You think Truman, our great President, would nuke people if he didn't have to? That the American people would have elected a man who would have nuked people if he didn't have to? No way!" It goes without saying that it had to be done, or else it wouldn't have been done. So you get a little fact like that, and you go, "Wow! Maybe if Arthur Vandenberg had had a heart attack and didn't have a chance to tell Harry Truman, 'Scare the hell out of them! Come on! Let's build up the USSR that just lost tens of millions of people in this war. Let's pretend they're about to conquer the Earth if we don't stop them.'" What if those people hadn't been there that week? Things might have been entirely different.

Van Buren: This is why we're talking about Hiroshima, which happened 74 years ago, because historians study history to avoid repeating the mistakes of the past, whereas politicians manipulate history in order to set up what they want to do in the future. There's really not a better case study of this than Hiroshima and the aftermath of the atomic bombing. You basically had, from a historical perspective, an action taken in the midst of war that happened the way it happened.

We can go back and look at it and say, "Was it truly necessary for the United States to annihilate two cities full of men, women and children?" Unfortunately, what the politicians did in America was to manipulate a set of facts and take advantage of some very skillful propaganda to repurpose those historical questions for the future in order to set up what they expected to be a nuclear conflict with the Soviet Union. Their example was going to be to use Hiroshima, not as a way of reflecting on our actions in World War II, but as a way of preparing the American people for the atomic conflict that they felt was imminent.

The way that people are manipulated by history by a constrained use of facts is very relevant. Most of us have lived through the post-9/11 years, where the events of September 11th were manipulated into a nearly endless series of wars that are continuing even through today. We have lived through the events of the so-called "Russiagate," which have been an attempt to manipulate facts in order to end the Trump presidency.

These lessons are not abstract. This is extremely relevant stuff, and by using a historical example, we gain the advantage of perspective, we remove some of the immediate emotion, and also, we take advantage of the fact that we have a fuller set of information to work with than we have on more contemporary events. Unfortunately, in our "transparent" democracy, it takes decades for information to slowly creep out of government hands because, of course, controlling information allows them to control the conclusions the public reaches.

Horton: There were some important truths and narratives that were coming out of the atomic bombings at the time. As you're saying, they weren't just covered up to protect the American government from losing face for doing such a thing to two places that were, in fact, not military bases as originally claimed, but more importantly, to set the American people up to get used to life in the era of atomic wars. Would you talk a little bit more about that, and specifically, how do you know that? What are the examples of where they decided that?

Van Buren: Let's set aside the question of "Were those bombs necessary to end the war at that point in time?" It's a difficult question, particularly when you're looking backwards and trying to place yourself in that room and think like 1945 people, not 2019 people. Let's instead skip ahead and use as a starting point the fact that the atomic bombs were dropped on Hiroshima and Nagasaki and decimated two undefended cities almost entirely populated by non-combatants. The war ended after that, and the American people initially were fed a line of that. Then we go back and look at the messaging as we talk about it today.

The original announcement of the bomb by Harry Truman was an absolutely vengeful statement of biblical proportions. He said, "We are now prepared to obliterate more rapidly and completely every productive enterprise the Japanese have above ground in any city." That's way beyond saying, "We're going to destroy their power to make war." He was pretty straightforward with his word choices. In that particular speech, he went on to talk about mass casualties and spoke of cruelty at levels previously unknown on this Earth outside of biblical times. That message of vengeance was the initial messaging, and it played well to a population that was weary of war. That's what stood for about a year and a half, to the extent that people even focused on it.

While writing what was a fictional book about World War II, I read through an awful lot of archive materials, and there wasn't as much conversation about the nuclear bombing of two cities as you might think would be engendered. The focus was on something else. The end of the war was the primary focal point.

It was John Hersey's 1946 article "Hiroshima" in *The New Yorker* that kind of began the change in conversation. When you study history, you pick out these events because it's a great way to talk about history. It wasn't that prior to Hersey's article no one talked about Hiroshima and suddenly it fell from the sky, so to speak. That isn't the case at all. GIs were coming back, photos were leaking, and this was a time when information spread very differently and much more slowly than today. The point is that John Hersey was one of the first reporters that went to Hiroshima to report on what happened there and was not censored. The war had ended, the military was less concerned about sitting on these types of articles, and there was certainly no tactical information that could be accidentally given away. Hersey wrote a brilliant article that later became a book, which explained in terrifying, horrific detail what happens to an undefended civilian population when they are nuked.

The timing of it, a year and a half or so after the war, Americans were anxious to forget as much as they could, but an economic boom was starting, and suddenly there was an opening for reflection about who we were and what we did. Hersey, almost by accident, stepped into that. The problem was that this period of reflection over how America made war was coinciding with what the government knew was the beginning of the Cold War. Inside Washington, they were no longer thinking about having defeated the Nazis and the Japanese; all eyes were on Moscow, and everyone was planning on how this next war was going to play out. We were a year or so away from what became known as the Berlin Airlift. There was the Iron Curtain, and all this stuff was percolating in Washington. They knew this was coming. The idea of softening the American people to the idea of nuclear war did not fit with the program.

John Hersey's article — and the introspection that it caused — was the wrong messaging for an America that Washington knew was heading into the Cold War. The Cold War was a nuclear conflict and was going to involve the annihilation of multiple Russian and/or Chinese cities. They didn't want any soft selling on that. Something had to be done, and what was done was essentially the last 70 years of propaganda. There's no real pinpoint starting point, but it would be fair to say that it started with a 1947 article "The Decision to Use the Atomic Bomb" by Secretary of War Henry Stimson, who actually wasn't the author. The author was McGeorge Bundy, who people will recognize as one of the best and the brightest in the John F. Kennedy crowd that drove America into Vietnam and the atrocities that characterized that war.

So McGeorge Bundy wrote this article, basically creating the talking points for the American government and the American people for the next 70-plus years — that the atomic bombs were necessary to end the war quickly, that the atomic bombs actually saved lives. The numbers they've used as it relates to that have jumped. Every time you go through a historical reference, the numbers get bigger, from 10,000 dead Americans invading Japan, to 100,000, to 1,000,000 on the beaches…

Horton: That's what President Bush, Sr. had said, that 1,000,000 American boys would have died invading Japan.

Van Buren: That's correct, but the interesting thing is that if you go back through the stories and articles and references and think tank pieces, the number of Americans that were going to be killed in that invasion increased by a factor of 100 in the five years after the war ended. More importantly, Stimson began laying out the idea that not only did we save American lives in the use of the atomic bomb, but we saved Japanese lives, too. You can't make this stuff up. Stimson's article, which is really the beginning of it all, was in *Harper's Magazine* in February 1947.

The idea is to say that we saved Japanese lives because, dang it, if we would have had to conduct a land invasion of Japan, we would have had to kill more of them little yellow fellers. So the people who died at Hiroshima and Nagasaki, they were the crumple zone for the rest of Japan that saved us from having to kill more of them. This was the creation of what I call the "Hiroshima Myth," which became the basis of the Cold War. The idea that annihilating cities was the way nuclear war was going to work, that Americans were ready for this, and that it's actually going to help them because the more Russians we kill in that initial blast, the less likely they are to continue the war, etc.

The Hiroshima and Nagasaki bombings were done as a proof of concept for the Russians. As in, "Look what we've got. You want to mess with us? This is what you're messing with." That Hiroshima Myth has survived mightily because in many ways, it became the core of convincing Americans that the Cold War was a noble cause and that dropping nukes was something we were going to have to do eventually. It was almost a certainty. If you read back through the archives of the 1950s, it was almost a certainty inside Washington that we were going to go to nuclear war with the Russians and possibly also with the Chinese. It really wasn't until the 1960s, when the

Russians started to get a little stronger and a little more time had passed since World War II, that the American government started to kind of imagine that maybe this wasn't so inevitable.

The bottom line is that the aftermath of Hiroshima was a myth that was specifically created for propaganda purposes by the United States to prepare the American public for the inevitable nuclear conflict to come. It served the interests of the United States government long past the Cold War and still kind of pops in around the edges, like when you've got to explain away accidentally droning a wedding party in Afghanistan to kill one terrorist. The Hiroshima Myth is underneath all that. You hear military people and politicians today saying things like, "Hey, if the people of Fallujah didn't harbor terrorists, it wouldn't have been necessary to destroy them." So there are hints and notes of the Hiroshima Myth that persist all through modern times. It's proven to be an extraordinary piece of propaganda, and it's still very much a part of us today.

Horton: You're right about that. It's like a big Milgram Experiment: "It's required that you continue to endorse this because everybody else does, and you don't want to look like a hippie."

Think about just how shallow and ridiculous the argument is, that to save American GIs, we've got to nuke women and children. Ralph Raico, in his article "Hiroshima and Nagasaki," says, "Well, what if we're just talking about the Nazis, who didn't have nukes, but instead they would just round up women and children in the town square and just machine gun them to death? Would it have been okay for the Americans to go into Hiroshima with our infantry, round up the population into the center of town and just machine gun them all to death like the Gestapo in order to get the government in Tokyo to surrender?" That's who we are now? But somehow, this is totally good enough.

In fact, look at Waco again. They say, "We have to get that David Koresh. I heard he deals drugs. Kill them all!" Yeah, but there's a bunch of women and children in there.

Van Buren: It comes down to a very core part of American policy, and that is that expediency always seems to trump morality. The right thing to do under difficult circumstances is how this stuff gets spun. Nobody is proud of killing women and children, but there's always an excuse like: "We have to get bad guys off the streets, and if there's some collateral damage, that's sad, but it's the unfortunate reality of things." That describes Hiroshima. It describes the atrocities that fueled the Vietnam War. It describes every civilian drone strike in Afghanistan, Iraq, Lebanon, Syria, Libya or wherever.

Let's look at Libya. The idea that we had to get rid of Gaddafi — because he was an evil guy — justified the destruction of the country and turning it into an ungoverned wasteland. We had to burn the village in order to save it. That's the Hiroshima Myth.

Horton: There's a new one for our time, which is George W. Bush's absolutely unprovoked, aggressive invasion of Iraq in 2003, where they just marched the Marines and the 3rd Infantry Division straight in from Kuwait. Now, anything less than that is fine. A great example of this was Jeremy Scahill on the show with Bill Maher on HBO a few years back, during the Obama years. Scahill was saying, "Look, Obama's murdering people." You've been mentioning wedding parties — I'm pretty sure that was one of the points Scahill made — and he said, "We're talking about innocent people who are being killed by these drones. A 500-pound bomb is not a scalpel. It kills innocent people." Bill Maher's response is, "Yeah, but compared to George Bush marching the 3rd Infantry Division into Iraq? That's nothing." He's making the argument that it is surgical in comparison to this other bigger thing, so we could drone strike all damn day.

Van Buren: This idea that evil scales — that something can be more evil or less evil, and that somehow that matters — I just have a hard time with that, but it rests at the core of America's actions. We like to imagine ourselves as nice people, and we want to have things like the Hiroshima Myth as a way to remind ourselves that we are nice people who make mistakes like everybody does. One of the things that has become so offensive to me as a thinking human being over the last three years is how all this gets garbled when it's run through the Trump filter. We now have revisionism about George W. Bush and how he did it, and we have completely whitewashed the Obama years of war to the point where every time I see an article about Yemen, my ulcer grows. I'm like the Grinch, my ulcer grows 10 times more when I hear people say, "Well, you know, Trump is doing this in Yemen." Yes, he is, but did we already forget how we got started in Yemen? We did. We did forget that, didn't we?

Horton: In a way, I almost wish they could just forget that Obama did it. That way, they'll finally raise their voice against Trump. Because you're right, they'll never turn against their savior, the Democrat from Illinois.

Van Buren: The idea that history is that malleable and we are that manipulable is why the Hiroshima Myth has survived so many years after being birthed in a magazine article in February 1947. We are willing participants in all this, and it's really quite shameful.

I had a really interesting experience being in Germany, and having lived many years in Japan. When you go to Hiroshima and the museum there, everything starts on that August 6th morning. There's no sense in Hiroshima that Japanese imperialism started 30 years earlier and inexorably led to — justified or not — what happened at Hiroshima. The Germans, to give them some credit, have a much, much deeper sense of this and are willing to talk about it. In Dachau, for example, the museum starts with the events of World War I and what happened in Germany between the wars, the stuff that made the rise of a strongman dictator as close to inevitable as history allows. You walk away from that museum with a much better sense of having been educated, of understanding things and how pieces fit together.

The Hiroshima Myth, I would be remiss not to say, has been strongly driven and supported by the Japanese themselves. They are thrilled to claim as much victim status as possible for those two terrible days. They work closely with us to make sure that those days stand in isolation from everything that happened before in their case, and everything that happened afterwards in our case. So, in a way, the politics of the Cold War contributed to the victims themselves supporting a false narrative.

Horton: It's sort of like what you were talking about with slavery and war here in America. The same kind of thinking exists in Japan, where people don't like looking back and being honest about what's going on. They identify themselves with the collective too strongly, and so it's an attack on their own psyche. I mean, that's what this all comes down to, right? Social psychology and whose side are you on.

That's the way everything is framed: You're either pro-America or you're not. So if you're willing to say that slavery was absolutely this bad, and the genocide of the Indians, and the end of the Filipinos and everybody else, for that matter, was this bad; if you're willing to do that, it almost always comes packaged with a full-scale anti-Americanism from the point of view of the right. It's hard to find people who are American patriots but who also are perfectly happy to explore all this to find out exactly what all it means for us then, now and in the future. In fact, if I really wanted to do this interview right, if we really want to persuade people, the first thing I should have done was lead with, "Let's talk about all the conservative Republicans and generals and admirals who opposed it." Because that's the thing that gets through to people the most. I've had people freak out — and heard a lot of stories, too — people freak out if you're against nuking Japan: "You must be the most anti-American, communist piece of garbage in the world!" Then you show them that, well, that's what MacArthur thought too, and that's what Eisenhower and Nimitz thought. Leahy, the Admiral Chief of Staff to the President, said, "Don't do this! I was not taught to make war against women and children!" Then they go, "Oh, my God! A bunch of admirals and conservatives and Republicans and MacArthur, of all people, were against nuking Japan?"

Van Buren: It becomes very difficult. The term "cognitive dissonance" is to admit you're wrong, and to admit that you made sacrifices for the wrong cause is humanly very, very difficult. So it's perhaps even unfair to expect people who made those sacrifices, or sent their sons and daughters to make those sacrifices, to ask them to see this as it truly is. But we're not them, and our job as historians, as thinking people, or as folks who are a step or two removed, is to try to look at these things more objectively. Honest to God, if I lost an arm in the Vietnam jungles and you tried to tell me that that was all a complete waste of time, that would be a hard thing for me to get myself around, right?

I see these Parkland kids on TV and the Parkland dads on TV, and I just feel for them because to try to get someone who's suffering the loss of a child to talk about these issues objectively is impossible. It's not even fair to expect them to. Unfortunately, there's plenty of folks who will exploit that for their own political goals. But when you pull back a step or two or three, we're supposed to be smart enough to do that. That's why, in talking about these issues and in the book I wrote, I set it in a fictional portion of World War II in the hopes of giving the reader that distance. The book revolves around a fictional struggle — and it's all true, but it didn't happen. The hope is that it can help us to look at these events as objectively as human beings are capable.

Horton: You were in Iraq War II, not as a soldier but as a State Department official, and you wrote the great book *We Meant Well.* Even while you were there, you were getting over it and started blowing the whistle to the American people about what was going on during that time. But you know what? You're not alone. There are a lot of people who served in Iraq War II, Afghanistan and all the other wars over there in this century who have changed their minds about this.

Van Buren: It happens, and I don't claim any virtue because I happened to see through it while it was happening to me. What I'm saying is that it's unfair to expect everyone to be able to do that.

Horton: No, I agree with that. I mean, I do *Antiwar Radio*, but the point is not to seek out the fathers of dead veterans and try to argue with them, or to confront the guys who were in the war. Because, as you're saying, these are people who suffered losses. Some guy who was in Iraq War II, I don't know how many of his friends died in front of him. I'm not trying to pick a fight with a guy like that. He's the implementer of the foreign policy. My fight is with the guys at the AEI and the Defense Department who got us into the war.

Van Buren: Just as an individual myself, there are parts of my life where I've been able to be introspective about, and there are parts of my life for which I haven't been able to reach that. I stumbled into it with Iraq, good for me. That I didn't do it with something else, bad for me. I don't expect that of other people, and it's unfair to ask it of other people. What is fair is to ask folks who do already have that distance to be a lot more thoughtful about it. Folks who have that distance because they're personally removed from a modern event or because we're looking back at history at events that none of us have more than second- or third-generational contact with. At that point, our excuses for not being introspective start to get a whole lot fuzzier.

The idea of historically looking back — for example, we've danced around the question throughout this interview about whether the atomic bombs were actually necessary. This is intellectually an interesting question, and it's something that is worth talking about because it informs the future. That's where the value always is, and the key is always to see if there is some way in the present for you to be able to pull in some of the perspective of the future.

In other words, you're making decisions in your life right now: "Should I buy a new car?" "Who should I vote for?" "Should I do this, that, or the other thing?" We're constantly doing that, and of course, in retrospect, when you look back at decisions, you have more information. So the key is always to try and bring that perspective in.

So when you zoom into the Oval Office in June 1945 and you hear them saying, "Mr. President, we're nearing the point where we need a decision on the atomic bomb," if you simply confine yourself to that window of time and assume that Harry Truman had a birth defect that didn't allow him to think ahead, then the decision to drop the bomb becomes as obvious as it seemed to be to Harry Truman in 1945. The thing is that you hope our leaders are bigger men and women than that, and that's what didn't happen in 1945.

I'll leave aside the question that this whole thing was a demo for the Russians. I don't think that was a driving force in the decision to drop the bomb. I think it was in Harry Truman's mind. You know, he was a diarist. I don't know how he had time to do anything else, but he kept incredible diaries. Stimson kept diaries as well, and those diaries are actually available in the Yale Library. I read through some of them. Truman has incredibly neat handwriting. It's really very easy to work with. So we knew a lot about what these guys were thinking — we're not trying to put words in their heads — and Truman, for his part, was not really thinking very far ahead. So while the Russian thing was certainly probably more on the mind of some of his generals, in Truman's mind he was a simple man, and he was told, "We've got a bigger bomb that's going to make a bigger mess, and we feel that the Japanese need just one to get them to surrender, and this bomb is going to do that. We've been working on it all these years, and it's ready to go. Sign here, sir." I don't know from my own reading that it was a lot more complex than that. We had been blowing up people for a long, long time in World War II. Civilians had been targeted from the opening days of the war, long before America was even involved, and I think, without fully understanding how revolutionary nuclear weapons were in July 1945, Truman just saw this as the obvious next step. That said, we hope that our leaders are bigger people than that and think further ahead than that, but Truman didn't.

The thing about bombing Kyoto and bombing other places — Stimson was the one who argued against Kyoto — there was a target list drawn up, and it was essentially a series of cities that had not been heavily bombed. For whatever reason, the military definitely wanted to blow up something that hadn't been blown up before, and they really were interested to see how this weapon worked. Keep in mind, they only had a lot of theory and a handful of small-scale tests out in the desert in the United States. They didn't really fully know exactly what was going to happen, but they wanted to know because this was going to be the weapon they were going to be going forward with post-World War II. Kyoto was on that list because it hadn't been bombed during World War II. Stimson said, "Hey, I took my honeymoon there. It's a World Heritage site," and a couple of the other people in the State Department said, "Hey, if we blow up Kyoto, it's going to piss everybody off because it's a World Heritage site," and Truman said, "Fine, I don't care what we blow up. What's next on the list?" There wasn't really an argument, per se. It was more of: "Fine, you don't want to blow up Kyoto? We'll blow up number two, what's that?" And that was kind of it. It's like you and I are walking down the street and there's three or four restaurants and I say, "Hey, Scott, how about this one?" and you say, "No, let's go to that one." I mean, it's really that kind of casualness. The idea was that Nagasaki itself was the secondary target, right? It was supposed to be a city — I think Fukuoka, I'm not sure now — but that's the idea. It didn't matter. As long as it was more or less a virgin target, it didn't matter which one it was.

So in 1945, the decision was, I think, seen as not that big a deal. I think the big deal came afterwards, when we truly understood the power of these weapons, particularly as we moved into the Cold War, where the way that nuclear war was envisioned was not as a capstone to four or five years of conventional conflict like World War II was, but that nuclear weapons and the destruction of whole cities was going to be the opening of the new war. That's where you get into the vengeance kind of thing. It's like, "Well, if the Japs hadn't bombed Pearl Harbor, and here's a list of Japanese atrocities throughout the war, and you know, we lost all those guys at Guadalcanal, and so they were really spoiling for this," and it fits. It's a very nice narrative that you know America's anger rose and finally our righteous smiting occurred. But it got trickier when you're talking about what was believed to be coming in the future, where on Monday morning Moscow was a happy little city and by lunchtime it was a smoking ruin, and that was the opening shot — and possibly the closing shot — of a war.

So the destruction at Hiroshima and Nagasaki, I think, caused people to start to think about this in ways that, honestly, I don't think Truman was considering in July 1945.

Horton: What year did you start working for the State Department?

Van Buren: I joined the State Department in 1988 when Ronald Reagan was president, and I left in 2012 when Obama was president.

Horton: So you joined right when the Cold War was ending, and the Soviet Union was falling apart?

Van Buren: Yeah, we were there for it. It was a heady time, as they like to say. I was actually in London from 1991 to 1993, which was a fascinating time because, you know, the Wall went down in 1989, but it took a while for things to really happen. We were being flooded with the Russian stuff during those years in London, because London was always one of those magical crossover points like Vienna or Prague, where the good guys and the bad guys saw some neutral territory. It was an amazing time. We saw, for example, this great exodus of Russian scientists who poured through London, guys who worked on all sorts of nasty weapons programs, who wanted to go to the United States.

Horton: What I want to ask you about is the threat of nuclear war now, because I have read things by people — for example, William Perry, who was Bill Clinton's Secretary of Defense — who now say that the threat of nuclear war between the United States and Russia is as great or even greater than at any time during the Cold War. That's because of, I guess, NATO expansion so that the buffer zone is now essentially non-existent, but also the advent of hypersonic missiles and the shortening of warning times and all these things. What do you think?

Van Buren: I can't subscribe to that theory. One of the things that was so predominant during those years was how the threat of nuclear war had diminished because of backing down from all of the hair-trigger stuff. The hair-trigger stuff, the idea that you might launch against me, so I'd better be ready to launch against you; and while you're ready to launch against me because I'm ready to launch against you, I'd better be extra ready. This kind of built up to the point where everybody had their finger on the trigger and was starting to apply just enough pressure to get the ball rolling.

That all kind of toned down, and in the aftermath of the Cold War, during those years when I was in the State Department, the stuff I was seeing and reading and being told was all built around accidents. The great fear was that the Russians would lose control of nuclear weapons, that they would end up in the hands of third parties, that people would stop caring, and that the Russian military might stop doing all the maintenance that they needed to do on the weapons, which meant that there was danger of accidents.

I think we have a firm enough understanding that the use of a nuclear weapon between the United States and any other nuclear-armed country means that we're racing to Armageddon. There's not a stop point easily found in this game, and this was always what restrained us during the Cold War, the idea that there was no halfway on this stuff, and I can't think of a justification to explain why anyone would think there's a halfway in it now. At the point where both sides have acknowledged there's no halfway, that once we start, we're both going to be more or less obliterated, that was an extremely powerful disincentive to war during the Cold War period, and I can't imagine it doesn't exist now.

There's plenty of reasons to keep building weapons, that's the military-industrial complex. But to use them, that's a different story. This is the whole idea about why there will never be a nuclear weapon-holding Iran. The day that Iran has a nuclear weapon will be the day that the United States and Israel are obliterated, because the idea that we could reach an Armageddon situation in the Middle East is simply not acceptable. There's only one, or one and a half, nuclear powers in the Middle East, that's the United States and Israel, and that's it. It will not be allowed to change. There can't be a nuclear standoff in the Middle East. It simply will not happen, and the United States and Israel will ensure that it won't happen. If you're not sure if I'm right, please go back and look at what the Israelis did to the Osirak reactor when Iraq was getting even vaguely close to developing something. Take a look at what they did to the Syrian reactor that supposedly the North Koreans helped them build, etc. The mistakes of the Cold War in that sense are learned. The Americans who wanted to destroy Russia's nuclear capability in 1947 or '48, whenever that was, their argument has resonated with the United States and Israel in the Middle East.

Horton: In other words, LBJ should have preemptively attacked China before Mao was able to get his hands on one?

Van Buren: Well, the problem with China was Russia; it wasn't a one-on-one deal. You couldn't have attacked China without bringing on nuclear war with Russia. You could have — and I'm not advocating — but you could have attacked Russia at its nascent nuclear stage, and nobody else would have done a damn thing about it.

Horton: Now, as for the standoff between the U.S. and China, or the U.S. and Russia, that nobody wants to lose their capital city. They'd be crazy to use these nukes because then they'd be dead. I'm not so sure that precludes the possibility of war. I don't know what the chances are, but when you look especially at, like we've been talking about this whole time, the level of just total nonsense, lies, false premises, ridiculous points of view and regurgitated propaganda talking points about what is even going on here from the American side about containing and defending the world from Russian aggression and all this crap, when that's just not really what's going on here at all. America is attempting to force our world empire even on Russia and China, who are powerful enough to be independent from us, and we can't stand it. But if you tell anyone in Washington, D.C.: "Come on, you guys know that the Clintons and Bushes picked this fight and caused this problem," they would deny it. They'd say, "No, it's all Putin's fault. The Harvard boys? That's ancient history. The war against Serbia? I don't know what you're talking about. The color-coded revolutions? I've never heard of those. What I do know is Russia, Russia, Russia, Russia, Russia!"

That's the kind of thing that makes me think that we can get into a war. It's that these people are, in a sense, insane, right? Their thinking is so off-base that I could expect anything to happen.

Van Buren: Well, let me reassure you in a way that I think will resonate. War is not as profitable as preparing for war. So if you look at it from the people who really run the United States, there is way more money to be made in making and maintaining nuclear weapons than in risking the use of them. In that sense, if you needed nothing else to reassure you, maybe hang on to that one.

Horton: I understand that, but say there's a naval confrontation in the Baltic Sea or in the Black Sea, and people start making ultimatums to each other, and then there's a coup in Belarus, or whatever it is, it could quickly get out of hand.

Van Buren: That kind of thing happened constantly during the Cold War. The Russians poked their army into countries all over Eastern Europe. The United States had its own little forward front all through Central America. We were constantly bumping into each other with submarines and surface ships and airplanes in all the coastal areas. There were little flare-ups on the inter-Korean border, whatever. It happened constantly during the Cold War, but it was in nobody's interest, financial or otherwise, for it to get too far out of hand. Militaries are good at following orders, and in each and every case, somebody was told to back off, and somebody backed off. That historical record should not be dismissed lightly.

Horton: Yeah, I understand because hey, who wants hydrogen fusion over their head? Nobody. Not even Bushes or Clintons. But here's my last best point about this then, which is what Pat Buchanan says about how we used to draw the line at the Elbe River. So when the Soviets cracked down on Hungary and on Czechoslovakia and on Poland during Eisenhower, LBJ and Reagan, the Americans said, "Hey, not my problem. Sorry, Charlie. But you Soviets had better not come into West Germany!"

But that was ceding a hell of a lot to them, as the right has kicked and screamed about ever since Yalta. Now, we've drawn the line at Russia's western border, we have a military alliance with the Baltic states, and we keep working on overthrowing as many governments as we have to in Ukraine until the situation is right for bringing them into NATO as well. That this is the kind of thing where maybe the Russians would do something that we would severely disapprove of, that Ike Eisenhower would have let slide, but that now we can't because now we'd be turning our back on Article 5 of the NATO Alliance.

Van Buren: Yes, maybe, but there are two conflicting arguments to answer that. Either the Russians were smart enough to realize that the rules are a little different and understand that they have to participate within different rules, or that the U.S. will recognize that the rules are made to be flexible. And this little sliver of Latvia that the Russians have always wanted their hands on is absolutely not worth nuclear war. I don't pretend to know enough about Ukraine to get into the nuts and bolts, but I can say that both sides have done an awful lot of weird, violent, nasty shit in Ukraine, and we've not gone to war over it, right?

Horton: Going back to the Cold War, when the CIA supported the Nazi right in Ukraine back then, too.

Van Buren: Well, that's a long tradition for us, but my point is that while the Russians let us do all the nasty stuff we've been doing in the Ukraine, we've made a lot of noise about what the Russians have done in the Ukraine, but we haven't dropped the paratroopers in to push them back or anything. The point is that I think both sides recognize the value of flexibility. Saber-rattling and fist-shaking serves everybody's needs, but actual war serves nobody's needs. And I think at the end of the day, as during the Cold War, that will be what keeps the peace.

Horton: Well, there you go, everybody. Humanity is going to survive after all! The optimistic take from Peter Van Buren today. I mentioned Ellsberg's book, *The Doomsday Machine*. Have you read that?

Van Buren: Oh yeah, of course. I have read everything Daniel Ellsberg has written. He's a hero of mine, and one of the highlights of my whistleblowing career was when Dan reached out and introduced himself, and I got a chance to sit down and have dinner with him.

Horton: That is very cool. I've gotten to interview him a few times, and he blurbed my book. That's not quite dinner, though. But anyway, that book will give you the heebie-jeebies. For people who haven't read it, he comes into power as the Deputy Assistant Whatever in the White House overlooking nuclear stuff in the Jack Kennedy years, and he comes to find out that the plan is that if anybody shoots anybody in Berlin, we're going to nuke every city in Russia and China, and there is no other plan. Also, the war could start accidentally about 150 different ways. It was his job to go, "Whoa, whoa!" and try to rewrite all of this stuff, try to get the military to even admit what their plans were and to try to make the nuclear Doomsday Machine a little bit safer to operate. But then, he kind of ends it with: "I haven't been there in a long time. I don't know how they do it now."

And see, the other thing I'm a subscriber to is what I think I'm going to start calling "The George W. Bush Theory of History," where all of our presidents are nothing but George W. Bush. Winston Churchill too. Nothing but George W. Bush. All of these guys are nothing but a bunch of Bushes. I mean, you would think Eisenhower would have been wise, but every bit of our mess in the Middle East is because of him and his predecessors and successors, you know what I mean? With that 1953 coup in Iran and the aftermath of all these things. So I could see them screwing up. I could see them getting us all killed accidentally.

Van Buren: The possibility of accidents happening remains. This has been my theory on the Koreas, by the way. Never mind that we're going to have a nuclear war with North Korea; I'm deathly afraid of a Chernobyl-like accident in North Korea that triggers the next thing.

Horton: That's a good point. And, you know, that Yongbyon reactor — I'm sure you do know — was built by the Soviets and is run off of pretty highly enriched stuff. It used to be run off of even weapons-grade uranium.

Van Buren: It's hard to imagine that industrial safety is a big thing in North Korea. So my great fear there is that an accident, a Chernobyl-level accident, causes the next thing. Whether that next thing would be, for example, for the Chinese to intercede, or a massive refugee flow into the South, or some kind of military coup? I mean, pick your favorite nightmare. But these kinds of things are of far more concern to me because they lack logic. At the end of the day, whatever you want to say about Putin, or whoever is the next Russian leader, these are still people that are trying to think these things through, albeit in a childish and ridiculously stupid way, but there is still a thinking process, which means there's still a chance to intercede. When you're talking about accidents that trigger spontaneous reactions, there is no logic there. It's just people running, and that's the most dangerous situation of all time.

Horton: There was that time the air force accidentally dropped a thermonuke on North Carolina, and eight out of nine safeties failed, I think they say, right? You have to wonder, if that H-bomb had gone off, would they have been able to admit that it was an accident? Or would they have had to say it was an attack and respond as though it was an attack?

Van Buren: Yeah. Accidents are right now what would worry me far more than any of William Perry's thoughts about how the Russians are going to launch a nuclear war.

Horton: Thanks, Peter.

Van Buren: Thank you.

Gar Alperovitz: The Decision to Nuke Japan August 27, 2019

Scott Horton: Introducing Gar Alperovitz. He's a historian who has written a ton of extremely important books, including *The Decision to Use the Atomic Bomb* and *Atomic Diplomacy: Hiroshima and Potsdam*. He's got this article in *The Nation* from a couple of weeks back, August 6th, the anniversary of the bombing: "The War Was Won Before Hiroshima — And the Generals Who Dropped the Bomb Knew It." Welcome to the show. How are you doing, Gar?

Gar Alperovitz: Great, thanks for having me.

Horton: Very happy to have you here. This is such an important article that you wrote. There's so much to talk about regarding the atomic bombing, and your article focuses right on what is probably the most persuasive and educational point for the average American: that all of the extremely credentialed criticism of the use of that bomb — contrary to what we're all taught in fifth grade, that this was an absolute necessity — everyone who you would think agreed with that back then, didn't.

Alperovitz: It's certainly true that the top generals and admirals, almost all of them went public immediately after the war, saying that the bomb was totally unnecessary because the war was going to end shortly without an invasion. And the bombing was almost entirely about civilians. Hiroshima was not a military center; there was that small installation there, but it was very marginal. It was basically a civilian target, and the generals and admirals, many of them old conservatives, weren't taught, as one of them said, to kill women and children unnecessarily.

General Eisenhower went public, saying it was outrageous. Curtis LeMay, the famous bomber pilot who bombed all of Tokyo, immediately after the war came out saying the war was going to be over within a couple of weeks, and that it was outrageous to bomb all these civilians. I can go on. The list is long. I counted twenty of them who are top people, five-star generals and admirals. It's quite an indictment. Most of them were conservatives, not liberals or radicals. They were old-fashioned conservatives who just didn't believe in killing women and children.

It's important to know that Hiroshima was not a military target; there was a small installation for training, but it was not very large. Which meant that with the young Japanese men off to war, the people left were women, elderly and kids. That was the population that was decimated by the bombing of Hiroshima.

Horton: These men that you've cited here — I'm asking for a generalization — would they have supported the atomic bombings if they thought that they were necessary? Would they have supported dropping the bombs if they thought it was necessary to end the war, or would they have invaded the place rather than drop the bombs to get that unconditional surrender?

Alperovitz: The Intelligence Estimates in April 1945 — the bomb was used in August — said that the Japanese are close to surrender and that when the Russian army attacks, that will precipitate an immediate surrender. The Japanese feared a Russian invasion more than anything else. They would rather have been invaded by the United States if they had to be invaded. So U.S. intelligence and British intelligence made this argument again and again starting in April 1945: that when the Russians entered the war, the Japanese would surrender as long as they could keep their emperor.

The U.S. had decided long beforehand that they were going to keep their emperor because we wanted to use him after the surrender to control the Japanese people. We were going to allow them to keep the emperor no matter what. They knew the war would end if you kept the emperor and the Russians attacked, and since we didn't know whether the atomic bomb would work — it was unknown until July 1945 — we were desperately begging the Russians to come in because that would end the war without an American invasion.

That was the plan up until August 1945. The Russians were scheduled to come in three months after the German invasion; it would take them that long to move their troops into position across Siberia. They were scheduled to come in on August 8th at our request because intelligence said that that would end the war, so long as we let them keep their emperor, which we were going to do anyway because they could then control the Japanese people.

In Potsdam, there was a big conference in Germany in July 1945 where a proclamation was issued by the U.S., warning Japan to surrender. It famously contained a paragraph saying they can keep the emperor, because we knew they saw the emperor as a god. If we didn't say that, if there was any doubt about the emperor, this god figure, they would not surrender. That paragraph promising them that they could keep the emperor as a figurehead with no power was extracted by the Secretary of State. The reason he extracted it, it's now obvious, is that he didn't want them to surrender. The reason he didn't want them to surrender was that he wanted to use the bomb rather than have them surrender, whether it be then or when the Russians came in, which is what the military intelligence said.

So, we kept them fighting by taking away assurances for the emperor. The Russians were scheduled to come in on August 8th. Hiroshima was bombed on August 6th. The Russians attacked on August 8th. Nagasaki was bombed on August 9th. That was the sequence. As I said, and as we now know from the endless documents that have come out, the military leaders, many of these old conservatives, were just outraged at the unnecessary killing of civilians with what Eisenhower called "that terrible thing."

Horton: On that question of the unconditional surrender there, when you say that the Secretary of State changed that in the official declaration there at Potsdam, you're also saying that inside the U.S. government, there was no change? It wasn't just that MacArthur decided for expediency later; they had decided long ago that of course we'll keep the emperor, but they kept that a secret and demanded total unconditional surrender so that they couldn't get it?

Alperovitz: Yes, that's right. They knew that they were going to keep the emperor in a powerless position in order to help control the Japanese people, in particular the Japanese army, once we invaded and were occupying Japan. That was the plan.

Horton: I'm just making this up, but the contrary narrative would be that there was a big debate inside the government whether to let them keep the emperor or not. In other words, this was a solid decision. There was no wavering on the decision. The only wavering was in public, for Japanese consumption.

Alperovitz: Exactly. The plan was that once we have an occupation of Japan, the best way to do it would be to keep the emperor as a figurehead because he would keep things under control for us. That was always the plan.

Horton: Let's go back, because you said you could go on all day about the names of these men. I'd like you to do just that. Please take your time, and go through and remind us. For instance, who is Admiral Leahy? Why would we listen to a guy like him?*

Alperovitz: Admiral Leahy was a fleet admiral with six stars. You can't get any higher than that. He was Chief of Staff to the President of the United States. He was chairman of the Joint Chiefs of Staff and the combined U.S.-U.K. Chiefs of Staff. He's about as high as you could get at that time, and he was a very good friend of President Harry S. Truman. Here's what he said after the bomb was used:

> The use of this barbarous weapon at Hiroshima and Nagasaki was of no material assistance in our war against Japan. The Japanese were already defeated and ready to surrender… In being the first to use it, we adopted

---

* See Appendix B: Who Opposed Nuking Japan?

> an ethical standard common to the barbarians of the Dark Ages. I was not taught to make war in that fashion. … Wars cannot be won by destroying women and children.

Horton: One of these that's very interesting to me is Curtis LeMay, who I don't know if you could call him conservative, he was certainly a right-winger, maybe more of a radical than a conservative. But we all know that he was perfectly willing to use H-bombs on China and Korea when the opportunity presented itself, so he was no humanitarian. He was the butcher of Tokyo, but he was speaking just in utilitarian terms that this was unnecessary.

Alperovitz: Here's a quote from General Curtis LeMay from a press conference on September 20, 1945. The use of the atomic bomb "had nothing to do with the end of the war. The war would have ended in two weeks without the use of the atomic bomb or even the Russian entry into the war."

A member of the press asks, "You mean that, sir? Without the Russians? Without the atomic bomb?" LeMay responds: "The atomic bomb had nothing to do with the end of the war at all. The war would've ended anyway." That's what most of them thought. That's what U.S. intelligence said.

Horton: You also mentioned Ike Eisenhower, who was the five-star commander of United Nations forces in Europe and later President of the United States.

Alperovitz: I've got a couple of quotes from him. Here, he's talking about before the atomic bomb was used, when he was told by the Secretary of War that it was to be used:

> During his recitation of the relevant facts, I was conscious of a feeling of depression, and so I voiced my grave misgivings first, on the basis of my belief that Japan was already defeated and that dropping the bomb was completely unnecessary. Secondly, because I thought our country should avoid shocking world opinion by the use of a weapon whose employment was, I thought, no longer mandatory as a measure to save American lives.

That's before he was president. After he was president, in a 1963 interview, he was quite blunt: "It wasn't necessary to hit them with that awful thing."

Horton: Can you also talk about Nimitz and MacArthur, who were the two major commanders of the Pacific Theater? They both condemned it as well.

Alperovitz: What we have from MacArthur is the recollections from two different people. This is one from Richard Nixon recalling that:

> MacArthur once spoke to me very eloquently about [the atomic bomb], pacing the floor of his apartment in the Waldorf. He thought it a tragedy that the bomb was ever exploded. MacArthur believed that the same restrictions ought to apply to atomic weapons as to conventional weapons, that the military objectives should always be limited damage to noncombatants. ... MacArthur, you see, was a soldier. He believed in using force only against military targets, and that is why the nuclear thing turned him off.

Then MacArthur's personal pilot, Weldon Rhoades, the day after Hiroshima, noted in his diary: "General MacArthur definitely is appalled and depressed by this Frankenstein monster. I had a long talk with him today, necessitated by the impending trip to Okinawa." What MacArthur was saying was the same as what almost all the top military felt. They just weren't taught to bomb civilians in general or in particular when it was unnecessary.

Horton: And Nimitz? There's an American class of aircraft carriers which is named after him. He was in charge of the whole southern campaign in the Pacific, and it was his forces at Iwo Jima, right?

Alperovitz: Exactly. He went public. This was an address two months after Hiroshima at the Washington Monument:

> The Japanese had, in fact, already sued for peace before the atomic age was announced to the world with the destruction of Hiroshima and before the Russian entry into the war. The atomic bomb played no decisive part, from a purely military standpoint, in the defeat of Japan.

Now you have to understand what's going on here, because these are the top leading military people of the United States Army and Navy fighting in World War II. These are not secondary figures. Immediately after the use of the atomic bomb, they're going public, attacking what the president had just done. It is amazing upon reflecting on it. The military is usually very quiet about the Commander-in-Chief's decisions, and here you find one after another of them doing the opposite. Not all of them, some of them waited until they had written a biography or their personal recollections, but a number of them went public after the war.

Horton: It really just goes to show the irony of American power in this age, where we're constantly having to rely on the military men to restrain the passions of the civilians, when it's the standing army that is the greatest threat. Apparently, sometimes they prefer not to be so misused, and they advise the civilians to be a little bit less worse than they would otherwise be. It's supposedly the civilians' job to rein them in all the time.

Alperovitz: It's quite interesting. I'm hardly a conservative myself, but these were virtually all very conservative generals and admirals who were very serious about defending the United States. They were also very serious about what I think McNamara once said: "Had we lost the war, we'd all been war criminals for bombing all these cities and civilians."

Horton: By the way, for anyone who is too young or doesn't know and has never seen that film *The Fog of War*, it's where McNamara finally comes clean. And he was guilty, not just in Japan, but also in Korea and Vietnam, of burning so many people to death. Then at the end of the day, he says, "Yeah, I guess I shouldn't have done that." Now, I know less about General Hap Arnold and Admiral Bill Halsey. Can you tell us about them?

Alperovitz: Hap Arnold was the guy who was running the air force at this time. He's another one who knew the war was over, that it was not in question. The intelligence was very clear. It was going to end when the Russians came in. The Japanese were desperate and on their last legs. Here's the kind of thing you got from him — this was in the *New York Times*: "The Japanese position was hopeless even before the first atomic bomb fell, because the Japanese had lost total control of their own air." In his memoirs, Arnold observed, "It always appeared to us that, atomic bomb or no atomic bomb, the Japanese were already on the verge of collapse."

And here's his deputy, General Ira Eaker. By the way, all these quotes are from a book that I wrote called *The Decision to Use the Atomic Bomb.* I just assembled all the quotes to find out what the military was saying at the time. This is in an internal military history interview. Inside the U.S. military, they very often do interviews, trying to find out what actually happened for historical purposes and for training purposes. Here's Eaker talking about Arnold: "Arnold's view was that the dropping of the atomic bomb was totally unnecessary. He said he knew the Japanese wanted peace. There were political implications in the decision, and Arnold did not feel it was the military's job to question them." Eaker reported further that Arnold told him:

> When the question comes up of whether we use the atomic bomb or not, my view is that the Air Force will not oppose the use of the bomb, and they will deliver it effectively if the Commander-in-Chief decides to use it. But it is not necessary to use it in order to conquer the Japanese without the necessity of a land invasion.

Again, it's virtually all of these admirals and generals at the very top who knew what was going on, who had exactly the same view: The war was just about over, so the bomb was totally unnecessary and shouldn't have been used.

Horton: So many people would take the position — and I have anecdotes like this — where they would say that the position you're taking is something that only the most horrible and hateful anti-American people could ever say. They'll react by saying something like, "Don't you know that millions and millions of Americans would have died trying to invade Japan?" That's an actual quote from President Bush, Sr. Not just one million, but millions and millions, he said. People really believe that stuff.

But what we've just gone through here, these quotes of these men, are the ultimate antidote to that, because it doesn't just say, "Actually, that's not quite true. Look at some of this criticism." This is the ultimate criticism, McNamara and Nimitz and Ike and LeMay and all these people. It's incredible, and it's shocking to people who have never been taught this. They don't ever teach us this. You don't get this stuff in school. So it's the kind of thing that can really make a light bulb go off: "Not only have I been wrong, but wow, maybe there's a whole other world that I was living in." If this many men, and as you said, conservatives, Republicans, right-wingers, lifelong military men, the heroes of World War II, they were the ones — the leaders, not secondary figures, not pencil-pushers back at the Pentagon, but the greatest men of the Greatest Generation — they were the ones who said we shouldn't have done this. So where does that leave us, then?

Alperovitz: It is quite shocking, because this material has been available — it is not secret material — it has been available for a long time. So it's been avoided, not looked at, and people have found it very, very difficult to believe what I think is true and what all these military leaders believe to be true: that the bomb was totally unnecessary, and it was aimed primarily at a civilian target. Hiroshima was not a military target. It did, as I said, have a very small training base there, but that's not why they chose it. They chose it because they liked the target. They thought it would show what the bomb could do. It was a very nicely designed city that would show off what the bomb could actually do, that's why they chose it. And they killed mostly old people, young kids and women who were left behind as the young men went to war.

So, it was clearly — and there's been research on this lately, some very good legal work — it was clearly a war crime. It is an international war crime to bomb any civilian target, which is essentially what Hiroshima was, and particularly when there's no necessity.

It is an outrageous story, the way Americans have not wanted to talk about it except for these leading conservatives. I'm hardly a conservative, but I have great respect for these old serious conservatives who were dedicated to protecting the United States. They were honorable army and navy and air force guys who were trying to do their duty. They just did not believe in killing women and children unnecessarily.

Horton: Every August, there's a lot of talk, every anniversary there's always some writing and that kind of thing, which we should be grateful for and which shouldn't take for granted. This time around, though, part of the discussion has been about more specific questions related to what Truman knew and when he knew it, what they told him, what he really believed, and whether he really thought he was nuking a military base. I saw this historian on Twitter talking about whether Truman really had any idea they were going to hit Nagasaki two days later. Apparently, according to one document, it seems like maybe he didn't realize that we were going to attack them again until the end of the month or something like that. What do you think about that stuff?

Alperovitz: I have not seen any serious documentation of that kind. I know this literature fairly well — I've done two big books on it. That doesn't mean I couldn't have missed something, but everything we know about the decision is that he was aware of the basic intelligence and that he understood what was happening. The story in human terms was that Truman was way over his head. He was a senator without any experience in this. He knew nothing about what was going on in the war. Roosevelt died, and he was thrown into this position in April and was kind of trying to get his sea legs. He was dominated at this point in time by the man he chose to be Secretary of State. He was really the person who was really involved, and he essentially guided Truman. His name was James Byrnes of South Carolina. He was a very important figure in the New Deal and was very slippery. His biographer titled the book *Sly and Able*, or something like that.

Byrnes was a senator at one point. He was also an Associate Justice of the Supreme Court, he was governor of South Carolina, and he ran the domestic economy for Roosevelt. He was a very powerful figure. He had also been at the Yalta Conference when Roosevelt met with Stalin. So he was on the inside of everything. But he had personally been in the Senate and was Truman's guiding figure when Truman came in and didn't know anything about the Senate as a young senator from Missouri. Byrnes took him under his wing and taught him the ropes.

Well, Truman made Byrnes his Secretary of State when Roosevelt died in April, and he also basically gave him control of the decisions about what to do with Japan's emperor and what to do with the atomic bomb. It's very much like George Bush was run around by his vice president, Cheney. Byrnes is a figure like Cheney, running around a young Truman who was just fresh in and didn't know what he was really doing at the outset.

Byrnes is on record saying at the time, when a bunch of scientists had come to him in April 1945 and said, "Look, we all know the war is over. You really shouldn't use this thing. It's very dangerous. You're going to start an arms race that could lead to the destruction of the world," and Byrnes said to them, "I know the war's over but," and this is a quote, "we need the bomb to make the Russians more manageable." There's a lot of evidence that Byrnes saw the bomb as a weapon against the Russians, not the Japanese. He thought using the bomb would scare them and they'd back off of Eastern Europe, although it was highly unlikely that they would relax there because the Germans had invaded Russia twice. Also, he thought that if we didn't get the Russians into the war in Japan, then they wouldn't come into Manchuria, and maybe we could have more influence in that area.

But Byrnes explicitly did say to these scientists who wrote it down: "No, I want to use the bomb because we'll make the Russians more manageable." And the scientists were shocked because they knew, as anybody who was on the inside and had thought about it for five minutes knew, that if we use the bomb to scare the Russians, they will have to get up on atomic bombs. The Russians would see it as a threat, and so that will mean that we will be in an arms race building nuclear weapons forever, which is exactly what happened.

So, I think that Byrnes is the key figure. Most of the discussions between Secretary of State Byrnes and Harry Truman were personal discussions held privately; they were old drinking buddies. We have very few records of what they actually said to each other. We do know that that's where the decisions were made. We know a lot about what they thought, and the best estimate of what was going on was that Byrnes really wanted to use the bomb, primarily because he saw its role in — not so much in ending the War, the war was going to end anyway without an invasion — but because of its implications: It might scare the Russians both in Manchuria and Eastern Europe.

That's very hard to pin down, but that's the best we can do with the available documents. Maybe someday — a good friend of mine, an investigative journalist, believes that there are documents — and if you really dig, you're going to find them in somebody's basement, and we'll get a better picture. But that's about as good as we can do at this point because Byrnes, for instance, he had a special system with a man named Brown. Brown kept a diary of what was going on, and Byrnes got ahold of the diary, rewrote it, and then filed it in the official documents. Then when historians would come up to him and say, "Well, what about what happened with the bombing?" He'd say, "Well, there's a diary from my assistant. You could go over to the archives and it'll probably give you all you need to know." Of course, this was the diary that Byrnes had privately edited and rewritten to be favorable to him. He was a very slippery character.

Horton: That sure is interesting. It's so important to question, like you say, this whole idea that we'll scare them. And then what? They'll leave power and go home? What's going to happen then? The obvious thing — as you detail, people had warned in advance — and how it played out, was that the Soviets embarked on a crash course to get an A-bomb as fast as they could. They got one within three years. Then within that same three years, America lost China to Mao and the Communists anyway.

So that whole thing about giving us an advantage in Manchuria by calling a premature halt to the Soviet invasion on Japan's western flank was null and void. Even the ulterior motive is some bogus garbage.

Alperovitz: At the time, the Secretary of War was a man named Henry L. Stimson, a very conservative Republican that Roosevelt had appointed, and he oversaw the generals who built the bomb. He understood the danger that if we use the bomb in a way that's irresponsible, the Russians will have to build a bomb because they will see it is necessary, and then we'll be in an arms race. He knew that these things could destroy the world, and so he desperately wanted Truman to approach the Russians with some kind of deal so that we wouldn't end up where we did end up. He made a desperate attempt to stop it but lost the battle to people like Byrnes. As Stimson put it: "The Russians will see us with this weapon rather ostentatiously on our hip, and they will know what to do." And now, here we are, with nuclear weapons everywhere.

Horton: Yep, and some Americans ended up helping the Russians with their bomb anyway. So they split the difference on the policy but without any of the trust-building, instead only with the atomic bomb-delivering.

Alperovitz: Yep, it's a very tragic story. Here we are, with nuclear weapons everywhere. It would have been very difficult to prevent an arms race, but it would not have been difficult to prevent the destruction of Hiroshima and Nagasaki. It would have taken great diplomacy to work out a thoughtful solution in advance. Oppenheimer and many conservatives within the government were trying to advance an arms control program at the outset, but Truman and Byrnes were having none of it. Their strategy was to push production and get ahead just keep getting ahead, which meant there would be a worldwide arms race. There were no serious efforts made. The efforts that were made about arms control and disarmament have been examined by many researchers, and they were pretty fraudulent. They could not have been accepted by Russians as put forward.

Horton: Well, I sure wish I had read both of these books that you've written about this. Are there any other major themes that are in there that we should know about?

Alperovitz: The first book I wrote on this was called *Atomic Diplomacy* because it shows U.S. policy towards the Russians in 1945. After the Germans surrendered on May 8, 1945, the bomb was used in August, and the Japanese surrendered shortly thereafter. There are now diaries available from that period which show how U.S. policymakers began to calculate what happens if we have this new weapon. They began talking about it as their ace in the hole, and they began seeing it as a very powerful weapon primarily against the Russians, but they were thinking more about diplomacy.

So I wrote a book about that, called *Atomic Diplomacy*, which was based basically on the diaries, particularly the Secretary of War's diaries. They came available when I was doing research on this. His diary is just extraordinary because he talks about all of this pretty much the way I'm talking about it.

There was nothing hidden. That book is all about the diplomacy around Hiroshima and the Potsdam Conference when Stalin and Truman met in Germany.

The second book I wrote is called *The Decision to Use the Atomic Bomb*, and it is a very detailed discussion with all of these military leaders. It goes beyond the first book in looking at the impact of the bomb on diplomacy, but also how precisely did it impact the decision, how was it made, and what we know about precisely who controlled the decision-making. The book goes through all of that step by step as best we can because a lot of it was done privately between Truman and his Secretary of State Byrnes, who had been Truman's teacher in the Senate and who kind of ran the show there in those first six months of the Truman administration. A lot of it is there privately between them. You can dig out people who understood what was going on, and you have to piece the story together, but the basic points are clear: They saw it as a weapon against the Russians, and Byrnes was a key figure.

Horton: I guess the good news is that there are only a couple of 10,000 of these on the planet now, whereas before the total was 60 or 70,000 of these things in the hands of the U.S. and the USSR at the height of the Cold War buildup, right?

Alperovitz: We're in a little bit better position, but it is not a very happy position that there's an awful lot of these things around.

Horton: It's funny to think of George H. W. Bush, despite the 30-year war in Iraq that he started, that in a way, from a quantifiable or measurable point of view, he may be the greatest man who ever existed in the sense of signing those last few disarmament treaties with Gorbachev and then with Yeltsin in just the last two years of his government there at the end of the Cold War. I mean, has any other man ever gotten rid of tens of thousands of nukes? He would share that, of course, with Gorbachev and with Yeltsin. I only learned recently that in January 1993, when he only had two weeks left in office, he went to Russia and signed a whole new treaty with Yeltsin. Since it was going to take too long for Bill Clinton to get started on his negotiations, Bush decided to go ahead and get rid of another couple tens of thousands of these things. I hate to give that ol' S.O.B. credit, but that's pretty great.

Alperovitz: On this issue, I think you're absolutely right that people don't recognize and credit him. Also, National Security Advisor Scowcroft, who was guiding this strategy there, was another former general.

Seymour Hersh, the great reporter, has also written — in a different vein, not about this particular subject — he has the same view of some of the generals and admirals, that some of them are really quite interesting and quite thoughtful. We usually think of the military guys as wanting to bomb everybody, and there have certainly been people like that, such as Curtis LeMay, who had insane views about using the bomb. But there also were a number of military figures who on the inside were very thoughtful about this. They had seen a lot of death, and they knew what bombs could do to people and their communities, and they didn't like war.

I came away with respect for some of these old admirals and generals. I was surprised once I learned what they really were talking about.

Horton: I talked with Peter Van Buren, and he said, "You know what, nukes are so horrible that we can trust that even the worst politicians know better than ever to really use them," but it seems like any of that kind of wisdom built up during the Cold War is spent. When I look at the idiocy in terms of America's position against Russia in Eastern Europe right now — no matter what fight we pick with them, we always pretend that *their* response is the start of the aggression — and you can tell that the people in D.C., they really believe it. Everything is Russian aggression. If it's raining, it's Russian aggression. They have terrified themselves into believing they're really on the defensive as they expand our military alliance right up to Russian borders, do coup d'états against all their friendly governments in the region, expand their military presence, arm up Ukraine and all these things.

I recall an NPR interview with Ivo Daalder, a think tank expert who was a former ambassador, during the worst part of the Ukraine war in the aftermath of the coup of 2014. He was saying, "Yes, we have to arm up Ukraine in this war. We should send them all these Lockheed products and all these things," and the NPR reporter says, "If we make the Kiev government stronger to win the war in the East, then to what end? What's next?" And the guy says, "Well, we think that if a bunch of dead Russians start coming home in body bags, that will cause the debate to go up." Like shades of Byrnes, "We're going to scare them." Yeah? And then what?

This guy says, "Yeah, we're going to cause the debate to go up in Russia." The idea being that the angry moms of the dead guys will just be angry at Putin for getting them killed, not angry at the U.S.A. for getting them killed; that they're going to want to back down rather than maybe getting really angry and doubling down. It's as though the negative possibility is outside of the realm of possibility, even though the whole war was the backlash of the coup not going their way. It was supposed to be easy, but it caused a war. This is a former ambassador, this is the brightest of the think tank world, and that's as far as they thought this through. "We're going to cause their debate to go up," when what he's really saying is that he might cause a world war. All these guys are just George W. Bush. They're just waiting to blunder us into another catastrophe at any minute.

Alperovitz: The tragedy is that these guys actually believe this is the way to help the world. They believe that this is going to do good, rather than seeing how short-sighted some of it is. The great historian William Appleman Williams wrote a lot about this. He wrote a book called *The Tragedy of American*

Diplomacy about how these guys actually believe what they're doing is going to be great for the world. What happens is the sort of thing you're talking about, because it is just so short-sighted that it produces a backlash, just the opposite of what you want.

Horton: That also helps explain why it's so hard to talk them out of any of it, because you're essentially asking them to stop caring about people when you're asking them to stop bombing people to death. They just think, "How can we abandon them now? We have to keep the war going." We hear this about Afghanistan daily.

Alperovitz: It's a genuine tragedy. It's outrageous, it's very costly for human lives, and it leads to destruction and arms races. So far, we've been lucky that these things haven't gone off, but what's going on in India and Pakistan may be the next concern.

Horton: Yeah, you've got that right, but I'll let you go now. Thank you very much for coming on the show, Gar. I really appreciate it.

Alperovitz: Thanks, I enjoyed talking with you.

Anthony Weller: My Father's Lost Dispatches from Nagasaki
August 10, 2007

Scott Horton: Yesterday was the 62nd anniversary of Harry Truman's nuking of Nagasaki, and our guest today is Anthony Weller. He's an author, musician, and son of Pulitzer Prize-winning reporter George Weller, from the *Chicago Daily News*. Last year, he published a book of his father's long-lost dispatches from Nagasaki, *First Into Nagasaki*, written shortly after the dropping of the second atomic bomb. Welcome to the show, Anthony.

Anthony Weller: Thank you very much for having me.

Horton: It's very good to talk to you. Yesterday, August 9th, was the anniversary, and every August, I like to cover the atomic bombings of Japan at the end of the Second World War. A couple of years ago, I spoke with Greg Mitchell from *Editor & Publisher*, and he mentioned some long-lost film that was found hidden in a ceiling. He also talked about the discovery of your father's dispatches back then, before they were published. Why don't you tell me, first of all, how it was that these dispatches were long-lost in the first place, and how did you find them?

Weller: Well, as you probably know, my father was a foreign correspondent for the *Chicago Daily News*, who had seen a great deal of the war, from Europe to Africa to the fall of Singapore and the war in the Pacific. He was one of 700 correspondents who were on the deck of the battleship *Missouri* for the signing of the treaty of surrender with Japan on the 2nd of September 1945.

Right after the signing, he and all the other reporters went into a meeting to figure out how we're going to spread out and see southern Japan, including the two nuclear sites. To their consternation, they were told that even though the war was over and Japan was no longer an enemy, they weren't going to be allowed anywhere south of Tokyo. In fact, they weren't even going to be allowed into Tokyo to cover the incendiary bombings that had gone on there for months in the spring of 1945. So they were all going to be ushered northward away from where the war had been concluded.

My father was furious about this. At this time, he was 38, he'd seen a lot of the war, and he'd suffered a huge amount of censorship at the hands of MacArthur throughout the war in the Pacific. So he decided that he simply was going to find another way, and there was a plane heading south to the very southern tip of Kyushu, the southernmost island of Japan, taking some American soldiers there. My father realized that he was going to be allowed to hitch a ride on this plane, so he volunteered to do so, even though there was really nothing much interesting to cover.

As soon as he got there, under cover of darkness he slipped away from his army handlers and made his way by boat onto the mainland and then, by a series of trains, up to Nagasaki. He got to Nagasaki and presented himself to the Japanese major general as an American colonel. Now, there were no American military there at the time, nobody had arrived, and the general said, "Well, how do I know? How do I know you're what you say you are? You look pretty rumpled to me." And my father said, "Look, if you don't believe I am who I say I am, just call General MacArthur, but I suggest before you make that call, you really think about the seriousness of your position." The general bowed to him and said, "What can I do to help you?" And my father took it from there.

Let me skip ahead a lot. Essentially, my father had three weeks in Nagasaki and in the nearby POW camps. He sent several dozen dispatches up to MacArthur in Tokyo to be sent to the States, because there was simply no other way to do it in those days. You had to get a passed-by-censor stamp, or your newspaper could not run your story. There were no cell phones. There was just no other way for my father to do it. So he was sending his stories up to Tokyo, and they were getting thrown in a wastebasket. He eventually realized none of them were getting through.

He left Japan, and in the course of a war-torn life from continent to continent, his own copies just went astray. After he died in 2002 at the age of 95 in Italy, I was fortunate enough, in going through this tumultuous archive of his papers, to find a crate with his original carbons in there.

Horton: And he had thought that they were lost before he died?

Weller: Yeah, he thought they were lost for decades. The story I always heard was not just "the scoop that got away," but "the papers that got away." I was just lucky to find them, but they were in pretty sad shape, they weren't in any kind of order, and they were all written in this weird telegram language that all the journalists back then were forced to use because they sent their stories by telegram. So all the obvious words were removed from every sentence, and I had to kind of reconstitute every line.

Horton: Now, you couldn't get this published in America, you had to go to Japan to have it published, is that right?

Weller: Well, it's a little more complicated than that. I mean, what I found was really kind of a mess, and I was finishing up a novel of my own at the time. I wasn't sure how to approach this because I could see it was a lot of work. I pulled together a version of some of the bomb stories — not any of the POW stories — and tried to see if an American magazine was interested in them, and they weren't, really.

To be fair, they were looking at something that was somewhat chaotic, but at least to me it seemed pretty full of all kinds of explosive material of its own. Remember, the existence of these stories was well-known. Greg Mitchell, your guest, had written about them in his famous book *Hiroshima in America.* They were well-known to students of censorship of the press. It's just nobody had ever seen a page of them. So essentially, some of the usual suspects of national magazines that might have been interested passed on this.

Fortunately, a very alert Japanese correspondent for the *Mainichi* chain, a correspondent named Sumire Kunieda, heard in an interview about a book of mine that I'd finally found these. She flew out to see me from Los Angeles, had a look at them, and her paper ran a long story about them as well as some excerpts. For those stories, she received the Japanese Pulitzer. As soon as they hit, in Japanese and then in English, on their website, essentially all hell broke loose, and I had ABC, the BBC, NPR and everybody coming to talk to me about it. Then, the book was possible.

Horton: The book is called *First Into Nagasaki: The Censored Eyewitness Dispatches on Post-Atomic Japan and Its Prisoners of War.* I actually want to ask you a little bit about the prisoners of war if I can, before we get too much into what he saw about the atomic bombing. My great uncle Bill was held as a prisoner of war by the Japanese and was worked almost to death.

Weller: Where was this?

Horton: I actually don't know specifically where he was. I don't think he was held at Nagasaki or anything like that. But I guess because it's part of my family history, I would like to know a little bit more. What can you tell me about what your father found at those POW camps?

Weller: It was a very bizarre experience for him because he spent 56 days in Nagasaki writing up all he could see of the bomb's destruction and the dying in hospitals, and of course this was utterly unlike anything anybody had ever seen. He sort of felt like he'd done what he could, and he had a sneaking suspicion that his stories were not getting anywhere up in Tokyo. He'd enlisted some Japanese military police, who believed he was a colonel, to ferry his stories up to Tokyo every day.

So, he decided to explore the Allied POW camps which were within a 30- to 40-mile radius of Nagasaki. He'd heard about these from speaking with some other American POWs who'd been held right in Nagasaki and actually had been about 300 yards from where the bomb went off, who had not been killed because they ducked into a trench. So he made his way north, and the first camp he came to was in a place called "The Mutha." It was Camp 17, and it actually was the largest POW camp in Japan, with about 1,700 prisoners — Americans, Brits, Dutch, Australians. This was about five weeks after the bombs had dropped, and these prisoners actually did not know the war was over. Some of their guards fled, but they weren't going to flee the camp themselves because they thought this might be a trap. They'd been worked in these long-shutdown coal mines that were terribly dangerous; they'd been tortured, they'd been starved, they'd been beaten, they'd been through a whole list of horrible experiences for three and a half years, and they were very weak. The death rates in POW camps in Japan were worse than they were in Germany.

Horton: I think I saw on the internet where you said that these men were people who had been left behind in the Philippines. They were men from the Bataan Death March. Is that right?

Weller: Yeah, many of them were among MacArthur's abandoned thousands in Corregidor and Bataan Death March, but some of them had been brought from all over the Pacific. It's amazing to think that some of these men had survived three years in a Philippine prison camp and two months in a hell ship while getting transported to their work in coal mines, they came up out of the coal mines, and saw both mushroom clouds explode, and not knowing what on Earth they were.

It's amazing what these guys went through, and my father essentially walked into camp and said to the commandant, "Look, the war is over. You've got to let these men go." And he spoke to the men and said, "I'm not an officer. I have no authority to tell you what to do, but here's what I would do if I were you: I would make my way south to the tip of Kyushu where there are planes landing American troops every day, and I'd just hitch a ride back to Okinawa." And at that point the men started to leave.

Horton: So your father, George Weller, the journalist, actually opened up and basically freed these men from their prison camps.

Weller: Well, he opened up the largest of those camps. He subsequently went to several others and encountered these rescue teams of American soldiers who were now starting to open them. But I think what's most valuable about his dispatches is that he's getting the eyewitness memories from literally hundreds of POWs, who have just heard that they're free men, about what their imprisonment was like. There's really no other document that exists out there like that.

Horton: All of these dispatches were censored by MacArthur as well?

Weller: Yeah, I think that those were censored almost out of spite. My dad and MacArthur had been at loggerheads for years. My dad won the only Pulitzer of any correspondent under MacArthur, but MacArthur had been censoring him and all the others left and right. Maybe we want to get into this, but there's a whole litany of reasons why the censorship would come down.

Horton: Let's go ahead and discuss what he actually found in Nagasaki, and the results of the bombing. I think that that would kind of make it evident why MacArthur didn't want anybody else to know. I've read quite a few excerpts on the internet, but I guess his first dispatches said essentially, "Well, best I can tell is that this was just a really big bomb. It sucks for you if you lived in Nagasaki, but that's the way things go. We had to end the war." And really, that was his impression of the atomic bomb at first — that it was just a bomb, only really big.

Weller: To put it in the context of memory, our vision of the atomic bomb comes in large part from what happened in Hiroshima, which was a flat city. It was just pulverized as if a giant bulldozer had hit it. That wasn't quite the experience in Nagasaki. Nagasaki is a hilly place, and most of the damage came not from the bomb itself, because the hill sort of bounced the blast around like a basketball; rather, most of the damage came from the fires that broke out as a result of the blast. It was a wooden city. The bomb hit at lunch hour, there were lots of cooking fires in people's houses, and the whole city became this inferno. So in a way, it was like a Dresden.

The other context to keep in mind is that my father was seeing it four days after he'd gone all over Tokyo, before MacArthur shut that down to reporters, and he'd written a long dispatch for the *Chicago Daily News* that wasn't censored but which the paper never ran, about the incendiary firebombing of Tokyo, which had gone on for months. There was one night of that in which more people died arguably than in Hiroshima. Certainly more people died there than at Nagasaki. When he was looking at it, he was seeing not a super bomb, but a bomb that had landed surgically and set off an enormous fire, so he wasn't yet seeing anything unusual. Then, by three days in, he'd gone around to the hospitals and was seeing people dying from radiation poisoning, and the tenor of his dispatches changed because it became evident to him that he was seeing something he'd never seen before in any theater of war.

Horton: It's pretty clear at this point that the military wasn't saying anything about radiation. They were basically letting the impression stand that it was just a really big bomb, even though they must have known better by then.

Weller: They had known better, from their earliest tests of the bomb in New Mexico, that there was something quite different, because a bunch of cows died because some clouds had drifted in the wrong direction. At that point, there was a major effort underfoot to quiet all discussion of radiation. Oddly, one of the things I discovered in researching this story was that there's an Australian journalist, Wilfred Burchett, who snuck into Hiroshima. He had written a story that was very unlike the story that a whole planeload of American journalists, who were flown into Hiroshima the same day as Burchett, wrote. They wrote, with only one exception, a series of stories discussing the power of this bomb but leaving out the issue of radiation. The whole point of their trip was to get them to write stories praising the bomb while not mentioning radiation. There was only one journalist among them who talked about it in terms of poison gas.

However, Burchett, who was an Australian and who had gone in by himself similarly to the way my dad got into Nagasaki, was absolutely clear on what he'd seen, in part because he was being shown around by a Japanese journalist who'd been there the whole time. When my father was filing his dispatches a couple days later from Nagasaki by himself, they were arriving in Tokyo the very day that the U.S. Army is holding press conferences denouncing Burchett's reports of radiation.

So, Burchett's in Hiroshima for one day. My father has been down in Nagasaki for several. He's sending out report after report. On one level, they're contradicting the story that the government wants to spread of how there was no danger from radiation. At the same time, he's sending up stories where he has interviewed American POWs, who all survived the bomb from 300 yards away simply by getting into a trench. So on some level, there is the danger of radiation which could be spreading to the POW camps miles away, but at the same time, it's not a super-bomb — you can survive it if you just get down into a trench.

Horton: So it wasn't just censorship. It was a cover-up.

Weller: I think the way to think about it, at least as I see it, is that there was absolutely no motivation on anybody's part to publish my dad's dispatches, because there's nothing in them that's good news. The real scandal that made my father furious for decades was that by now, five weeks have passed since the bomb, nearly four weeks have passed since the Japanese surrender, and the Japanese are no longer our enemy, but not only are reporters not being allowed in, but no medical assistance has arrived at all. These people are dying, and of course nobody knew at that time that maybe nothing could be done to save the ones who'd received radiation poisoning, but there were no doctors or hospital ships or nurses or anybody there to help them. This made my father furious, and he felt that this was the scandal that MacArthur really wanted to keep quiet.

Horton: The fact that America was not providing any medical care to the people.

Weller: Absolutely! Because they were no longer the enemy, so why shouldn't we be helping them? And indeed, President Truman, four days after my father's first dispatches, circulated a secret memo to all the print and broadcast media in the United States, asking them as a favor not to talk about the atom bomb. The only discussion of the atom bomb in the media for months afterward that was permitted was by Atomic Bill Lawrence, the *New*

York Times science reporter who was being paid under the table by the Manhattan Project to write PR about the bomb.

Horton: Amy Goodman and her brother just ran a great article in the *Boston Globe* about how the Pulitzer Prize committee ought to strip the *New York Times* and Mr. Lawrence of the Pulitzer that he won for his propaganda efforts at that time.

Weller: Sure. I mean, he actually wrote a speech for Truman announcing the Bomb, and it was so insanely enthusiastic that even the White House said, "We can't use this. This is over-the-top."

Horton: And for Truman to find something to be over-the-top is quite a statement.

Weller: Absolutely. Lawrence was actually in one of the planes that dropped the bomb on Nagasaki, and he claimed that he saw a giant Statue of Liberty emerging from the mushroom cloud. What's interesting in all this is that if you look back at accounts of the press, right up into the early 1970s, this relationship between Atomic Bill and the *Times* is seen as something to boast about. It's only after Vietnam that accounts start to criticize it, and of course now it is quite a large scandal.

Horton: Of course, a big part of that is that after all this time and all these different historians going back, it's now clear that the Japanese had been trying to surrender for a long time. They only wanted to keep their emperor. We know now that Truman, after both atomic bombs, finally dropped the demand for unconditional surrender and agreed that they would be allowed to keep their emperor.

So, the whole argument that this was done to save American lives because "we would have to do a land invasion of Japan" is just a bunch of bunk. The only reason Truman wouldn't accept their surrender is that he wanted an opportunity to use his new bombs on them.

Weller: I think it may be slightly more complicated than that, to the degree that when you read the machinery of all this, it looks an awful lot like an extremely complicated contraption that's running down the hill, and without the Japanese actually saying, "We surrender," nothing could have stopped it. You can also see that once they start imagining they're going to test a new weapon of war, it's very hard to stop that. I mean, the original plans for one bomb, in fact, included the second. So I guess if there had been better weather, the original plan was that the second bomb would be dropped five days, not three days, after the first. But, in fact for Truman, the order for one was really a two-for-one offer. All the machinery was so single-minded in making this happen.

At the same time, there was a Japanese major, I believe, who went down from Tokyo to inspect Hiroshima the day after the bomb dropped, and he reported to the emperor that it wasn't really so bad. If he had reported, you know, the truth, then who knows? Perhaps the Japanese would have absolutely thrown in the towel, not after Nagasaki, but after Hiroshima. And Nagasaki would have been spared. But it's very sad to read all of that.

The question of whether the emperor should have been allowed to remain is also a complicated one, because for the Japanese — who regarded him as a kind of godhead — it ensured a sense of national exoneration for a war which was, after all, of their creation. I don't think that this has been, in the long run, profitable for the Japanese. The Japanese now don't learn that anything in the Second World War was of their doing; they simply learned that they were bombed mercilessly by the Americans. Obviously, the truth is a little more complicated than that.

Horton: That's a good point, the national vindication that came with being allowed to keep their emperor. Now, let's get back a bit here to when your father first came to Nagasaki. One of the things I read here was that the fires were still burning a month later when he arrived?

Weller: Yes, that's right.

Horton: And we mentioned how he didn't understand all the implications of the bomb at first. He just wrote, "Wow, this is a hell of a bomb," but that's about all. But now I'd like to read a little bit here from one of his first dispatches, from Saturday, September 8, 1945, at 2300 hours:

> In swaybacked or flattened skeletons of the Mitsubishi arms plant is revealed what atom can do to steel and stone. But what the riven atom can do against human flesh and bone lies hidden in two hospitals of downtown Nagasaki. Statistics are variable, and few records are kept, but it is ascertained that the chief municipal hospital had about 750 atomic patients until this week, and lost by death approximately 360. About 70 percent of the deaths have been from plain burns, but most of the patients who are gravely burned have now passed away, and those on hand are rapidly curing.
>
> Those not curing are people whose unhappy lot provides an aura of mystery around the atomic bomb's effects. They are victims of what Lieutenant Jacob Vink, Dutch medical officer and now allied commander of prison camp 14 at the mouth of the Nagasaki Harbor, calls "Disease X." Vink points out a woman on a yellow mat in the hospital who has just been brought in. She fled the atomic area but had returned to live. She was well for three weeks except for a small burn on her heel. Now she lies moaning with a blackish mouth stiff as though with lockjaw and unable to utter clear words. Her exposed legs and arms are speckled with tiny red dots and patches.

And it goes on like this. These are the people who survived the atomic bombing and then weeks later came down with symptoms like these. I think one of the excerpts I read talked about a woman who was in perfect health, it seemed, but who had gotten a small cut on her finger making dinner for her husband and had never stopped bleeding and bled to death from the small cut on her finger.

I guess if you can just fill in any more gaps about what your father saw in these hospitals, and maybe if you can tell me what were the numbers, total, of people who died of the radiation after the bombing, if you know?

Weller: Well, you hear numbers in all kinds of directions. The number that seems certainly safe is that at Nagasaki, about 70,000 died. The original numbers my father was given were in the low 20s, but that was in the first week, and those numbers had probably gone up to the high 20s by the time he left Nagasaki. He left Nagasaki on the 26th of September, and the bomb had dropped on the 9th of August. So we're now seven weeks after the dropping of the bomb? That would suggest that 40,000 more died.

He tried his best, he told me, to be an inquiring visitor, not a conquering visitor, when he went to the makeshift hospitals. One of the things that impressed him most was that the Japanese doctors, who were very alert and very expert, had already catalogued all the effects of the radiation on the various organs of the dead. And the big effect that really did not account for the bulk of the death at that moment — the bulk of the deaths at that moment were from fires — but in time, the bulk of the deaths from radiation were a result of the death of the platelets, which is the element in the blood that allows it to clot.

So, there were innumerable instances like the woman you described there, where a simple cut just kept them bleeding and bleeding and bleeding. Interestingly enough, my father in Tokyo, before he even snuck away from MacArthur's handlers, had interviewed some friends of a famous Japanese actor and an actress who were in Hiroshima, who survived the bombing, thought they were fine, went to Tokyo, and then one of them felt his insides were burning out. Then the woman had, I believe, cut herself, and she did not stop bleeding. That dispatch was actually passed by censors, but again was not published by the newspaper. So there was another kind of censorship, sort of an unwitting censorship, that wanted to hear not about how the war had been won but about the peace that was coming along.

In any event, my father saw this endlessly, and I think that was part of the medical outrage. Of course, he didn't know at the time — nobody knew — whether there might be some way of stopping the death of these platelets. It was like a leak in a boat. If there was a way to stop it, then maybe you could save an awful lot of people, but there weren't any doctors on hand then.

Horton: Do we know now, is there a way to treat someone who's been blasted by gamma rays?

Weller: I don't think that anything could be done about the platelets, but these days, all the atomic discussions that lean on the horrors of Hiroshima and Nagasaki are kind of missing the point. I mean, present-day nuclear weapons just don't have much resemblance to these.

Horton: Oh, right. These were what? Nagasaki was 10 kilotons or something?

Weller: I think it's 15, but these are pop guns compared to what they do now.

Horton: Yes, that's right. They have hydrogen bombs now that could kill Houston in one shot.

Weller: Right. They're really not very similar, unfortunately.

Horton: Well, don't worry, though, because we have a whole new generation of "usable" tactical nuclear weapons so that our DOD can find places where it's okay to go ahead and start nuking people again. They've got to get rid of that nuclear taboo, you understand.

Weller: You can imagine my relief.

Horton: Yes, exactly. Now, there were a bunch of pictures too, right? Your father had a photographer with him who had taken hundreds of pictures?

Weller: No, it actually wasn't like that. He, like many other writing journalists, was made by his newspaper to carry a little Leica along with him. My father was no photographer, but he was often in situations where there was no photographer. I had never, ever heard from him about the existence of any photos. All he ever talked about was, as he would put it, "I lost my war at Nagasaki." All I ever heard about was the dispatches that got destroyed and then lost.

When this Japanese journalist was coming out to interview me back in 2004 or 2005 about having found these, she asked, "Are there any pictures?" And I said, "No, there are no pictures, my father wasn't a photographer." Then I went rummaging around, and I found about 100 photographs that he'd taken. There are probably 25 or 30 that are actually superb pictures, and the book contains the best of them. These were pictures of Nagasaki looking pummeled and the POWs looking starved. It's pretty dramatic material, I think.

Horton: Pictures from inside the hospitals as well?

Weller: No, there's no pictures in the hospitals. I think it felt wrong to him to do that. In all his descriptions in the dispatches, he speaks of the scenes in the hospitals as piteous and dignified, and I think he felt it would be wrong to photograph there. So they're mostly photographs of the destruction in Nagasaki and of the actual camps.

Horton: In your afterword to the book, you talk about how your father would often mention the firebombing of Tokyo in comparison to Nagasaki. You make the very important point that he wasn't trying to downplay Nagasaki; he was just trying to point out that, "Hey, let's not forget the humanitarian catastrophe that is Tokyo as well, whether they used nukes or not."

Weller: I think it upset him terribly that the atomic bomb rapidly became a kind of cliché, a kind of super-bomb nothing could stop. Of course, he'd seen that, in fact, quite a few things could stop it. If you ducked into a trench, you could be three football fields away and emerge unscathed. If you were dropping it on hilly terrain rather than the flat plain of Hiroshima, its effect was radically diminished. I mean, the destruction in Nagasaki would have been pretty minimal if a) the Japanese observed their own air raid warning system and b) the city was not made of wood.

So, it annoyed him that this kind of caricature was purveyed and also that nobody mentioned the incendiary firebombing. For him, that was more destructive than Nagasaki. I mean, one night in Tokyo certainly killed 120,000 people — just fried them on a platter, as he so poignantly put it. For people to talk about the atomic bomb and not even to mention what these conventional weapons could do seemed to him absurd and a weird kind of self-flattering myopia. Remember, the firebombings went on for months and never made it into the press. My father's dispatch and, I'm sure, other people's dispatches about them didn't get mentioned at all. That's why MacArthur immediately shepherded all the correspondents who'd arrived for the surrender out of Tokyo and forbade them to go into Tokyo.

Horton: I think probably to this day, most Americans don't know about the firebombing of Tokyo. The censorship has lasted.

Weller: I think there's very little in that story that people want to know about — from the firebombings, to the atom bombs, to the POW camps — which were horribly brutal. Notice that the Nazi POW camps have entered the kind of cinema mythology, but the Japanese POW camps, which were seven times deadlier, have not. The hell ships are virtually unknown, the war trials are virtually unknown; and the fact that the Japanese carried out biological experiments just like the Nazis and that MacArthur enabled a lot of those scientists to come to America and become pharmaceutical magnates. They traded their freedom for the secrets of their experiments in case we needed help with tortures of prisoners in the future. All this stuff is unknown.

Horton: See, I knew about them bringing in Nazi scientists. You're telling me they brought in Japanese torture experts as well?

Weller: There was a whole unit called Unit 731 that was based in Manchuria that carried out biological experiments every bit as grisly as anything the

Nazis thought up. After the war, MacArthur felt that their discoveries might be of some use to us in the future, so they were allowed to go free in exchange for what they learned.

Horton: Talk about becoming your enemy, huh?

Weller: There's very little about the Japanese war, on any side of it, that's really known in a clear-minded way, here or in Japan.

Horton: From all your experience growing up with this and finding your father's long-lost dispatches, what are the most important lessons of the nuking of Nagasaki for people to take away from this?

Weller: Well, I can't pretend to be an expert on the nuclear side of things. My side of it, of course, is very personal. I found his dispatches, which meant the world to him and which he thought he had lost, and I managed to save them. For me, the lesson in my father's story is the importance that a rogue reporter can have. You know, he wrote in different places that all censorship is fundamentally propaganda, and that all military propaganda ends up being political eventually, and that's what we have.

It's not difficult to transpose the lessons of that to the present. I think the means of censorship today are different. There are cell phones. You can't stop a story, but you can stop access to a story. That's the way it's done today. I think that's why Walter Cronkite, in the forward to my dad's book, was at pains to point out that these are the lessons we need to take away from this saga.

Horton: Very important, and particularly now when, as you say, there's cell phones everywhere and wide-open access to information. Yet, still we see that the lie carries the day.

Weller: Sure, and if you can't get into a place, you can't get the story out. Also, if journalists are not the bulldogs that they ought to be, then it can take several years for the story to get out.

Horton: Long past the point when it really mattered.

Weller: Yeah, absolutely.

Horton: Alright, everybody, Anthony Weller. He's an author and musician. Son of George Weller, the Pulitzer Prize-winning reporter for the *Chicago Daily News*. The book is *First Into Nagasaki: The Censored Eyewitness Dispatches on Post-Atomic Japan and Its Prisoners of War*. His website is AnthonyWeller.com. Thanks very much for your time today, I appreciate it.

Weller: Thank you so much for having me.

Greg Mitchell: The Real History of Hiroshima and Nagasaki July 31, 2020

Scott Horton: Introducing Greg Mitchell. He is the author of the new book *The Beginning or the End: How Hollywood and America Learned to Stop Worrying and Love the Bomb.* Welcome back to the show, Greg. How are you doing?

Greg Mitchell: Thanks, very happy to be here.

Horton: Good to talk to you again, and what a great book. I learned so much about the atomic bombing itself and about the making of the docudrama *The Beginning or the End.* It's great, and I know everybody's going to love it. So tell us, what is this movie, and who's got anything to do with it?

Mitchell: Well, we are marking the 75th anniversary of the atomic bombings of August 1945 over Hiroshima and Nagasaki, which killed, by most estimates, at least 200,000 people, 90 percent of them civilians. About two months later, Donna Reed, the actress, was contacted by her former high school chemistry teacher back in Iowa, who had disappeared into the Manhattan Project two or three years earlier in Tennessee. He urged her to get Hollywood to make a big-budget movie warning the world about the dangers of nuclear weapons, the coming bigger, more powerful weapons such as the H-bomb, and a nuclear arms race with the Soviets. She had a husband who was an agent, and he went to Louis B. Mayer, head of MGM, a famous and legendary figure. He loved the idea. He said that it would be the most important movie he would ever make, and so they started down that path to write a script and so forth.

The thing to remember is that the movie was inspired by the urgent warnings of the atomic scientists raising questions about the use of the bomb and the potential use of more of them. The book basically tells the story of what happened in the next year, where the movie was revised and revised, cuts were made, and falsifications were added, basically because MGM gave General Leslie Groves — who was the head of the Manhattan Project — and President Truman himself script approval. So the script, as I charged in the book, over the course of those 12 months went from raising severe issues about the bomb to being basically pro-bomb propaganda.

The movie was released. It did not do that well, so it's not surprising that you hadn't heard of it or seen it, but it did find an audience. It got sort of mixed reviews, and it lives on today on DVD and sort of regular showings on Turner Classic. People might stumble upon it, or they may even look for it. So the book is really a story about this turning point of that whole first year or two after Hiroshima, when the U.S. had a chance to turn away from nuclear weapons, or at least to keep them in check, and thanks to official intervention, censorship, suppression and propaganda, we did not do that.

Horton: Yeah, well, ain't that always the way with stories from the world of nuclear weapons! There's a long line of suppressed stories, like that of George Weller. I've interviewed his son Anthony a couple of times. They've got that great book called *First Into Nagasaki.*

Mitchell: That's right, he was the first reporter into Nagasaki. His stories were suppressed by the MacArthur censorship office in Tokyo, and they actually did not surface for, what, 60 years? There are many other serious examples of suppression in that period, of which this movie was one, but it was a very prominent one.

Horton: It's really sad the way you explain that this really started out as one of the scientists who was in on it having real doubts, especially after Nagasaki, and said, "Man, we've got to make a movie so that people can understand how horrible this is." Then, by the time it came out, it was just a bunch of "Oorah!" and "What a great job we did protecting ourselves from a bunch of myths."

Mitchell: It was the development of what I call the "Hiroshima Narrative," which still holds sway to this day. Some people would say, "Well, why write about or care about a movie that's 75 years old? Why even write about what happened to Hiroshima and Nagasaki? You can't bring back the dead. You can't change what happened then." But the key thing is that this narrative about the use of the bomb — allegedly why we had to use it, that we didn't have other options, and it saved American lives and so forth — is still the main narrative today. We saw it this year. I mean, the main book, the main publicity to this date for this year, the 75th anniversary, is Chris Wallace's book — the Fox News host, who wrote a book that became a bestseller basically endorsing the official narrative.

Those of us who have questioned the official narrative for decades are always up against it. I'm trying in my own way, with books and numerous articles and interviews, to put a dent in it. I may be partly succeeding, but we'll see how this ends up. When August is over, if your listeners have seen a flood of contrary articles that raise questions, then I guess maybe we're making progress. But if it's the usual, almost blanket endorsement in the media, then I don't know what you can do. This is the 75th anniversary. You'd think it would be a time for reconsideration.

Horton: I was taught the story for the first time in fifth grade in social studies class, and the teacher actually did say something like, "Hey, there are people who disagree about this. It's not an open-and-shut case kind of thing." Yet he pretty much let us know that the narrative here is that President Truman did what he had to do. *Of course* he had to, or he wouldn't have done it. *Of course* he was the right guy to make that decision, or he wouldn't have been there. He was put there by the democracy, and the democracy knows what to do. So this is a thing that happened. It's like you're saying that Moses shouldn't have come down from the Mount with the tablets. This is how it goes, and that's it.

Mitchell: As you know, Truman became president only because FDR died, and of course, there's speculation, including in my book, on whether FDR would have gone ahead with the bombing. Truman became totally wrapped up in this movie, in the sense that he and his aides monitored the scripts. Maybe the two key chapters in the book come fairly near the end, where finally they have a screening of the near-final movie in Washington, and a top White House aide sees it and then waves a red flag to MGM, saying they must re-shoot the most key scene in the entire movie where Truman explains why he had to use the bomb, and so they re-shot the entire scene. They fired the actor playing Truman, again at White House request, and re-shot a scene in which they were able to lay out "the Hiroshima Narrative" — that it was the only option that Truman had, it saved a million American lives, and going forward, the bomb is something we're going to have to embrace. That's how the movie ended up. That is the final key scene in the movie.

Horton: Whether the movie was a big smash hit or not, that certainly was the narrative, and the movie certainly helped to establish the narrative from that time that that's just the way it is. That's the way it happened because it had to happen. The implication is that the only people who would object or say, "Okay, it's 75 years later, and now we should rethink it and look at it seriously," are essentially just hippies or somebody who is just taking the contrary position to what everybody knows. You can do that all day, but it doesn't change the narrative of the majority.

It would have to be something huge. In fact, I think one of the most important points that you make in the book, one of the things that I know absolutely blows the minds especially of right-leaning people, was when you start naming admirals who opposed the attacks. Then once you start naming them, then you get to the generals, you get to Leahy, the chief of staff, you get to McNamara and Eisenhower and Nimitz, they all opposed it. You quote one of these admirals as saying, "Oh, well, the Americans, *ahem*, the Americans, they have their new toy, and they want to play with it." Oh, my God. They didn't teach me that in fifth grade.

Mitchell: The contrary narrative makes use of what a lot of these military people said. General Eisenhower was probably the most vociferous and actually told Truman and his Secretary of War, Stimson, in advance. It wasn't Monday morning quarterbacking; it was actually in advance.

Horton: I'm sorry because I think I mistakenly gave McNamara credit there, I meant to say MacArthur, which is even more to the point — MacArthur is the guy who wanted to nuke North Korea over and over again, he didn't care about using nukes, but in this case, he thought, why bother?

Mitchell: It's funny that you mentioned McNamara. Of course, Robert McNamara famously later admitted that even the regular carpet-bombing of Japanese cities was a war crime. So he kind of was admitting he was a war criminal.

Horton: That's in the documentary *The Fog of War*. Everybody, if you've never seen that, you absolutely have to see it. It's just remarkable, and that is just one of the many important revelations in there.

Mitchell: Just to complete one thought I had, which is why does this matter today? Succinctly, it's simply because many people don't even know the U.S. still has a first-use policy, as it's officially known. What that means is that any president — we especially have to worry about the current one — has absolute control, ability and permission to initiate a nuclear war — to launch nuclear missiles not in retaliation, but in a preemptive attack. It could be in any conventional war. It could have been, for example, when we invaded Iraq; if Saddam had threatened to use chemical weapons, we could have nuked Baghdad.

It can happen even with threats. If Iran, North Korea, Pakistan, or whoever started raising threats in a crisis, it's absolutely U.S. policy to be able to launch a first strike. This goes back to 1945, and the fact that while people in the media and officials kind of wring their hands and say, "Well, we must never use nuclear weapons again." I mean, how many times have you heard that? But in fact, they endorse the two times we have used them, and the same arguments — about saving American lives, how we had to use them first to end the war — are all very much in play.

Polls show that at least 30 to 40 percent of the public, if not more, explicitly endorse scenarios for us using the bomb first against North Korea or Iran, for example. So this is not just a story of the past. I know, if people watch this movie, it may seem kind of hokey, and they might ask, "Why should we even care about this?" But in fact, it's part of the whole scenario where America came to embrace the bomb. Although we don't think about it much today, it still accepts and endorses the bomb today.

Horton: Ray McGovern was on the show earlier and was very much putting the media in the military-industrial complex with the rest of them. So it is a hugely important story because it just goes to show how easy it is for the media just to defer to the government. I think you have the quote, "We owe it to our government to defer to its opinion," or something like that.

We do? Says who? Why? And, when? Especially when they're wrong, justifying a monstrous war crime. Their whole argument is: "What if the

Nazis had gotten it first?" Okay, let's run with that. What if the Nazis had gotten it first, and the Nazis had used a bomb? Would the story be that, "Hey, they had to do what they had to do to get unconditional surrender," or would it be part and parcel of the entire narrative that Nazis do what Nazis do because that's who they are, and dropping bombs is a Nazi kind of thing to do?

Mitchell: The specific revisions in the movie, which I document in the book for the first time anywhere, reveal that the U.S. was, I guess you might say, defensive about this story at this turning point. For example, it's almost dark humor. As the revisions went on, they injected specific references which were totally false and totally based on myth, where the president or General Groves warned that the Japanese might have atomic weapons, and if we don't use ours first, any invasion will be met by the Japanese with atomic weapons. They even had a scene where they have a German submarine surface off the coast of Tokyo, and a German scientist is brought to shore. He meets with the top Japanese scientists and gives them the secret of the atomic bomb that Germany had. The Japanese scientist then heads off to their big laboratory to create this atomic bomb.

Horton: These people are just shameless.

Mitchell: And where is that lab located in the film? I'll give you five seconds to guess where this laboratory was depicted to be located. Well, it's located in Hiroshima. You can't make this stuff up, but they did. That scene was cut from the final film, rather surprisingly, I guess, but there are so many other changes that reflect this kind of meddling. I'll just mention one more. There are many people who might not be sure where Hiroshima ranks, but they are pretty sure that Nagasaki was a war crime. So, recognizing that, over the course of those months, they cut every mention of Nagasaki in this movie. Anyone who watches the entire movie would never know that we even used a second bomb. Of course, this was one of the more sensitive and vulnerable parts of the official narrative. People accept one bomb, but the second bomb seems like torture, so they had to cut that.

There are numerous other things, but like I said, these revisions in the script really reveal the American defensiveness or thinking, and it worked. I had a major article published at *Mother Jones* last week about how people think about how John Hersey's "Hiroshima" article in *The New Yorker* — which has been called one of the great journalistic feats over the last century — had a tremendous impact and changed minds and so forth. It's still read today in book form. In reality, it did have a big impact for a couple of months, but then Truman and his allies rallied to have the Secretary of War Stimson write a cover story for *Harper's* defending the use of the bomb. The press, which had started to raise questions after the Hersey article, then did a 180. It was like they said, "Well, okay, this settles the question forever. We don't have to think about this bombing again, maybe forever, and the *Harper's* cover story settles it for history."

So the Hersey article was in the same period as this movie, summer and fall of 1946. The bomb is coming under scrutiny. It was a possible turning point. Many of the atomic scientists like Albert Einstein were totally against it, but all sorts of questions about the Hiroshima bombing were laid to rest by early 1947.

Horton: Just to go back to the first-use policy real quick. I wanted to point out that Obama, at one point, had put out his Nuclear Posture Review, which said that we no longer threatened first strikes against non-nuclear weapons states. Except Iran — we still might use a first strike against them.

Then Trump rescinded that, and so now we reserve the right to launch a first strike against Australia or anybody. That's back on the table. So it's worse than you could even describe. Really, the whole thing is just insane.

Mitchell: I realize there are a lot of other serious issues today, including the pandemic, of course, and climate change, which in some ways has usurped a lot of the interest, especially among younger people, about the dangers of nuclear weapons, which is understandable but in some ways unfortunate. In the Republican or the Democratic debates, the primaries, all the many debates over the last year and earlier this year, nuclear policy came up almost never. One question somehow came out of the blue for Elizabeth Warren in one of the debates. She was the only one asked, "Do you support the first-use policy?" Of course, I rubbed my hands and said, "Wow! Someone's actually being specifically asked about the U.S. first-use policy," and she bravely said that no, she was for no-first-use. I was hoping they would then go down the line and ask people to raise their hands, but they didn't do that.

Horton: Gabbard did bring up nuclear issues a couple of times herself, but they certainly never gave her the opportunity to do so deliberately.

Mitchell: So she said this, and sure enough, the next day, Republicans and conservatives in the media just exploded at Elizabeth Warren. Liz Cheney blasted her. This is one person who dared to question this first-use policy, and she was smashed by even Democrats and, of course, the Republicans.

Horton: In 2008, Obama during the campaign was saying, "Yeah, I'm going to start bombing Pakistan. That's where al Qaeda is. I'm going to Pakistan." Then one of the reporters asked him, "Would you use nukes on Pakistan?" And he says, "No, I wouldn't use nukes on Pakistan." And then Hillary Clinton attacked him saying, "Don't ever say who you won't use nukes against! Barack Obama is so naive and unprepared to be the president. Did you hear that? He just promised not to nuke Pakistan."

Mitchell: Well, I'll give Obama credit. I know he was disappointing in many ways, but he did raise questions about first-use. He raised questions about cutting back.

Horton: Meh. He gave them a trillion dollars for a whole new arsenal too, but he did talk a good game sometimes.

Mitchell: As I chronicled at the time, he became the first president to ever visit Hiroshima and Nagasaki while in office. In fact, he was the first to send the U.S. Ambassador to their memorial ceremonies — that was Caroline Kennedy, of all people. Obama then visited it, and of course, the right wing had a storm of this in advance. He was going to go and apologize! He was going to say we did something wrong! And of course, he did not. But he did go there briefly. He hugged one of the survivors. So I'll give him at least credit for that, how it took 70 years for a U.S. President to even step foot while in office in one of those two cities.

Horton: Let's talk about the premises of the nuking here. You go through this real well in a book about a movie. You really tell the story of the bombing and the decision-making here so well and how it all came out. You have this ever-inflated number of the expected American invasion force casualties, you have ever-changing dates of when it was going to happen, you have ever-changing claims about how certain it was that the invasion would be necessary. You always have the begged question that "of course the surrender's conditions must be unconditional" and all of these kinds of things that almost always just go without saying, without discussion, when this subject comes up. You break this down really well and give people the opportunity to see this a different way.

Mitchell: I appreciate that because in a way, I like to think of the book as one of the better breakdowns of the decision to use the bomb, disguised within a book about a Hollywood movie. Rope them in with a Hollywood story, and then convince them. It's important to break it down a bit because I suspect even people in your enlightened audience, and certainly there are many liberals and even lefties, who still cling to this official narrative. I get it all the time in interviews or in comments where, you know, the audience is more towards the left, and they're still raising the same issues. So it's probably important to briefly break it down just in this way, and to say that when we got to July 1945, we had taken all these islands in the Pacific, we had already leveled most of the Japanese cities, at least in part, with the carpet-bombing, Japan was surrounded, so the question was: How can this war end now? How do we get this war to end?

Of course, there was absolutely a plan, as there had to be, a detailed plan for an invasion of Japan with millions of American soldiers, but it was not set for August 1945; it was set for November, December, January — many months off. There was no pending invasion, but there were certainly pending serious plans for it. They were going to do a partial invasion, and if that didn't work, there would be a landing near Tokyo. If that didn't work, there would be more. Of course, you game out if that happens, if Japan still is capable of fighting, how many Americans would be killed? At the time, there were no estimates of a million American casualties, let alone death. The number was 200,000 or 400,000 or 100,000.

Horton: And that's when you're talking total casualties, not just killed, right?

Mitchell: That's right. And that's bad enough. However, the invasion was many months off, and Japan was in an incredibly precarious state with their economy, their ability to fight, they were surrounded, they had no oil, and so there already were peace-feelers from Japan. There were even people within Truman's orbit, aides and advisers, who were saying, "Japan is ready to throw in the towel. You just have to find the right terms here." The most serious thing probably was when the U.S., after Germany was defeated, demanded of Joseph Stalin that Russia now declare war on Japan. When Truman was at Potsdam in July, Stalin confirmed that he would be in the war around August 9th. Truman even wrote in his diary, "Fini Japs when that occurs," and he was referring strictly to the Russian invasion, not the use of the bomb. He has a second reference in his diary along the same lines. So Truman and others recognized the importance of the Russians coming into the war, with Japan already on the verge of collapse, and many did think that that alone would cause a quick surrender.

The other issue was that the U.S. demanded unconditional surrender, which the Japanese rejected. However, there were many people close to Truman, including Stimson and John J. McCall, who are not doves by any means, saying that if we tell them that if we add one condition — which is that the emperor could remain on the throne, even symbolically, and would not be executed, and the Japanese who cared would still have their royalty — that alone could prompt a quick surrender. Of course, we rejected that. We demanded unconditional surrender. Then, after dropping the two bombs, we suddenly accepted the emperor remaining.

Horton: I like the anecdote that you've got of Allen Dulles of the OSS, who got a message from Japan saying, "Hey, we want to talk," and they shut it down.

Mitchell: There are very respected historians, even some I know well, who attribute the decision to bomb mainly to wanting to send a signal to Russia, mainly to stop them from getting closer to Japan, and as the beginning of the Cold War. I've always believed that to be a factor, but it's just one of several factors. An important factor, perhaps, but as you say, there were top people that were far from doves who felt that we'd have to be crazy not to modify our surrender demands and not to wait a few days for Russia to declare war.

The Japanese feared the Russians far more than they feared the Americans. They had a long history with Russia, not with the U.S.

So, I think it's plausible — as many of these Truman people recognized at the time, and I think fed why Eisenhower and Leahy and all these other people were against the use of the bomb — to believe that the combination of the Russian declaration and the allowing the emperor to stay on the throne most likely would have ended the war, if not in the same day, then within the same week that it did end. We would have saved all those civilian lives that were lost. I know Americans don't feel it's such a black mark on our U.S. reputation, but most of the world sees this as a war crime, sees this as unnecessary. It does not speak well for us, so that probably could have been avoided — not certainly, but probably, if only these other options had been tried.

Horton: Back to some of the lies that made it into the movie. The one that seems the most important is the thing about the leaflets, but also the flak and the rest of it. They really reframed the whole thing where, even though Japan was completely beaten, we were still the Rebel Alliance and they were still the Empire, right?

Mitchell: Well, I mentioned them allegedly having atomic weapons themselves.

Horton: Right yeah, like "Oh, they're going to meet us on the beaches with their own atom bombs," that somehow they haven't seen fit to deploy up until this point.

Mitchell: You mentioned a couple examples, but this was part of the need of this movie to picture this attack on Japan within the bounds of American decency. Because, as you know, we were very happy to denounce Japan for so long for their sneak attack on Pearl Harbor as cowardly. They got no opposition, at first, flying over Pearl Harbor, so here we were, attacking these two defenseless cities.

Now, in real life, our bombers reached Hiroshima and Nagasaki without any opposition. There was no anti-aircraft fire. They just dropped their bombs from an incredibly high altitude and got the hell out of there. But of course, the movie had to picture this as part of the usual U.S. valor and bravery and risk, so they had to introduce in the script the bombers coming under attack by light flak. Then it was turned into heavy flak and anti-aircraft fire, and then finally, Japanese fighter planes in the distance, ready to fire on them. It had to show this as incredibly risky.

You mentioned the leaflets. There's another good example. The film claims at least twice, maybe four times, that the U.S. dropped millions of leaflets to warn the people that this new weapon was coming, and so this was not a sneak attack with an atomic weapon. In real life, we did drop leaflets in advance, but it did not warn of any new weapon or say anything beyond the ordinary. They did prepare millions of leaflets specifically warning of the atomic bomb — or specifically saying, after Hiroshima, that we had used the atomic bomb — but these leaflets were not dropped over any city, including Nagasaki, until the day after Nagasaki. So if you were unfortunate enough to be in Nagasaki and survive the atomic bombing, the next day you might have found drifting down from the skies a leaflet warning you that the atomic bomb might be coming. It didn't do a lot of good, and that's another myth that's introduced more than once in the movie.

Horton: In fact, they even create the whole line of dialogue too, where someone's like, "Oh, we've been dropping leaflets on them every day for 10 days." Which makes no sense, right? You're claiming that this is a military base. Why would you have been warning them for 10 days to put all of their weapons on railroad tracks and get them the hell out of there? So are you saying it's a city full of civilians? Well, that's kind of different.

But anyway, then the guy responds, "Jeez, I only wish they dropped leaflets and given us warning at Pearl Harbor." Yeah, like that time they wiped Honolulu off the map and killed all the civilians, right? Oh, but that's not what happened. They sank a couple of military ships. That's different.

Mitchell: You basically have that accurately, but again, it shows the revisions. That's what makes it fascinating. That was not in the original script, that was not in the first round of many rounds of revisions ordered by the military. That snuck in towards the end, when they were fine-tuning this.

Horton: In other words, the more they watched it, the more they were like, "Jeez, we really look like cold-blooded murderers here, man. Maybe we need to propaganda this thing up a little bit."

Mitchell: Again, that's what makes it interesting, these specifics. It's not just like, "Oh, it started this way, and it ended up that way," and isn't that a shame?

Horton: The same thing they say about how *Star Wars* was saved in the edit, where the first time, once Han Solo comes in and Darth Vader's ship goes spinning off, then Luke goes all the way around again, and he's got all this time to take the shot. There was no tension at all. Then a good editor came in and said, "No, Darth Vader's right about to kill him right up until the point that he takes the lucky shot, and then it works out perfect, see?" Same kind of thing.

Mitchell: I will bow to your knowledge of *Star Wars*, because mine is almost non-existent.

Horton: It's a perfect parallel to what's going on here. Like, "Jeez, they're not in danger. Let's put Darth Vader right on their tail," and that way, they're defending themselves as they nuke this city.

Mitchell: You mentioned the whole question of a military base. In fact, Truman's initial announcement to the world about this bombing was simply a press release from the White House from Truman. In the very first sentence it said, "X number of hours ago, we bombed. We used this revolutionary new weapon to destroy Hiroshima, a Japanese military base." No mention that it was a city, of course. There was a military headquarters in Hiroshima, but that was not the target of the bomb. The bomb was dropped in the center of the city. Now, that Japanese headquarters was destroyed, and there are estimates that perhaps as many as 10,000 Japanese military personnel may have perished, but that's a relatively small number when there's 125 or 150,000 others.

In Nagasaki, there are estimates now that maybe 150 to 200 military personnel died in total in the bombing, out of 100,000 dead. There's a naval base in Nagasaki, but it was on the edge of town, and it was basically untouched. The bombing was over the city, but again, that's part of the myth that we *had* to do this, that these were legitimate military targets. Of course, we'd been firebombing the cities for months.

There is the argument that any city with industry is a military target. You can imagine, if the U.S. was ever bombed, if they just bombed Detroit or Pittsburgh and said, "Well, you know, they don't have big military bases there, but there's a lot of industry."

Horton: The fact that McNamara and LeMay had not already burned these two cities to the ground with firebombs was proof of what low priorities they were. That was why they were chosen, because "We'll get a good test out of this thing."

Mitchell: You're right. They wanted a pristine city. That's all part of the official narrative with the legitimate military targets and so forth. There are people who raise very interesting historical evidence and points. There's evidence of how difficult it was to enact the Japanese surrender. They raise images of the Japanese fighting on, and women and children with spears, and everything like that.

There certainly are, as I've already said, reasons not to be certain that Japan would have surrendered in that near term without the use of the two bombs. You can never prove that. However, it has to bother one that when people raise the possibility that they weren't necessary, raise moral issues and raise human issues like the dead civilians — women and kids were the majority of people who died — that's supposed to be kind of off-limits. You really shouldn't raise those issues.

Again, just to say one more time, the reason it matters so much today is the example it sets, the precedent it set, and the media coverage we get that endorses the use. And why? If you endorse it then, you're giving a signal to

Trump or anyone else that these weapons are usable, no matter how often we read that they're not really usable, and that they're only for deterrence.

Horton: You quote somebody in the book here saying, "Look, we hold the Japanese people collectively responsible for the crimes of their government, and that's our excuse for doing this to them." Same thing for carpet-bombing German cities and saying, "Well, they should overthrow the Nazis for us," or something like that, as an excuse. But then, our country is a democracy. That means that the civilians of our country would then be that much more responsible for this particular act. The truth is that you couldn't hardly blame the average Japanese citizen when the emperor and his military command have total power over the society and the people have essentially none whatsoever in comparison to it. He makes a good point about the ones who are dropping this bomb on these helpless people who were in no way responsible or able to determine their government's policy on any issue whatsoever.

Mitchell: You referred to this earlier. You mentioned Ayn Rand in passing, and people might say, "What does Ayn Rand have to do with this?"

Horton: This is hilarious, everybody. For libertarians, there's a longstanding beef, and some overlap, between libertarians and objectivists, but we all like making fun of her because she's a clown. Anyways, go ahead.

Mitchell: I imagine Ayn Rand fans may know this, but most people don't know that she was for a period in the early to mid-'40s a Hollywood screenwriter. Like so many, she went out there to make her fortune, and she did write two or three screenplays. Then, practically the same month that MGM went ahead with this movie, Paramount decided to do the same thing, and instead of a race for the bomb, this was a race for the first bomb movie.

So, Paramount launched their major effort, and they hired Ayn Rand as the screenwriter to write their film. I was able to go through all her papers, scripts, outlines and so forth at the Motion Picture Academy Library and chart her involvement — why she got involved, how she got involved, what she said out front the film was going to be like and then her early scripts. Even the fact that she managed to do two interviews with Robert Oppenheimer, the so-called "father of the atomic bomb." MGM wasn't able to accomplish this. She sat down with Oppenheimer twice, and ultimately, the Paramount film fell through, I think largely because the producer Hal Wallis recognized that Rand's script was kooky, and so he sold out to MGM and the competition ended.

But Ayn Rand went on to write a little book called *Atlas Shrugged* with her free time, and she used Oppenheimer as her model for one of the key characters, the scientist Stadler. So she did get something out of this. Quite a bit of this is in the book. Some people may find it amusing, some people may find it horrifying, but Ayn Rand in Hollywood is quite a sight to behold.

She was coming off big success with her first book *The Fountainhead*, which had been sold to the movies, so she was a well-known quantity. People were interested in her doing this.

Horton: It is hilarious, too. For anyone who is not too familiar with her character, I suggest you Google "Mozart Was a Red," which is a really funny play that Murray Rothbard wrote about his time being introduced to the Rand cult in New York back in the 1960s. Think about a very kind of young Ayn Rand but still with the cigarette holder and the passionate self-interest and all that.

But the way she decides to frame the whole thing is that this is all the free market and individualism that did this, the creative genius and individual men triumphing over goals, trying to spin it as essentially a heroic tale of individualism, when you're talking about building a giant *bomb* for the *state.* This wasn't merely a government program, but the worst government program of all government programs that had ever been devised in world history.

Mitchell: Yeah, well, thanks to industry. Thanks to DuPont and all the rest.

Horton: [Laughing] Yes, that's right. "We never could have done this without DuPont, young man!"

Mitchell: Her biggest fear was that FDR would get any credit. In one of her outlines to Paramount, she said basically that if such a film attributed any of the success to Franklin Roosevelt this would essentially be a crime against humanity and she was going to quit. It had to be individual scientists, the industry, the flash of genius from free men, and the only reason we could create this was because we're America. The Nazis could have never.

I think most people would say the Nazis could have. If the Jewish scientists had stayed there, the Nazis would have beaten us to the bomb, but history turned out different.

Horton: Same kind of thing happened with their doctors: "Oh, you kicked all the doctors out of your society? Let me know how that works out for you." So fortunately, they did not. And we know now that they went off on bad dead-ends in their nuclear research, yet that's when you hear quite a bit of, "Oh, the Germans were about to get it." No, they weren't. They had gone off on rabbit trails and given up on them not too long into the war. There wasn't even progress being made in Germany at that point.

Mitchell: As I talk about in the book, some of these scientists who were against the use of the bomb against Japan have signed off on the use against Germany. They, the Jewish scientists particularly, had a special and reasonable grudge and fears about a Hitler takeover. But two months later, Japan's surrounded, Japan's not as big of a threat, and so they said, "Okay, Germany, maybe. But Japan? No."

Horton: That's really kind of a common thread throughout the whole thing, all these scientists thinking better about what they had done, and trying to do something about it, right?

Mitchell: Well, it's too long a story — there was only one real whistleblower who really tried to get something done, Leo Szilard, the famous scientist. He launched a petition campaign and got many top scientists to sign this petition asking Truman to hold off, but it never even got to Truman's desk.

While all this was going on, MGM had to chase Oppenheimer, Einstein, Szilard and others to get their permissions in a contract to be portrayed in the movie. At the very same time, J. Edgar Hoover and the FBI were surveilling these same men because of their alleged communist or left-wing credentials. So in the book, we see Einstein having his mail opened and telephone calls monitored, Szilard being followed in the street, his mail opened, and Oppenheimer's phone being tapped. There's a whole scene in the book where Oppenheimer talks about how terrible this movie is, thanks to an FBI transcript. Ultimately though, they all signed their contracts. So all these things come together — MGM, FBI, the atomic scientists — it's all in the mix there.

Horton: The book is called *The Beginning or the End: How Hollywood and America Learned to Stop Worrying and Love the Bomb*. That's the title of the movie as well, *The Beginning or the End*. Alright, you guys. That's Greg Mitchell. Thanks again.

Mitchell: Thanks, Scott.

Josiah Lippincott: Wholesale Slaughter of Japanese Civilians in WWII September 11, 2020

Scott Horton: Introducing Josiah Lippincott. He wrote this really interesting thing for *The American Conservative* magazine, called "Wholesale Slaughter of Japanese Civilians in WWII was Evil." Welcome to the show. How are you doing?

Josiah Lippincott: I'm doing well. Thank you for having me on, Scott.

Horton: Really happy to have you here. You really advanced my understanding quite a bit, telling me what you read in a very important book I'd never heard of. It's also interesting that this article came out not right around the time that we write our Hiroshima articles in the beginning of August every year, but at the end of August or beginning of September. Your article is actually a rejoinder to a hawkish piece in *The Federalist* making the typical conservative case for why the Red, White and Blue had to do what it did on the 6th and the 9th when they nuked Hiroshima and Nagasaki back in 1945.

So, if you want to start with that, who wrote it, and can you rehash the case he makes? It sounds like his take was the pretty conventional one, right?

Lippincott: The author of that piece is a guy named Josh Lawson. He was actually a graduate student at the same program that I am attending currently. I think he's an editor at *The Federalist*, and he makes the case that I've seen made elsewhere, to the effect that we had to drop the bomb, right? We were confronted with Japan, they're evil, and if we don't nuke them, then the civilian casualties of a siege or from an invasion of the island would be just orders of magnitude greater than anything we experienced prior in the war. So, confronted with that choice, the argument that Lawson puts forward, much like many other conservative commentators, is that we just had to drop the bombs. We had to engage in strategic bombing in order to bring the war to an end sooner. That meant targeting civilians, but it's justified because we're saving more lives in the aggregate.

Horton: By the time I was just getting into high school, President H. W. Bush said, "Look, we had to save a million lives, and so we saved a million lives" — by ending a couple hundred thousand in these two cities, but that was what it was. That was the loss that they were trying to prevent there, although I don't think that's really the history of the numbers that the military estimated at the time.

Lippincott: Well, of course you can talk about things that didn't happen and say, "Well, this would have been an abject disaster if we hadn't done X." This book that I referenced in my piece is by this guy named Paul Ham. It's called *Hiroshima Nagasaki*. I highly recommend it to anyone who is interested in the atomic bombings.

I had begun to question the narrative around strategic bombing on military grounds about a year before I actually ended up being stationed overseas in Japan. I was on Okinawa. I served as a Marine. So reading that book, just seeing the facts, really changed my perspective on what had happened, and I realized this narrative that we would have had to invade the island doesn't make sense. Why would we have wanted to fight the Japanese to the death on their home turf? Was there another way? Well, it turns out that there was. You had the possibility of negotiating a surrender. By 1944, the Japanese had basically been defeated militarily. It was over. By Leyte Gulf in October 1944, all of their capital ships were either damaged, destroyed or sunk, and completely outnumbered and out of position. They had lost territory, they had lost their navy, they had really no air force to speak of, so at that point, the Japanese government knew that they were defeated.

What they were not willing to accept was total submission to the United States, and that was what was being demanded — unconditional surrender. You can see in the notes, which Paul Ham helpfully put in the book, that the Japanese high command did not want to see the emperor removed from his throne. That was the issue upon which the continuation of the war hinged. If he was not willing to step down, there were those within the Japanese government, especially the war minister, Korechika Anami, who were willing to continue the war indefinitely. They were willing to commit national suicide (*gyokusai*, or "shattered jewel").

It turns out, they wanted to negotiate. They wanted to negotiate a surrender or some sort of peace treaty, and the United States government was unwilling to do so. That, for me, was what really put it over the edge because then that argument falls apart; you don't have to invade and sacrifice millions of lives. In fact, then you start asking, "Why did we attack Okinawa? Why did we invade Iwo Jima?" An amphibious landing means charging straight into enemy gunfire. That's one of the first things they teach you at Basic School: "Don't charge at the enemy's strong point. Go around if you can. Attain your objectives at minimal cost by attacking the enemy where he's weak." Attacking a fortified island is not attacking the enemy's weak point; it's attacking his strong point. So the idea that has gotten into people's minds that, "Oh no, we would have had to attack the Japanese mainland and fight to the death of the Japanese." That is crazy because it ignores that there were options here, and we should have pursued those options. That's what really changed my mind about the bombing more than anything else — actually knowing some facts about what happened.

Horton: Especially, as you recount, the chronology about when their navy had absolutely had it and all the time there was to spare. And just like with almost everything, it's a giant presumption at the beginning, and the rest of the discussion takes place after that. The question of unconditional surrender is never debated. Everybody knows that these guys and their Nazi allies were so evil that we demanded absolute surrender because we had the right to because of how bad they were, and that's just how it is. Maybe they explained that to you one time, maybe not even that, but everybody knows that of course that was the right thing to do. People would say, "Oh, this is Monday morning quarterbacking stuff. If only they had known what you know now, but at the time, they thought, 'How can we ever accept anything less than total surrender, and how can we ever approach their islands without taking these other islands?'" So what do you say to that?

Lippincott: I think you've hit it on the head there. Which is to say, you have to reject the false binary. It is not "Unconditional surrender, or fight to the death." That's an infantile way of looking at history. I know good people with good character. I've met Josh Lawson, the guy I was responding to. This is not an evil person. At one point, I made the same arguments. My problem is that we have got to get outside of the idea that you're either on the side of the devils or the angels. It makes it hard to think, and we have got to think about what we're doing, especially about military matters. I say this as someone who served in the military, who cares about this country. I want to win wars. I do not want to waste the lives of the young men who go out and fight, and I don't want to kill civilians either. I'm not a barbarian, but I get it. Sometimes in war there's collateral damage. People die. It's not good.

The problem is that we've gotten in this mode where suddenly, anything goes. That was not the way that it was in the West prior to World War I. People did not have this idea in their minds that it was okay to obliterate civilians intentionally in order to attain your objectives. There were limits on what war was supposed to be like, and a lot of that has been forgotten. The old international law has basically been shredded up, despite the fact that it represents civilization. I like civilization, I like winning wars, and it turns out doing the right thing morally corresponds to the good.

I'll make one point here. The strategic bombing campaigns took up enormous resources. You had to take these islands like Okinawa and Iwo Jima so they could base our fighters and provide landing strips for bombers. And if you begin asking, "What can Japan actually do militarily?" It turns out not very much, as they had basically been defeated. So why are you pouring resources and lives and money into annihilating cities?

Horton: I forget whether it's in your article or it's one of the things that you linked to, where you tell the story about how they decided to even go to Iwo Jima. It was basically two against one among the Navy, the Army Air Forces and the Marine Corps. Three men had a discussion in a room and said, "Okay, fine." Just like out of *Catch-22*, this is how they decided to go to Iwo Jima. It was a decision that very well could have gone the other way. We could have just gone around them. What are they going to do? All we've got to do is leave one battleship offshore, and what are they going to do, swim? They can't do anything. They're stranded there, so we can just move on right past them at that point. Unless, as you say, "Oh, we need those airstrips for our long-range bombers that we've already committed to seeing through the use of."

Lippincott: I am totally willing to be wrong. If someone wants to come out and make the argument that we had to do what we did in terms of taking Okinawa or Iwo Jima, then people need to make that argument. That's what I'm getting at here — reject the false binary. We have to think about what was done. Right now, it's just assumed. We just have these narratives, right? You see the photo of the Marines planting a flag on Mount Suribachi, and that's what the Battle of Iwo Jima becomes. So the logistics, the decision to go to battle, the individuals involved, it all falls away, and what you're left with is a photograph or a slogan or a term.

This is a huge problem within the military now, that we continue to live off of the fumes of what was accomplished 75 years ago. We continue to have bases overseas, and we continue to pour millions of dollars into this. I don't think most Americans are tracking, but there are 22,000 American personnel on Okinawa. There are a ton of people. You have to ask, "Why are we still there? What are we trying to do? Where does this strategy come from?" It turns out that a lot of it is the momentum from World War II with us today, and there's just no discussion about how maybe our fundamental premises aren't right here. And we need to have that conversation.

Horton: Boy, ask the people of Okinawa. They'll tell you.

Lippincott: It's controversial there. They get money from the American government. That's good for some people, but they give up sovereignty, too, which isn't so great.

Horton: Well, it's not really up to them, right? Okinawa is sort of separate from the rest of Japan. They're kind of like Japan's Puerto Rico. They're going to put up with what's decided for them.

Lippincott: Yes, the politics there are very complicated. There's been talk of moving the base from Okinawa to Guam, but I think what really needs to happen, and what I'm working on now, is to say, "What do we need to contain?" The argument that is made now is that we have to contain China, we have to do X, Y and Z to make sure that we can continue to trade in the Far East. Then you have to ask, "How is what we're doing now relevant to that end? How relevant are the weapons systems we're using? How relevant are the arguments?"

I saw people making arguments that the way we were going to defend from invasion is that we're going to have these islands and defend them, and it looks a lot like what the Japanese did in World War II, but in reverse. That's not going to work.

Horton: I'm sure you saw the big *Reuters* piece about how you're right — our entire Navy is obsolete and we can't use it because if we get within range of them, then they're in range of us, and then all of our ships are at the bottom of the Pacific. But you know what we're going to do, we're going to use our B-1 bombers. So they're already fessing up that the entire Navy in the Pacific is obsolete for anything other than keeping Filipino pirates at bay or something.

Lippincott: I think that gets to the core of what's happening here. Which is to say, strategic bombing, the B-1, that's going to save us. The whole reason we have a separate air force from the army is so that we have these bombers which are designed to hit targets deep within enemy territory. This isn't close air support; this is going hundreds of miles deep to attack these nodes, these centers, so that you can disrupt them, halt manufacturing and wreck their supply line.

Well, it turns out that that mode of fighting an air war, that kind of strategic bombing, I have come to question all of that. Part of it is because of World War II, when it didn't work. Bombing all these cities in Japan, we bombed five dozen cities; over 60 cities were burned to a crisp by American bombers. On August 7th, the day after the bomb fell on Hiroshima, they had the discussion, "Should we surrender?" No. On August 8th, the emperor is still insisting, "I'm going to stay on my throne."

Horton: In other words, two days after Hiroshima has been obliterated in a single shot.

Lippincott: You know what changed their minds? It was the Soviet Union invading Manchuria. They took over Manchuria, and when that happened, the Japanese understood they cannot negotiate their way out of this. Then they're willing to consider surrendering.

Horton: And they knew that Stalin was willing to send his infantry and lose them. They would rather be occupied by the Americans than the Soviets, and that was the choice they were making at that point, right?

Lippincott: Again, one thing you do not see — at least I have not seen — is anyone who's really explicitly connected the invasion of Manchuria by the Soviet Union to the Korean War, which happened just several years later. That was a bloody war. I think it's something like 30,000 dead Americans in Korea, fighting the forces of communism. You have to ask, "How did the communists get into Manchuria and North Korea?" It's because we let them invade. If we had negotiated with the Japanese sooner, then the Japanese would have owned Manchuria, and we might have been able to work something out with them.

So, this is where the narratives get really strange. Here we are on the side of the angels in terms of fighting the evil Japanese. We're also fighting with the Soviet Union, with the communist powers, and then we end up with a Cold War where we're fighting them again. Whose bright idea was this? And was this the best option?

You can stretch this back even further, back to the early 1900s, when the Russians had owned parts of China; they invaded all of Manchuria by 1900. China was being divided up by the European powers, the Japanese were trying to occupy territory as well, and yet the narrative that gets presented is that it's bad for the Japanese to have an empire in China, but it is not bad for the Russians to do that.

And by the way, what about the United States? We had troops inside of China in 1940 and 1941. The 4th Marine Regiment was stationed in Shanghai and got moved in November 1941 to the Philippines. We had troops in China.

Horton: America also backed Japan's puppets in South Korea in that war, the people who had fought as the quislings of the Japanese Empire against those who had fought for independence, who were led by the communists in the North.

Lippincott: Again, you can just see how complicated this is. We have to think about these things, and we need to think about them more seriously than we have. That's really the point of me writing this piece, not just to say the bombings were immoral, which I think they were; but to say they also were driven by bad policy, and we have to think about that clearly.

Horton: It's got to be mentioned that the American entry into the Pacific War was not altogether on the up and up anyway, if you're familiar with Robert Stinnett's great work in *Day of Deceit* and the rest of that. So this whole conversation presumes that we have got to get one or another kind of surrender here, but that's already assuming a lot to start.

Lippincott: This is something I'm interested in. I haven't read Stinnett's book, but I'd like to. Once you start asking one question, then you start asking others. That's liberal education. That's thinking, to say, "I have this view, I want to see if it's true."

So, Pearl Harbor, why is the fleet at Pearl Harbor? Turns out a lot of those ships had been based out of Coronado, near San Diego, California, in the continental United States, but had been moved in the summer of 1940 to threaten Japan because of their actions in China and then in Indochina, which we know as Vietnam. So if you ask, "Why was Pearl Harbor attacked in December 1941?" In June, the United States had embargoed oil, rubber and, I believe, scrap metal to Japan based on the fact that they had an airbase that they were trying to get in Indochina. They had conducted negotiations with the Vichy French regime and were attempting to base out of Indochina. So because of that, the United States cuts off the Japanese supply, putting the regime under pressure to decide if they can maintain their position in China, which they saw as necessary for their national survival. Or, do they have to seize more territory to get the resources they think they need?

Horton: How do you like that for Chapter One of the Vietnam war? I like it.

Lippincott: I think all of this goes back to how the United States does all these things to restrict the Japanese, to contain them, and then we end up fighting in the same places to contain communism. By the way, when we embargoed the Japanese in the summer of 1941, what were they then not able to do? Fight the Soviet Union on their eastern front.

Horton: You could say that Japan had no right to Manchuria or any of China, we needed them out of there, but we might have negotiated with them. We could have said, "Look, do your best not to leave Mao and the Communists in the best position," but the way that they withdrew, they left all their equipment and everything behind for the Maoists to get, which helped turn the tide in favor of that side in the civil war that followed. Then Mao, internationally not so much, but domestically, was the greatest monster in all of world history in terms of his regime's forced famines and all the rest of that. Containing communism in Indochina meant that America was the one that kicked over the sock puppet royalty in Cambodia and spread communism and the nightmare of Pol Pot and the Khmer Rouge to Cambodia. It wasn't the Chinese and the Vietnamese who did that; it was the Vietnamese who tried to stop them, but then Carter and Reagan ended up taking the Khmer Rouge's side against the Vietnamese.

Anyway, I want to go back to this great thing that you wrote in your piece, this really important note where you say how it was only after the Soviet invasion that Hirohito agreed to a full cabinet meeting specifically to discuss surrender to the U.S. That meeting began three hours later, at 10 a.m., an hour before the atomic bombing of Nagasaki. So it was not just this horrific war crime in Hiroshima which, as you say, was not decisive, but Nagasaki was completely superfluous at the same time, murdering approximately 100,000 people in a moment.

Lippincott: ...While also annihilating the largest Christian community inside Japan at the time. And this shows you how things are really done. Truman did not know about the bombing of Nagasaki. He was not told, and he did not sign off on it. I want to make that clear. The people making that decision were inside the military bureaucracy, inside the Pacific. In fact, Leslie

Groves, the guy who's in charge, did not know when the bomber took off that they were going to hit Nagasaki. They made that decision in the air because of the weather over Kokura, which they were actually planning on bombing.

Horton: In other words, they had devolved the orders so low down the chain of command like this was just another bomb, and we're going to keep dropping them until somebody makes us stop.

Lippincott: The managerial elite are the ones who get to make the decision as to what actually gets hit. You can see the "deep state," the administrative functionaries which extend hither and yon, they're the ones who are driving policy and making these decisions without the consent of the people and without reference to politics. That's not good. You can see what a nasty effect this has on the American regime, and that runs throughout the Cold War.

Horton: The other ugly irony here that you write about with regards to the timing of all of this, is the "Byrnes Note" written by the very hawkish new Secretary of State under Truman. He writes the Byrnes Note that's issued on the 11th. And it says what?

Lippincott: What Byrnes ends up putting in that notice, he says for the first time that the emperor can be allowed to stay on his throne.

Horton: Two days after Nagasaki.

Lippincott: Right. It's only after the bombings that we finally clarify the position of the emperor. That blows my mind, how they couldn't get that the Japanese view the emperor as the embodiment of the regime, regardless of what regime he's actually supporting.

Horton: Was that the date of the actual notification? I know they signed the document later, but that was the day that they notified the Americans that they were surrendering, the 11th?

Lippincott: Yes. It took them a little bit. They had to have a debate, there was some question over the English translation of the orders, but I think August 15th is the day that they surrendered. They agreed to the armistice, and the emperor came out and told the troops to stop fighting, and then September 2nd was the USS *Missouri* signing in terms of the official surrender.

Horton: The common narrative is just that MacArthur decided to let them keep the emperor anyway, but what I'm getting at is that they were told, "You can keep your emperor," before the Japanese had really even notified anybody that they were surrendering.

Lippincott: The note itself says something to the effect of: "The Emperor, subject to the Supreme Commander of the Allied Forces..." The phrasing is such that it implies that the emperor will remain pursuant to the command of the Supreme Allied General, or the post-war ruler of Japan. So there is this sense that, you know, MacArthur was the one who let him stay, and by the way, that may have been MacArthur's prerogative to decide.

Horton: I guess my point really was that they really were notified after the second nuking, "Okay, okay. You can keep your emperor." Then, for all practical purposes, the surrender came after that. Not just after the Russians, but also after the Byrnes Note. Is that correct, or not?

Lippincott: That's correct.

Horton: That's what I really learned here that I did not know before.

Lippincott: Yeah, they were willing to keep fighting even after the bombings.

Horton: And after the Russians crossed the line.

Lippincott: Potentially it was those two events. When they realized they could no longer negotiate, there was a willingness there that wasn't there before, where they wanted to figure out how to surrender and keep the emperor. So when they get the Byrnes Note on August 11th, they say, "Okay."

Horton: And that shows how Truman and Byrnes are desperate, right? "Here come the Russians, we'd better go ahead and give them a concession so that they'll surrender to us instead right now."

Lippincott: Right. Again, talk about a disaster. And the whole history here is worth getting into. On one hand, they're publicly asking the Soviet Union to help them in the war effort, and on the other hand, they very much do not want them to help because already they're seeing the outlines of the Cold War. It was very unclear in the minds of the people who were making policy what they were actually trying to do, and so that bleeds over to where you're saying, "Unconditional surrender. We can do anything to you after you surrender unconditionally. We're going to govern you. We're going to institute your form of government."

Then in the Byrnes Note, they finally give them something where it sounds to them that the emperor will be allowed to stay. That's what gives the impetus to the emperor to say, "Okay, I'm ready to let the war come to an end."

Horton: And like you say, if then, why not last November?

Lippincott: Yeah, there's a great book by Edward Luttwak called *The Grand Strategy of the Byzantine Empire*. I recommend it to anyone who's interested in foreign policy. What he points out is that when Attila the Hun was conquering, when he would invade a country, he would immediately make a peace deal with the people he was conquering. What this did was create a peace faction within his opponents by offering, "If you want this to stop, you can end it right now." It was usually pretty mild in terms of what Attila the Hun was demanding of his opponents. So in this way, Attila the Hun, the barbarian, is more humane and better at international politics than these modern, 20th-century foreign policy supposed experts who are fighting this massive global war. Attila the Hun. That, to me, was striking.

Horton: Part of it is because of the economics of democratic politics, where everybody is such a weak coward that they're afraid of being called a weak coward. So they all act completely crazy.

Lippincott: Part of this too is that you have very small cliques of people who end up taking power, and then they end up driving policy. Look at something like the atomic weapons program. Vice President Truman did not know it existed. You had 20,000 people working on a super-weapon, and Congress didn't know, the American people didn't know, and no one ever voted on it.

That, I think, is what's so striking: that you'll have these small groups of people making these policies that dramatically affect our lives, and they often do things which in hindsight are questionable and worth probing deeper. There are many such cases, such as the history of the CIA or the history of the lead-up to the war in the Pacific, and this can open people's eyes to thinking more clearly about foreign policy and about our own regime. That's the goal here.

Horton: Yeah, for sure. And it's a really great contribution to that. Anything that is speaking with reason on this issue under the name *The American Conservative* magazine is doubly important in value. Who knows about what liberals think nowadays? But if you're going to get an emotional, super-patriotic, how-dare-you-question-this-decision reaction from somebody, that typically tends to be like a talk radio listener and a Republican voter type.

I've seen this in my own life, and I've heard many stories by other people having the same exact experience, that it could be so productive to actually pick a little bit of a fight and then explain that all these generals and all these admirals who were in charge at the time all opposed it. From Eisenhower to Curtis LeMay — for God's sake — to Nimitz and everybody in between. Even MacArthur, and he wanted to drop extra-radioactive cobalt-laced nukes on the border between Korea and China because he was the biggest barbarian in the history of East Asia. Yet when it came to Hiroshima and Nagasaki, he was like, "Why would we do that? We already won the war."

Lippincott: I'll say this in defense of people who, like me, grew up Republican and conservative, the reason we ended up defending the bombings is that it's considered that if I don't do this, then it means I don't believe in my country, and I have somehow betrayed our fundamental principles or betrayed something that I love. I would say that's not true. The

American people were always very hesitant to go to war without really good cause.

Horton: It's almost like because it's all black-and-white pictures of Truman that just sort of makes it so historical that it's beyond question. But if you frame it like this — where Harry Truman was nothing but Bill Clinton — if Bill Clinton dropped an atomic bomb on someone, you wouldn't say, "Hey! That's my country that did that! You'd better shut up, dude!" You would say, "That's Bill Clinton, of course, he's the worst person in the world, and that's just one more proof."

Lippincott: I think the way is to make it clear to people that America can win wars, we can defend our rights, and that does not necessarily require us spending hundreds of billions of dollars. Then we can think about this. America is blessed in so many ways, we have lots of natural resources, we have oceans separating us from potential enemies, we have the ability and technology to establish border controls to keep ourselves safe without needing to be involved everywhere. That was what really changed my mind on some of these things, by asking: "What are we really attempting to do overseas? What were we attempting to do historically? How do we go from here? How do we make policy?"

I think the reason to do some of these historical writings and reading is to try and think about these things so that we can make better policies for our kids and for their children. I want America to be safe, but I also want America not to sacrifice the virtue and the best of our young men for bad policy made by people who made obvious errors. And we should at least acknowledge them as such and say, "That was wrong. Don't do that again."

Horton: The point here too is that if you're questioning the decision to nuke Hiroshima and Nagasaki and the end of World War II in the Pacific, then anything is fair game after that. That is what is really dangerous. Because George Washington and Abraham Lincoln are dead, that was too long ago. Really, it's FDR and Truman who are the Founding Fathers of the United States, of our modern American empire, and you're sitting here saying that they lied and tried to deny it, and you're challenging the very kind of basis of our civic religion at this point. FDR made the bomb, Truman used it, and it was great. That's how you know it was great, because they were the ones who did it. So if you could question that glorious decision that saved a million lives, then what does that mean for Korea and Vietnam and Iraq and their legitimacy today?

Lippincott: I think the most important thing Americans can take away is to walk away, being able to ask questions about policies made in their name, that they then have to pay for. I'm a big fan of the American people. I like them. I am an American citizen. Like I said before, I want my country to be safe, but then I have to ask, "What is the value of the things we're doing, and where do we get these ideas from?" When you look at someone like FDR, I think you really begin to see that that was a re-founding of the American regime. What he wanted to do with American power was different from what someone like George Washington thought we should use our power and resources for. I mean, George Washington said we want to be friendly to all, right? If we have to fight wars, then we need to win them, and then we need to go back to having peace and securing our rights, and I'm all about that. But there needs to be more of a discussion on this topic as we go forward, and that's the goal. That's the hope.

Horton: Well, this is a great contribution to it. I really appreciate your time on the show to talk about it, too.

Lippincott: I'm very grateful. Thank you for having me on.

Part 5

Protesting Omnicide

"While the bow and arrow and even the rifle can be pinpointed, if the will be there, against actual criminals, modern nuclear weapons cannot. … These weapons are ipso facto engines of indiscriminate mass destruction. … We must, therefore, conclude that the use of nuclear or similar weapons, or the threat thereof, is a sin and a crime against humanity for which there can be no justification. …

"Therefore, their very existence must be condemned, and nuclear disarmament becomes a good to be pursued for its own sake. And if we will indeed use our strategic intelligence, we will see that such disarmament is not only a good, but the highest political good that we can pursue in the modern world."
— Murray N. Rothbard, 1963

Paul Kawika Martin: The New Treaty to Ban Nuclear Weapons March 29, 2017

Scott Horton: Alright, introducing Paul Kawika Martin. He is the political and communications director for Peace Action. That's PeaceAction.org. The subject here today is this big United Nations meeting on the abolition of nuclear weapons. So the big question for you, Paul, is: What all did and did not happen at this meeting?

Paul Kawika Martin: Yeah, the conference is actually still going on, so we'll have to see exactly what pans out. It's the first part of negotiations in a convention to ban nuclear weapons. This is one week long. They start the negotiation process, and then there will probably be talks behind the scenes — working groups and whatnot. Then they come back again in June and actually try to finalize some sort of a treaty.

Now, realistically, is this going to happen? Probably not. December is when 113 countries voted to have these negotiations. As you would imagine, most of the nuclear weapons countries voted against it, but with some exceptions: China, India and Pakistan actually abstained from the vote. The U.S., of course, has been driving countries now to be against this negotiation process. So far there's been, I think, about 40 countries that are boycotting these negotiations, which is very sad.

Some of the listeners may not remember, but at one point in the world we had tens of thousands of nuclear weapons, and those weapons were 10 to 100 times more powerful than the ones that were used in Hiroshima and Nagasaki. As a reminder, those killed several hundred thousand people instantaneously, and several hundred thousand people more over the lifetime of those victims. So it's important that we try to get to a world without them. We don't need a world with nuclear weapons. This was one step in trying to negotiate that.

So that's what's currently happening in the UN. We'll have to see how this week plays out and continue to push on countries through the whole process, which should end sometime in June.

Horton: Alright. Well, lots of stuff to go over here, but first of all, China, India and Pakistan abstained. Does that mean that they're sending us a signal that, "Hey, if the other major powers were willing to go ahead and forgo nukes, then we would be for that?"

Martin: I think in some ways, yes. I mean, let's take a look at China. The best intelligence says that they probably have about 200 nuclear warheads. Compare that to the U.S. We have approximately 1,500, give or take, that are strategically deployed — that means on missiles, bombers and submarines — and another approximately 3,000 in reserve.

Again, these are not the same weapons used in Hiroshima and Nagasaki. These are 10 to 100 times more powerful. If they're ever used, you can just not even imagine the devastation they would have on any major city and how that would affect the world. It's also clear that if a small nuclear war were to happen, let's say, between India and Pakistan, it's very possible that it would cause what's known as nuclear winter, which is putting up enough dust in the air that it would significantly change the climate of the Earth and how we live on it.

So there are countries, we think, that are against it, like China, who would be probably for moving towards not having nuclear weapons. They only have 200, and some intelligence says that their warheads are not even on missiles right now. So that means they wouldn't even be able to launch them immediately, to respond to any kind of attack. They think nuclear weapons are just kind of an insane thing to think about using.

India and Pakistan is a more complicated issue. It's unclear exactly why they abstained. I think that both countries realize that any kind of nuclear exchange between the two would affect both countries. Even if a small nuclear weapon were to be used on Pakistan, you would have radiation that would come over into India, and vice versa. So they hopefully are thinking about détente and that it is certainly not a wise thing to have these weapons around, since they may fall into the wrong hands.

It is a huge concern, especially the command and control of these weapons that could change the face of the planet. It's one of the concerns we have here in the United States. There is not really the "red button," but the president does have access to the nuclear "football" and can launch nuclear weapons at any time without necessarily obtaining congressional approval, or anyone else's. So these are concerns not only here in the U.S., because you might have some of the same dynamics in Pakistan and India.

Horton: So what's the deal with the Non-Proliferation Treaty? Because the non-nuclear weapons states who sign it promise never to get nuclear weapons, by hook or by crook, and to sign a Safeguards Agreement with the IAEA to allow them to verify that they're not participating in any non-peaceful uses of nuclear energy.

But then the nuclear weapons states have already agreed — this excludes India, Pakistan and Israel — but the rest of them are members of the NPT, and they have all sworn to get rid of their nuclear weapons. But is that just written in there as the vaguest of promises, as sort of a fig leaf to get the non-nuclear weapon states to promise to stay that way? Do we need a new treaty that's much more specific in saying that the U.S., U.K., France, China and Russia have to give up this amount of nukes by this timeframe and that kind of thing?

Martin: Clearly we do. And as you mentioned, you exactly laid out basically what the NPT did, when it entered into force in 1970, so many, many years ago. Article Six of that treaty clearly stated that countries were supposed to move deliberately, or whatever the wording is, towards getting rid of nuclear weapons. The interesting language there is where it says "and general disarmament," which some people say actually means slowly getting rid of conventional weapons as well. So the nuclear weapons states argue that they have been doing that. I mean, look, we did have 70,000 nuclear warheads, and I think now we're down to about 14,000, or something like that, combined, in the world.

Horton: Yeah, I like how you keep emphasizing that. I think that's the important thing, because on the one hand, it sounds like some hippie, utopian thing. Where people say, "Yeah, Jane Fonda thinks we shouldn't have nukes," or some crap that's so easy to write off. But then, "Oh, yeah? How come the Republicans reduced them by the tens of thousands?"

Martin: Yeah. Let's not forget that it was Reagan who finally decided that this was what we needed to do. It was Reagan who built a lot of this stuff up and then finally realized what had to be done. History will have to show exactly why, but the best guess is actually because of his daughter Patti, who was bringing in people like Helen Caldicott and other folks to have conversations with him, that he finally kind of realized that.

Horton: Hey, I didn't know about that part of it.

Martin: This led to, "Surprise, diplomacy does work," in the discussions with Russia. The Cold War ended when we started signing treaties with Russia on nuclear weapons. That's how we went from the 70,000 when we started these great treaties, and moved ourselves away from the brink of world war and complete disaster to where we are today.

Horton: Yeah. Well, and Bush, Sr. too, did his part and pulled them all out of Korea. He continued the reduction by the tens of thousands before Clinton ever came in. In fact, it may have all mostly already been done before Clinton ever got there, if I remember it right.

Martin: Yeah, and this is completely a nonpartisan issue.

Horton: But it's a great way to argue it, right? Because the frame of reference is so often that, "Oh, yeah, hippies are against nukes." But then if you can outflank the right from the right and say, "Look at all the greatest Republican heroes who say that this is absolutely unnecessary, that it threatens our civilization, and that it costs too much money; that it's this dangerous for all these reasons," then that confuses the usual narrative that this is kind of an issue to dismiss, and makes it all of a sudden a very serious one. You know what I mean? Citing military officers and that kind of thing.

Martin: Yeah. You're exactly right, Scott, and even if you want to believe in the idea of deterrence, which I don't, but if you want to believe in this idea that you need to have nuclear weapons to deter another country or actor from attacking you, then you should remember that under the Obama administration, the Department of Defense said, "Oh, we think we only need a thousand at the most for that."

There have been even air force generals who came up with a way to have deterrence with only 300. We have 1,500, plus 3,000 in reserve. It costs billions of dollars a year to maintain them, to secure them, to keep them going. There's currently a plan, actually, to escalate our nuclear weapons program to the tune of about a trillion dollars over 30 years, that would be replacing all these warheads with newer ones, replacing —

Horton: All new factories, right?

Martin: Oh, yeah.

Horton: A whole new infrastructure for the industry, everything.

Martin: Well, yeah, it's not necessarily completely all new, but just a lot of money. They want new ICBMs, new submarines, new bombers, replacing the current arsenal. A trillion dollars, which is hard really even to comprehend that amount of money. Like you said, even the most conservative of conservatives realize that we can't afford this. It's at the cost of other security needs, whatever that might mean to you. Maybe you think we need more tanks, or we need more planes; we're not going to be able to afford those if we're going to spend this kind of money on nuclear weapons, which in my opinion actually make us less safe rather than more safe.

Horton: Well, that's the thing of it. If you stop halfway through the thought, then nuclear deterrent makes a lot of sense: "Nobody's ever going to mess with us. We've got H-bombs."

Okay, but then what if they do anyway, and you've got to follow through now? Now you're talking about using H-bombs. I mean, this is crazy. Here's the way I think of it: Are you saying that for the next 700 years humanity ought to have this equal balance of hydrogen bomb, fusion terror, so that nobody had better mess with anybody ever? Does it make sense to think that this is a long-term solution to the problem of atomic power? Or are we going to have to come up with something else?

You know what I mean? In other words, "unthinkable" doesn't mean "impossible," as much as we would like it to mean "impossible." Look at the world wars. That was all unthinkable, yet it happened anyway.

Martin: Yes, all kinds of things have been unthinkable — Three Mile Island, for instance. You can just go down the list of things that we could never imagine would happen, but which have happened or may happen.

So it's certainly our responsibility to attempt to mitigate these risks. That's why this convention was brought up, because most of the countries who don't have nuclear weapons are opposed to the world having nuclear weapons. It's almost this kind of weapon classism, where you have the nuclear haves and the nuclear have-nots; and the nuclear haves want to keep their powerful weapons as a way of putting pressure on other countries with their military hegemony. Then you have the have-nots, who either are trying to oppose, or are submissive to these other countries and try to get them to keep them under their so-called "nuclear umbrella."

This is Japan and other countries who we are supposedly protecting with our nuclear umbrella. At the end of the day, these geopolitics just don't work, and it needs to be something else. That's why this convention is moving forward. Whether one happens or not, it'll be clear at least to nuclear weapons states that countries are getting more and more serious about taking their NPT obligations more seriously, demonstrating that they're moving towards a world without nuclear weapons.

Horton: You know, Paul, I think maybe somebody needs to do a remake of *The Day After* or something like that. Because you know what we have here? I know you know this a lot better than me, but you have this Global Zero campaign, and there are a lot of people, usually old guys who are retired now, who have a lot of experience with this issue — people like William Perry, Bill Clinton's former secretary of defense — who really know. Dan Ellsberg is another great example. People who really know about nukes and what they really mean, who don't have to resort to vague imaginations of nuclear detonations the way most Americans do, who are saying that something has got to be done about this.

Right at the start, as you're saying, of this brand new trillion-dollar project — I know all the faux controversy over Putin and Trump right now could poison this a little bit — but just seems like a perfect time to go ahead and say, "Actually, no, instead of doing the trillion-dollar thing, we are going to go the other way. We're going to go ahead and get all the way down to 300," or whatever, something more reasonable — as though that's very reasonable at all — than the stockpile that we have now.

It seems like we need something to really galvanize people's attention. But we don't want the smoking gun to be a mushroom cloud, Paul. We want to get people's attention some other way. I guess an ABC prime-time TV movie wouldn't do it in the year 2017, but something.

Martin: Yeah, it's interesting. It's really the fault of the educational system here in the United States. I think there's a couple of things happening. First, as you mentioned, a lot of the old people who actually remember what it's like in the duck-and-cover times, where people really did fear for their lives... You know, people were building so-called underground bunkers thinking that would protect them, even though they probably wouldn't have in a complete nuclear war. People were really fearful for their lives at every moment that there was going to be a nuclear war. That generation is starting to die off, unfortunately, so that memory is being lost. Young people, if you ask them, think that maybe we have a couple hundred nuclear weapons, and they also don't even understand the impact, as you mentioned. So the educational aspect of this is very important.

That's part of the reason that this convention is important, because it is getting the attention like you do, but you talk about these issues all the time. You've been an excellent person on foreign policy for a number of years, but there are a lot of people who don't talk about this issue. So when this issue comes to the UN, it generates stories about it. Hopefully we can educate people about these risks and costs, so they can think twice about whether our country should have so many and should spend so much money on them, or whether there are better ways to spend our tax dollars and better ways to protect the planet from possible annihilation.

Horton: Yeah. Alright. Well, make some suggestions. How do listeners get involved in this issue?

Martin: PeaceAction.org is the best way. We have 100 chapters of affiliates around the country, so you can look there to see if you have a chapter to join. You can get on our email list. We send an email about once a week about various issues where you can take action. That's a good place to start. PeaceAction.org.

Horton: Alright, good deal. Thank you very much, Paul. I appreciate it.

Martin: Thanks for having me, Scott.

The Treaty on the Prohibition of Nuclear Weapons entered into force on October 26, 2020. The United States of America is not a signatory.

Frida Berrigan: A Childhood Ruined by Nuclear Weapons April 26, 2019

Scott Horton: Alright, you guys. Introducing Frida Berrigan. I'm very happy to have you back on the show. How are you?

Frida Berrigan: I'm doing well. How are you doing?

Horton: I'm doing great. Listen, this is such a great piece that you wrote here at WagingNonviolence.org. It's called "Nuclear Weapons Ruined My Life, and I Wouldn't Have It Any Other Way," and that is because you are quite famously the daughter of Phil Berrigan and Liz McAlister, a priest and a nun. Is that right?

Berrigan: Yeah, Liz McAlister was a nun at the Religious of the Sacred Heart of Mary and was up at Marymount on the Hudson River in New York, and Phil Berrigan was a Josephite priest. He was originally stationed down in New Orleans, and they eventually met at a funeral of a Catholic Worker, a mutual friend, in the midst of the Vietnam War. They were both involved in resistance to that war. One thing led to another, and here we are.

Horton: Well, good. I'm glad you exist. So these two have quite a career, and this is no hyperbole when you say this ruined your life. Your family was broken into pieces like a car wreck; only it was all over nuclear weapons.

Berrigan: Right. So I was born in 1974, and I have a younger brother who is a year younger. We have a sister who was born in 1981. My parents estimate that they spent 11 years of their marriage separated by prison, which means that all three of us grew up with either our father or our mother missing pretty much every major milestone you could think of, because one of them was in jail or prison for acts of anti-nuclear, disarmament, or different acts of resistance throughout our lives.

Part of my reason for writing this piece and sort of reflecting on our upbringing in that way is that my mom is now 79 years old. My father died in 2002 at the age of 79, but now my mom is 79, and she is in a county jail in Brunswick, Georgia. She's been there for more than a year. Now, instead of being the child who is making her way in the world with either parent in jail, I am sort of explaining this experience to my own children, trying to explain why Grandma is in jail, why nuclear weapons exist, why police officers kill black motorists, and just trying to explain our whole messed-up, crazy world to a five-year-old, a six-year-old and a twelve-year-old.

All of that makes me think a lot about how our crazy, messed-up world was explained to me forty years ago — this pretty brutal education I got in U.S. foreign policy, U.S. imperialism and U.S. racism from these two peace activists and all of the friends and fellow travelers that they attracted along the way.

Horton: Well, it's a really great account, and I like the way you're using, as a narrative device, this history of the hardship of your upbringing, being completely frank about it like, "Yeah, it did hurt not to have my dad around." But you're using that to tell the story of: Where was he? "Well, he was in the pen." And what was he in there for? "He was in there for spilling blood on a nuclear missile." So go back and tell us some of these stories, and what this is really about?

Berrigan: I start the piece with being born in 1974, and then my first real awakening to this issue right around my fifth birthday, when the partial meltdown at Three Mile Island happened, which was about 90 miles from where we were growing up in Baltimore, Maryland. There was a lot of concern about radiation and contamination throughout the mid-Atlantic region, and my parents took my brother and me to West Virginia. That was the beginning of my young education into the havoc that's been wreaked on our land and on everything by nuclear weapons and nuclear power.

I talk in the piece about coming back to a changed diet. We were forced to drink miso every morning. My mom had read that workers in Hiroshima and Nagasaki who drank miso suffered fewer of the ill effects of radiation exposure by drinking this fermented soybean paste, and it was really awful. I kind of like it now as an adult, but as a child, we just held our noses and drank the salty, brown, icky stuff every morning because our parents were so concerned about us being exposed to radiation.

For many parents, that would have been enough, right? But just a couple of years later, in September 1980, my father and seven other people gained access to this nuclear weapons production plant in King of Prussia, Pennsylvania. They had cased it out, and they prayed over it, and in the course of planning, conspiring, thinking and learning about one another — they're all Catholic, and this call from the Hebrew scriptures to turn "their swords into plowshares and their spears into pruning hooks" resonated with the whole group. It became this metaphor, and actually a lot more than a metaphor, as they went with household hammers into this production facility, found Mark-12A nose cones, poured blood on them, hammered on them with their little hammers, and then stayed and took responsibility for what they had done.

They then had this incredible trial, where a jury of all very staunch Republicans and hard-nosed, right-wing people had to grapple for the first time in their lives with nuclear weapons and with this power that at the time was just held by a handful of nations — the power to destroy the world. At the time, both the United States and the Soviet Union were not only building nuclear weapons and testing nuclear weapons but were articulating a first-strike posture towards the other and thinking that not only could we fight a nuclear war, but we could win a nuclear war. The United States thought they could win, the Soviet Union thought they could win, and there was this heated arms race with our Cold War superpower rival.

Into this big conflagration step these eight very simple people, and then this movement took off. Since that time, there have been more than 100 Plowshares actions, all over the world, mostly in the United States but throughout Russia, throughout Europe, in New Zealand, in Australia and in a handful of other places where citizens take personal responsibility for what's being done in their name, with their tax dollars and in the name of their security and safety.

All of this is not only bankrupting our nation and poisoning our land and water, but it is holding the whole human family hostage. Some of us are particularly afraid of Donald Trump having his finger on the nuclear button, but every president since 1945 has held all of our future in his hands, and it's a terrifying prospect.

Horton: Dan Ellsberg shows in his book *The Doomsday Machine* how there are thousands and thousands of men in all of these countries who have the ability to launch these weapons. You don't need the president to open his football and type a special secret presidential code; they have fail-safes for that. What if somebody killed the president and he's not available, Colonel Hapablap has to have the ability to launch a cruise missile — and it's already there. That whole *Dr. Strangelove* scenario is exactly right, and Ellsberg talks about how when that movie came out, he and his friends at the RAND Corporation went and saw it, and when they left, they were kind of laughing and saying, "You know, it's not funny that that really is accurate about the way things are set up here."

Anyway, so now I have to tell you, through my whole childhood in the 1980s and my teenage years in the 1990s, I've always known that there are these Catholic priests and nuns who go and get themselves arrested breaking into nuclear weapons facilities. I never knew anything about it because they never did a prime-time live TV show where they explain the whole thing or anything like that, where they give you guys a fair hearing, but there's always a headline about it here or there.

It's really powerful, because it's not some hippie in a tie-dyed shirt that does it; it's a Catholic priest and a Catholic nun, and they break into a military base, take a hammer, and they start banging on a missile? Wow! That's an interesting thing, and that's certainly to their credit, and it certainly brings an inescapable narrative of who these people are and why they're doing it; it's that they're religious people, very serious ones, and they're doing it for the most serious reasons. This is not vandalism; these are people who are looking straight at the New Testament, and when they believe it, they really believe it, and they're trying to hold their current society to those kinds of standards. That's the kind of thing that you don't have to be a Christian to see, that these are serious people of serious faith sacrificing themselves for what they think is not just right, but the most important thing.

Berrigan: Right, sometimes it really resonates. As I said, there have been more than 100 of these actions, and some of them have kind of bubbled up into popular media. There was a Plowshares action in 2012 that was organized by Sister Megan Rice, who was a nun in her eighties, and Catholic Workers named Greg Boertje and Mike Walli, who were both in their mid-sixties. They gained access to the Fort Knox of uranium in Tennessee. There were Congressional hearings about it, and it was front-page news. It was very embarrassing to the National Nuclear Security Administration that these amateurs, these random people, were able to gain access.

The real question of why we have a Fort Knox of uranium and why we're continuing with the production and continued improvement of nuclear weapons was really the issue that Sister Megan, Greg Boertje, and Mike Walli wanted to draw attention to. But you had this odd spectacle — which is documented in a movie called *The Nuns, the Priests, and the Bombs* that was done by a filmmaker named Helen Young a couple years back — of members of Congress thanking Sister Megan for showing the weakness of our security and telling her, "Thank God you weren't a terrorist." That was sort of laughable.

One of the powerful things about Plowshares' witness is this question of what is proper about these nuclear weapons? What makes it okay for them to exist? International law says that they shouldn't exist, religious law says that they have no right to exist; yet here they are, enshrined and protected behind all of these fences and this huge security apparatus. Meanwhile, so many other of our institutions are so threadbare in this country. So these Plowshares actions really get at the question, or hope to get at the question, of what we are doing with these weapons. Why do we have them?

Given the opportunity, the activists do the action and then do the time, and are really able to speak a very powerful truth before judges, juries and prosecutors, and aren't looking to get off. They aren't looking for leniency but are really looking for truth and conversion and transformation not only of nuclear weapons — which is the metaphor out of the Hebrew scriptures, to turn a sword into a plowshare and a spear into a pruning hook — but the transformation of hearts and minds as well. Where people who never even considered that we have thousands of nuclear weapons on hair-trigger alert and that we spend $90 billion a year on these weapons of mass destruction, have never even thought about that before, are confronted with people who are willing to spend the rest of their lives in prison to get them to think about it. That is what my mother is willing to do. She is willing to spend the rest of her life in jail so that you and I can have this conversation and so that the average Jane and Joe can really think about this.

Horton: Alright, just a little bit of internal education for my movement here. The great libertarian sage Murray Rothbard said that, according to our Non-Aggression Principle, all nuclear weapons must be banned forever for the reason that they are impossible to use with discrimination. They're only a weapon of indiscriminate killing and therefore cannot be used in any way that could possibly be justified, even in so-called defense. To kill civilians en masse in another country just because their government had even nuked your cities still makes no moral sense whatsoever; it cannot be tolerated whatsoever, and there's no question about that. I like to show people that because it doesn't necessarily have to be any kind of religious principle at all; it just has to be a recognition of what Sheldon Richman calls the Non-Aggression *Obligation*, that what we owe to each other as social animals is that we don't attack each other. Then you extrapolate out from there.

Anyway, so what you're saying about the accepted permanence. Everybody knows that we have nukes, but everybody forgot about it. Or, like you say, they never really had to stop and think about it for a minute. But, as you say, there are thousands and thousands of these things, enough where if a general nuclear war broke out under whatever scenario, you're talking about losing a couple of billion people right off the bat. Billions!

Berrigan: That's right.

Horton: So yeah, there's something to make people kind of recognize: Did we ever resolve that thing with the nukes? Maybe we never did? We're friends with the Russians now, aren't we? Or maybe we're not again.

Berrigan: Yeah, Putin's over there shaking hands with the Chinese prime minister, so maybe we're not so hot on him right now. But right, all of this is so mercurial, which is not how you want to think about the end of the world.

One device that I used throughout this piece is the Doomsday Clock. My brother and I joke a lot with our sister about how our social life as children was tied to the Doomsday Clock, which is a device that the *Bulletin of the Atomic Scientists* developed in 1954 to graphically demonstrate the nuclear threat and how close we were to nuclear midnight, the Doomsday scenario — billions of people dying in an instant, in a heartbeat. When I was born, the Doomsday Clock was at 12 minutes to nuclear midnight, and throughout my life, except for one point in the early 1990s, after the end of the Cold War, it has been moving closer and closer to the point where we're now at about two minutes to nuclear midnight according to that Doomsday Clock.

Our father would always bellow at us when we would ask, "Dad, can we go to the movies?" He would be like, "You want to go to the movies? We're four minutes to nuclear midnight, and you want to go to the movies?" And we'd say, "I don't know if going to see *Jumping Jack Flash* is going to tick the clock any closer to nuclear midnight, Dad." But it really just kind of tick-tocked all the way through my upbringing, and it's sobering as the daughter of these two people who devoted their whole life to nuclear disarmament. We haven't had a nuclear conflagration, and we're obviously grateful for that, but the danger is still here, it's even more present, it's even more existential and palpable, and yet it really is off the radar screen for the majority of Americans.

There was a made-for-TV movie that was shown in November 1983 called *The Day After*, and 100 million people sat down together all across the country to watch that movie. It was a little over-the-top in some ways, but it told the story of a nuclear attack on the United States — really a nuclear tit-for-tat between the United States and the Soviet Union. It focused on a town in Indiana and the effects of this nuclear attack on the people of this community. One of the striking things about it was, who are the lucky ones? Are they the people who died right away from the fireball and the explosion? Or are they the people who survived but then had to deal with the impact to the infrastructure, the nuclear radiation, the blocking of the Sun, the anarchy that followed the attack, and the uncertainty, the instability of the human interactions that we take for granted? We watched this made-for-TV movie with our parents, as a nine-year-old and an eight-year-old — our little sister was spared this — but it was this indelible moment, and it really drove home this stark message that a nuclear war is not survivable. And even if you're lucky enough to survive, it's not the world you'd want to live in. There was a trope at the time that the scariest part of the movie was your children asking you afterwards what you were going to do about it.

Ronald Reagan, President of the United States at the time, watched the movie, and he wrote in his diary that it depressed him deeply to watch this film. When you think about his public persona as this Cold Warrior, it's kind of laughable.

Horton: They say that really helped kick-start a change of heart, that his brinksmanship policy ended there, and the beginning of his *détente* and willingness to negotiate in good faith with Gorbachev really began right there. That he thought, "Man, I cannot let them ever nuke St. Louis," or whatever it was in that movie, that it gave him nightmares. Which I guess is plausible because I was an elementary school kid at the time, and we did all watch it. That same experience you described where everybody watched it, and that's what it really would look like if an H-bomb went off over your city, everything bleached white by the heat afterward, all the rubble, Steve Gutenberg's face falling off — and poor John Lithgow, they cut the movie short before he got to finish his arc.

Berrigan: Everyone watching the final season of *Game of Thrones* is not comparable, right? It is not asking us to do anything with our own lives, and it's not asking us to demand anything from our leaders. Our culture and our media have changed a lot since then, but what hasn't changed is the amount of resources we devote to nuclear weapons as a nation and the way in which those weapons still operate within U.S. foreign policy and economic policy. They are the fist inside of the glove, although we don't have much of a glove these days.

Horton: The felt's worn thin, that's for sure. Mr. Burns on *The Simpsons* complains of the expense of these do-nothing nuclear missiles, which is funny because we would rather not get our money's worth if it's all the same, you know? But it is a huge expense. But that's the thing, it seems like we had this opportunity at the end of the Reagan and Bush era, and instead, Bill Clinton was elected. In fact, Reagan and Bush had done far more to dismantle the nuclear weapons than Clinton ever did. Bush, while he was the lamest of lame ducks in January 1993, when the USSR was gone and it was just Russia, he went over there to make one last deal to get rid of another few thousand nukes. It's funny to think that probably the most heroic thing anybody ever did is the dismantlement that H. W. Bush ordered. But they didn't complete it. There was no mandate by the American people saying, "Hey, remember that movie from 1983 when we decided how we really want to get rid of these H-bombs?" But there was no consensus, and nobody cared enough. There's no organized interest group against it, but there are plenty of organized interest groups for keeping them in the military industry and in the manufacturing industry.

Berrigan: So then you have President Obama standing before a throng of thousands in Prague in 2009, saying that we're going to rid the world of nuclear weapons and how he had this vision of a nuclear weapons-free future. But that steady work of these arms control treaties and this kind of latticework tapestry that was moving us quite slowly but surely towards disarmament really didn't advance under Obama. Obviously now it's not advancing under Donald Trump.

Horton: It's funny, with or without Russiagate, the best move for him would have been to invite Putin straight to D.C., take him to a rock concert and out to dinner, sign a new nuclear arms reduction pact and then dare the Democrats and demand that the Democratic Party leadership explain to their constituents why they're against this new nuclear arms reduction pact. But this goofball, he didn't know anything about the art of dealing anything — some ghostwriter wrote that book — and so he let them persecute him with this fake narrative the whole time. He could have turned the tables on them right from the very beginning and just said, "You know what? I'm inviting Russia into NATO, how do you like that?" I mean, he really did have those kinds of instincts, but there's no way he could pursue that now, even if he really wanted to, and I guess he didn't ever really want to.

Anyway, so let me ask you this: What do you know about Global Zero and all that? We see Henry Kissinger and George Shultz and William Perry and these others saying, "Oh yeah, you know, Berrigan's right. We've got to get rid of these things," but they don't ever seem to have much effect even though they're the greyest of greybeard, foreign policy, centrist, expert CFR types up there.

Berrigan: That was an interesting development. One way of looking at all of that is that they were comfortable with the Cold War parity, with that Mutually Assured Destruction paradigm and the rationality that they thought undergirded that. But then when nuclear weapons democratized, with India and Pakistan having their own tit-for-tat, when other countries got into the mix as well, when Iran started developing their nuclear program, when North Korea starts developing nuclear weapons, when the plans for dirty bombs are available on the internet, all of a sudden, they kind of have stepped in and said, "Whoa, whoa, whoa. Wait. We do not want a world where every country has nuclear weapons, and yet we want to decouple nuclear weapons from being a global power, a global influencer and a fully formed state."

Right now, the five permanent members of the Security Council are all nuclear weapons states, and so there is this aspiration where, if you want to be a real country, you have to have nuclear weapons. That was the model that came out of the Cold War. So now we have these Four Statesmen of the Apocalypse stepping in and saying what we have to change. There's something very colonial about that, there's something very paternalistic about it, but this effort really begins and ends with the United States and Russia, who together have more than 90 percent of the world's nuclear weapons. We can get all bent out of shape because Iran has a nuclear program, or Pakistan has a handful of nuclear weapons, or China is modernizing their modest nuclear weapons arsenal, but we as the United States can't really say, "You disarm first," when we are stepping so hard on the scale of the balance of power and when we still have a first-strike policy in our Nuclear Posture Review. So the United States and Russia really have to set the pace, and other countries are waiting for us to do that. I think it was a good moment when you had somebody like George Shultz saying that he was wrong, or Henry Kissinger sort of reconsidering the past, but the impetus really is still on the United States and Russia to disarm first.

Horton: And especially just to back off, too. Because after Iraq and Libya, Korea would be crazy to give up their nukes, and Iran would be crazy to give up their mastery of the fuel cycle and their ability to enrich weapons-grade uranium. Not that Iran ever has moved toward making nukes, but they have a civilian program that has essentially a latent nuclear weapons capability, and they'd be insane to give that up, with the precedent of how we treat their neighbors.

The thing is — just taking the exact same thing that you said but back the other direction — if America really took the lead on this, if you just had *Mr. Smith Goes to Washington* does the right thing, decides to lead the world toward nuclear disarmament and says, "Yes, exactly; beginning with us and Russia first, because we have the most. Our allies in Europe, in Israel and our good friends in India and Pakistan, we expect them to disarm, too. We're doing this. Everybody, come on." If America really did that themselves and insisted that they could do it, that would be the end of that. That's it.

Berrigan: Yeah. And then we really would have a peace dividend that would repair our roads and bridges, take care of our schools that are falling in on the heads of our children, and the whole country would look really different.

Horton: Right. That shouldn't be magical thinking. That should be, of course, the consensus. How in the world are we sitting here, tolerating the existence of these things for one more second? That's the real question. The whole burden of the argument is on them.

Let's wrap up with this then. Tell me, what was it that your mom did this time to get her locked up in Georgia?

Berrigan: So on April 4, 2018, she and six other Catholic peace activists gained entry to a naval base in Georgia, the Kings Bay Trident submarine base, which is home to a fleet of six Trident submarines, each of which can carry 250 Hiroshimas in its belly deep below the ocean blue. They went deep into the heart of the base, they spread banners, they poured blood on the entrances to three different facilities there within the base and were arrested. They came with signs that quoted Dr. Martin Luther King. It was the 50th anniversary of his assassination. Martin Luther King had said, "The ultimate logic of racism is genocide." They were so bold as to update that thinking with a banner that said, "The ultimate logic of Trident is omnicide" — the killing of all Creation.

My mom, a Jesuit priest named Steve Kelly and a Catholic Worker activist from New Haven named Mark Colville, the three of them have been in jail ever since. Their four co-defendants were released on bond, are wearing ankle monitors and have been out since late last summer. They're still awaiting trial. There's no trial date set yet, despite the fact that it's been more than a year. They've launched this novel defense for themselves using the Religious Freedom Restoration Act, which is a piece of legislation that was used by Hobby Lobby and by the homophobic cake baker, so they're trying to frame their opposition to nuclear weapons, their need to respond to the burning fire, the crisis, the terror of nuclear weapons by trespassing onto the base, as religious speech and a religiously motivated action.

As far as I know, the Religious Freedom Restoration Act hasn't been used as a defense in a criminal case before — those were civil cases that were previously brought — so things are moving very, very slowly in their case. People can learn more at KingsBayPlowshares7.org. They have a really great website with a lot of information, their action statement and all their motions related to the Religious Freedom Restoration Act there, too.

We're hoping for a trial by the end of the summer, a trial by jury, and an opportunity for them to tell their story, to share their motivation, their intention and their vision with the people and with the jury as they speak to why they carried this action out.

Horton: Well, I don't know, it just seems like a great opportunity for a bunch of well-meaning hypesters to make a cause célèbre out of this thing. What a great bunch of poster children, these elderly religious figures who are obviously willing, essentially, to risk dying in prison in order to try to make this point. I know that defending the CIA and FBI and the military is a big popular thing among liberals these days, but there are still lots of good leftists out there who aren't falling for this new Cold War with Russia stuff, and these people are doing the work. It's up to the rest of us to draw some attention to what they're doing, to show how important it is and how brave that is. Seriously, a little lady, 79 years old, sitting in jail, awaiting a trial that might not come? This is crazy! It's a great opportunity for something on Twitter to get retweeted. Somebody do a TV show about it. Somebody put a thing on Netflix about it. You never know who's listening.

Anyway, you say there is this movie, *The Nuns, the Priests, and the Bombs.* Where can people watch that?

Berrigan: At NunsPriestsBombsTheFilm.com. The filmmaker Helen Young is really trying to get it out there, traveling with it, speaking about it, etc. It's a new film. It tells the story of two of these Plowshares' actions and speaks to a lot of people about the history of this kind of resistance to nuclear weapons.

Horton: Cool. It is absolutely brave and heroic work, and for the very best of reasons, it really deserves all the attention it can get. So that's great, I'll see if I can get that director on the show then, too.

Before I let you go, did you want to say a word about your uncle Daniel? He just died a couple years ago, right?

Berrigan: Yes, that's right. He died in 2016, just short of his 95th birthday. He was part of the first Plowshares action in 1980 with my dad and six other friends. He was a Jesuit priest, and he really brought the art and the poetry and their precision of thinking to this work and to this question of what kind of property deserves to exist. What is proper is that which gives life and protects life, and nuclear weapons do none of that. People keep turning back, as they're inspired to carry out this kind of witness and this kind of action.

They keep returning to that question which really kind of began with Catonsville and the draft board raids in the late 1960s, which caught fire throughout the anti-Vietnam War movement, and this is maybe an evolution of that action. That great statement, "We're burning paper instead of burning children," that was articulated by the Catonsville action. This is just a new facet of that same work.

Horton: That's great. I'll throw this in parenthetically that it was great when I read in here that you used to work for Bill Hartung back in 2000. I was going to make some kind of joke about "No wonder both you guys are so good on everything!" Because I'm a huge Bill Hartung fan, and I just thought, "Oh, doesn't that make perfect sense that you two worked together?"

Well, listen, I will let you go. Thank you so much for coming back on my show, Frida. I really appreciate it.

Berrigan: Scott, it was great to talk to you. Take care.

Elizabeth McAlister: Nuclear Winter and the Kings Bay Plowshares 7 October 18, 2019

Scott Horton: Alright, you guys, a very special guest today. I'm joined on the line by Elizabeth McAlister from the Jonah House and the Plowshares, and of course she is one half of the most famous anti-war couple in American history, really, with the late Philip Berrigan. She's the mother of our friend Frida Berrigan, a great friend of the show we've interviewed many times here. We're very happy to welcome you to the show. How are you doing, Elizabeth?

Elizabeth McAlister: I'm doing just fine, thank you. And Frida will be in town for the trial.

Horton: Okay, well, that's what we're here to talk about, the trial. When does it begin?

McAlister: Monday [October 21, 2019].

Horton: And this is regarding an action that your group, the Kings Bay Plowshares, committed on April 4, 2018. Could you please tell us about it?

McAlister: Well, we went to the Kings Bay Naval Weapons Station, and we went to three distinct sites in small groups. One group went to the shrine and made their presence felt there. Three of us went to a section up on the hill and close to their nuclear weapons. The third group went to the administration building and left many messages there. So we were pretty present to different moments and different works that the plant is about. Of course, it's about nuclear weapons, and the nuclear weapons that are carried by the Trident submarines would be enough to destroy everything on Earth, which is what we've learned over the period of study leading up to the action.

Horton: And you guys chose April 4th for a very specific reason as well, is that right?

McAlister: That's correct. The anniversary of the assassination of Martin Luther King. We wanted to remember his witness and to celebrate it.

Horton: And that's an important point — his "witness," as you call it. This all is very religious to you. Your motivation here is your Christianity, is that correct?

McAlister: That's correct. It's Christianity, but it's humanity, and it's Earth, and the reality is that everything is up for grabs with these weapons. People, places, history. There is enough firepower on the base right here in Kings

Bay to make it that — if that were ever used, we have read — that's the end of life on Earth. It would spell the end of life on Earth if the firepower on each of the Trident submarines home-based here were ever used. Now that's sobering.

Horton: Because of the nuclear winter, you mean.

McAlister: Well, the effect of the weapons themselves, yes. They affect climate, they affect everything. They affect the air we breathe. They affect the soil. And it's just a very, very dangerous weapon.

Horton: Well, I have to tell you, I have no idea when I first learned of you and your husband, but I've always known about these Catholic priests and nuns who go and protest the nuclear missiles and make a controversy out of America's nuclear arsenal when most of America just never pays attention to this issue at all. I think you and your groups that you have done these actions with over the years deserve the lion's share of the credit for people even thinking of nuclear weapons as controversial at all. Most of the time they still go without saying anyway, as if it's just part of our security. But I don't think that people can imagine what it would be like to really see an H-bomb go off over an American city.

McAlister: No, the thought is horrendous, and that's why most people don't think about it. But I do want to say that we are far from alone. We are surrounded going into our trial by some extraordinary human beings who are attorneys and who are just the most thoughtful, prepared, intelligent people — and human, they're beautiful human beings. And the privilege is to be able to work with them and to have them share their powers and their training with us. So the trial should be very, very strong, and very fine, thanks to them.

Horton: And they've already said in the news that they mean to mount a religious freedom defense, is that correct?

McAlister: Well, I'm sure that's going to be at least part of it, if not the whole thing, yes. That's correct.

Horton: That's very important because, of course, right next to how much American society cherishes our nuclear weapons, we cherish our 1st Amendment, and even out of all five protections in the 1st Amendment, the freedom to worship as one sees fit — or not to worship, or otherwise — I think, is the one that everyone can agree on as the single most important right of all to be protected.

McAlister: That's very, very true.

Horton: And they're trying to put you in jail for 25 years here — in prison, for 25 years, is that correct?

McAlister: I thought it was 26, but I will not argue over a year. I have already served 1 year and 8 months for this in the local jail.

Horton: So does that mean they refused to bail you out, or you refused to be bailed out?

McAlister: I refused the conditions of release until they finally just released me without conditions.

Horton: I see. I had read that a few got out and were on house arrest with ankle monitors.

McAlister: Yes. Clare Grady just had her ankle monitor removed the other day, and she's been out for months. I had it for 3 or 4 days, and they took it off. My release was very, very weird and an utterly unexpected and unprepared-for thing. They just wanted me out, and so I'm out.

Horton: You've been through this quite a few times. I know you and your husband have served not just time in jail, but time in prison over the years for these kinds of actions. How severe are the charges here compared to previous actions?

McAlister: I think that the charges are not different; I think the consequences are a little more steep than what we've faced before, but not that much. What is it, 20-some years in prison? This is a figure I've heard, I haven't explored it, though. We'll see. [Laughing] But if it's 26 years, I probably won't make it.

Horton: And how many times have they sort of backed down and let you go with time served in the past, versus really seeking to press the charges as hard as they can?

McAlister: Well, never.

Horton: They always really try to put you in prison for as long as they possibly can each time?

McAlister: Well, for a significant period, put it that way. And for what they feel they can get away with without looking too bad.

Horton: In other words, they don't treat this any different than if you were some kind of foreign spy sabotaging their precious missiles, rather than a former nun protesting in the name of Jesus and the human race.

McAlister: Yes. I think the way you're putting it is quite correct and faithful to what we have experienced in the courts, yes.

Horton: Well, I'm glad to hear that you've got such great, talented lawyers. We would hate to hear that the worst happened here. But it goes to show the level of your commitment, especially having been through this before, that you are willing to continue to put your life on the line in this way. It's incredible.

McAlister: Well, I've had some very, very good examples of people who will not be silent. I have children, I have grandchildren, and it's impossible for me to think of them and not want to do what I can to make their future a possibility. Those children that are close to me are ways of also feeling for all of our children — all of our children whose lives are really up for grabs, given the way this culture, this society, is being run. And it's not getting better.

Horton: I have to tell you — I'm sure you already know, but in case no one's reminded you lately — that tales of what you have done have really meant a lot to a lot of people throughout this society, and even people who don't know your name have heard of the Catholics who break into the nuclear weapons facilities to protest before. Everyone knows that's a thing. And who is it that's doing it? Priests and nuns are the ones doing it. It's immediate. It grabs people's attention. It makes people want to find out more and at least to try to grapple with the situation here, and I think you have had such an important effect on so many people. I hope you know that, regardless of how this thing plays out, it has been worth it in a way to those of us who are paying close attention to you.

McAlister: Thank you for that. I hope for the fidelity to keep on keeping on, so we'll see how it all goes. But I think that those of us who were part of this most recent action are feeling ready to get to court and to make what presentation we can make. So we'll see what comes of it.

Horton: Great. Well, we'll be following it as close as we can at Antiwar.com. Again, that's you and six others, the Kings Bay 7.

McAlister: Thank you for this, and thank you for your own work, which is really important. Keep on keeping on there, brother, okay?

Horton: Thank you so much, Elizabeth, and best of luck to you.

McAlister: Thank you.

Francis Boyle: In Defense of the Kings Bay Plowshares 7 November 7, 2019

Scott Horton: Time to welcome Francis Boyle back to the show. He's an international law expert and a would-be expert witness for the Kings Bay Plowshares 7, who were convicted at the end of October for their protest at the Kings Bay Trident submarine nuclear facility. Y'all might remember that we had Elizabeth McAlister on this show just a couple of weeks ago, three days before the trial started. Well, they were convicted, and so Francis Boyle is here to talk about the trial and what comes next. So welcome back to the show, sir. How are you?

Francis Boyle: Thank you very much for having me on.

Horton: I'm very happy to have you here, sir. Please remind us about the Kings Bay Plowshares 7, who they are, and exactly what happened.

Boyle: Well, these are religious people affiliated with the Catholic Workers and also the Plowshares movement that was founded around 1980 by Phil Berrigan, Dan Berrigan and Phil's wife, Liz McAlister, whom you've already talked to. They went on the Kings Bay Trident II nuclear weapons site that has the Trident II nuclear submarines there on the East Coast. They got in there, they did what they called "symbolic disarmament" — they prayed and things of that nature. They were arrested voluntarily. They were peaceful, nonviolent. They did not resist arrest, and the United States government indicted them for depredation of government property, which comes with a sentence of 10 years; destruction of Naval property, 5 years; destruction of U.S. government property, 5 years; conspiracy, 5 years; and trespassing, 6 months.

So, I work for Liz McAlister. She was represented by my friend, Bill Quigley. Liz and I have been friends for many years, as well as her late husband, Phil, and I've represented them in other contexts. The other Kings Bay Plowshares decided to represent themselves. There were basically three approaches to defending them. The first was the Religious Freedom Restoration Act. This is the first time ever, to the best of my knowledge, that such a defense was made in one of these Plowshares cases. The second — Dan Ellsberg put it in an affidavit on the basis of his book of *The Doomsday Machine* — outlining the necessity defense. And then I had a very extensive declaration, which was about nine pages long, that you can find on the internet at the Institute for Public Accuracy, going through all the international criminal law violations by Trident II. It also included the argument that they really did not have the criminal intent necessary to constitute these crimes, that the government had to prove that criminal intent beyond a reasonable doubt, that my declaration had created a reasonable doubt and that the charges should be dismissed.

Now, on Friday evening at 10 p.m., in the dead of night, before the trial was to open on Monday morning, the U.S. federal judge — who was a Bush, Jr. appointee — issued an order that struck them of all these defenses. In addition, she muzzled them as to their capability to raise these types of issues in their own testimony. She also threatened them and their lawyers with contempt and other harsh sanctions if they disobeyed her order. So after reading that, I did put out a press release at the Institute for Public Accuracy, saying that this is a kangaroo court with a rubber stamp on a railroad towards their conviction, and that's exactly what happened in court. You could read the transcript if you want, at the Kings Bay Plowshare website. They were repeatedly cut off, interrupted, threatened with contempt, etc., etc. So not only were they stripped of all their defenses illegally and unconstitutionally, but they were muzzled and threatened on the witness stand. So of course they were convicted.

Horton: And of course, the jury was full not of the peers of the accused, but of the peers of the accusers, just like always. Meaning government employees from the local area probably. But now, I guess, if you give them the benefit of the doubt, maybe if they'd gotten to hear the defense's side of the story, the jury would have at least had a chance to do the right thing.

Boyle: That's correct. But not only that. There's a good chance they would have been acquitted. I've worked on these kinds of cases before. The first anti-nuclear protest case I ever did for Pax Christi was in 1982. The first Plowshares case I did was in 1985. You could read about those in two books I've written: *Defending Civil Resistance Under International Law*, published in 1987, which I wrote for lawyers to use in these cases; and then *Protesting Power: War, Resistance, and Law*, published in 2008. When we've been able to get these arguments to the jury, we have been able to get outright acquittals or hung juries, at least.

Horton: Really? I didn't realize that. That was going to be one of my other questions, how often it is that judges are this total in banning the defense from even being allowed to put on a case in this way.

Boyle: It's typical in Plowshares cases, yes. They don't really care. Federal judges, they're just part of the federal system. You have to understand that the Plowshares go directly at the heart of the American empire and against the Pentagon. Indeed, the U.S. Department of (In)Justice has a special task force to deal with Plowshares cases, to monitor them, to direct the persecutions — as I would call them — and always to go for the maximum charges and the maximum sentences. That's been the consistent pattern that I've seen since doing these cases, starting in 1985, for the Plowshares.

Horton: Let's break this down a little bit. When you say "go after," they're doing symbolic protests; they're not truly threatening to disable any nuclear weapons. They're not cutting open the fence so that al Qaeda terrorists can sneak in. They're not doing anything other than creatively protesting and committing acts of civil disobedience to bring public attention to the danger of these weapons. It goes to show just how important nipping even that in the bud is to the state. They don't want it to look like any old nun can protest a nuclear weapons facility and get away with it, because somebody else might copy her. That's a pretty total stand to take against what on the surface is just left-wing direct-action protest tactics, which aren't really dangerous. It's not like they're breaking windows at Starbucks or something terrible like that, you know?

Boyle: Well, everything you said is correct. I worked on the case of three nuns in Denver who went out to an ICBM silo site, cut the fence, went in there and prayed. And the federal government — it was again a total kangaroo court proceeding before a Bush, Jr.-appointed judge — found them guilty of depredation and sabotage, and then the United States government demanded the maximum 30 years for both of them. And they were elderly, they were 66 and 67 years old. This would have been life in prison. At the end of the day, they got short of three years. This is typical because the United States government understands that this goes right up against the heart of what the Pentagon is all about. When you go into these trials, you know you're up against the Pentagon. That's who is really behind the Department of Injustice persecuting these people.

Now, I would only make one correction to what you said, Scott. This is not a case of civil disobedience going back to Dr. King and the American Civil rights Movement, the courageous African Americans and others who supported them. This is a case of civil resistance in that the crimes here are being committed by the United States in violation of international criminal law and U.S. domestic criminal law, including, but not limited to, the Nuremberg Charter judgments and principles, the U.S. Army Field Manual 27-10, the Geneva Conventions and other basic sources of international criminal law. That gets incorporated into United States domestic law — the U.S. War Crimes Act, for example. So these people aren't disobeying anything. They are obeying the rules of international law, which are part of United States law, and also criminal law, in trying to prevent the ongoing commission of crimes. So they are the sheriffs, and the people running the Trident II system, they're the outlaws.

Horton: I think I just figured out why the judge won't let you testify.

Boyle: That was all set forth in my declaration. When we get these arguments to the juries and I have enough time to present these arguments and we can bring in an expert witness on U.S. nuclear weapons systems — such as Dan Ellsberg like we wanted to do here — we can usually turn around a jury and get an acquittal or a hung jury.

Horton: I'm so interested in the topic of nuclear weapons and all the politics surrounding them, and I'm just ceaselessly fascinated by how little anyone cares about it except when it's a news story like this. But for the most part, these things are just out of sight, out of mind, even though everyone knows enough about them to know that one good one can kill a whole city and that, at best, we've got a bunch of lawyers in charge of deciding when they should be used. The same kind of soulless monsters that would put an old lady in prison for life for praying at a facility are the same kinds of people who decide whether to use these or not.

So far, we've been lucky, but it just sort of goes without saying in our society and around the world that we're just all going to have H-bombs pointed at each other from now on, the Non-Proliferation Treaty and the rest of these things you cite notwithstanding. But nobody cares, really. They care more about TV shows than about H-bombs, and I'm not sure why. I guess we all feel powerless to do anything about it, that's a big part of it, but what an emergency! To think that these guys are holding on to bombs that are measured in the megatons when they detonate.

Boyle: In the case of the Trident II, its use would probably destroy all of humanity. These are 150-kiloton bombs, maybe 10 times the size of Hiroshima and Nagasaki. It would definitely set off a major nuclear war between the United States and Russia and/or China, and I don't think anyone would survive. Human life would be extinguished. According to a scientific study, the only form of life that would survive would be the cockroaches, because their shells would resist the radiation. That's what we're dealing with.

Horton: I should mention that there are less drastic scenarios for nuclear war that still include the deaths of billions of people, widespread famine and essentially the cancellation of human civilization back to bronze-age levels. Absolute catastrophe beyond anyone's imagination. But, you know, in the Southern Hemisphere some humans might be able to pull through. So maybe it wouldn't be down to the cockroaches, but without question, a war between America and Russia would mean absolutely the end of all northern civilization on the planet and certainly the destruction of billions and billions and billions of whoever's left after that.

Anyway, sometimes people hear things stated so totally and maybe dismiss it, like, "Oh, I don't know, you could have a war where hundreds of millions or billions die, but maybe not all eight billion of us," you know?

Boyle: We could have brought on Dan Ellsberg with his new book, and I think Dan's opinion would have been that using the Trident II nuclear weapons system would have extinguished, basically, all forms of life on Planet Earth and turned it into a radioactive wasteland for cockroaches. We could have had Dan address that precise issue, that's what we were going to do.

Horton: Hey, I'm with you. Everybody needs to read *The Doomsday Machine*. That thing will absolutely blow your mind. I actually have another friend, Gordon Prather — you might remember him from the George W. Bush years — who wrote so much great stuff for Antiwar.com debunking all the lies about Iran's supposed nuclear weapons program, and he used to make H-bombs for Uncle Sam during the Cold War. He knows a hell of a lot about it and says that if you survive the initial dose of radiation and you're far enough away from the actual explosions, the radiation eventually dissipates from that, and so you'd survive that. Then, obviously, the weather patterns changing, crop failures and all those things, I don't think he discounts nuclear winter altogether, but I think he also said that some of those things might be absolute worst-case scenarios.

Then again, if this is the low end of the sliding scale, where we're still talking about billions of deaths, we're still talking about catastrophe beyond anyone's imagination. That's hardly an argument for going ahead and hanging on to these things.

Boyle: Well, Dr. Helen Caldicott has written books on the dangers of radiation, and I think she would respectfully disagree with this. Also, Dr. Sternglass has written a book on radiation. So I don't think we would have had much difficulty, if we'd had the chance, establishing in court that the use of Trident II nuclear weapons systems would have extinguished life on Planet Earth as we know it.

Horton: Yeah, maybe I should get her and/or Dan back on the show to talk about that. The Trident, we're talking about *one* missile with multiple warheads, launched from one submarine, that alone could extinguish all life in France or something, maybe. But if you're saying that the use of one of those would mean an unstoppable chain of events that would lead to a full-scale nuclear exchange between America and Russia, or America and China, then that's a different question.

Anyway, I'm sorry. I don't mean to bicker with you here. I'm absolutely on your side, and the idea that humanity can even entertain the possibility that we would hold on to weapons that, if used once, would kill an entire city's worth of human beings is just absolutely, unforgivably mad.

Boyle: As I said, the systems on both sides, the United States and Russia, are on hair-trigger alert that could go off at any time. It could be a computer malfunction, a flight of geese, bad satellite images, or things like that. In other cases I've worked on, we have introduced evidence of repeated malfunctions of warning systems. So it's far more dangerous than most people realize.

Horton: Yeah, there have been at least 20 extremely close calls to the absolute brink of war. I'm not really good with math and statistics and stuff, but a cat only has nine lives. We got away with 20 almost-H-bomb wars over mistakes, over a bomb that accidentally fell out of a plane, or a misunderstanding between the Americans and the Soviets over their intentions at different times. That's cutting it way, way, way too close. In fact, my dad was at UCLA in 1962, and his professor disappeared for two weeks — evidently advising the government in Washington on some level — and came back and said, "I'm here to tell you kids that this Cuban Missile Crisis is as close to nuclear war as we could possibly get without having one," that this is the absolute brink of the crisis. The fact that we survived it is a miracle.

Boyle: As a matter of fact, if you study all the literature on the Cuban Missile Crisis, that's exactly correct. We came within a hair's breadth away from nuclear war with the Soviet Union. Indeed, I argued to that effect in a court in Greenock, Scotland, where three women went out and damaged a tender for the UK Trident IIs that we gave them. I argued that exact point. You can see my argument testimony on the website for Trident Ploughshares. At the end of the day, we got a directed verdict from the judge of acquittal on behalf of all three of them, four different charges each on various different counts of destruction of property. It even made the British press the next day say, "Our nuclear deterrent is illegal," which it is.

We have to understand that despite all this nonsense about deterrence, it's a joke and a fraud. All our weapons, our strategic nuclear weapons, are designed for offensive, first-strike, strategic nuclear attacks on Russia, China, or now other targets that we allege have nuclear weapons. Indeed, Doctor Michio Kaku wrote a book, *To Win a Nuclear War: The Pentagon's Secret War Plans*, on this, and he goes through all the times we threatened or prepared for a first-use strike of nuclear weapons. So these weapons, they're not there to deter anything; they are there to be used in a first strike.

Indeed, in a Plowshares case I worked on in Wisconsin for Tom and Howard Hastings, they had damaged the ELF extremely low frequency towers that communicate with Trident II submarines. They were both facing charges of sabotage, and we brought in a retired naval captain who used to command submarines, and he said it's well-known that the ELF Trident II system will be the bell ringer for the start of an offensive, first-strike, strategic nuclear war. With his testimony, along with my testimony and my arguments, we got outright acquittals for both of them. So again, it goes back to the point that when juries actually hear the destructive capabilities — I mean, we're talking about weapons that far exceed even the wildest fantasies of Hitler and the Nazis. These are Nazi weapons to the umpteenth degree. And when juries get to hear that, and that this is ongoing criminal activity involving conspiracy, planning, preparation to commit Nuremberg crimes against humanity, war crimes, crimes against peace, and outright genocide, typically they'll acquit.

By the way, the United States government knows that the Department of Justice does that, which is why they fight so hard to keep my testimony out of there. Putting aside my testifying to the court — why I was standing by to fly out there — even my two declarations they did not get to see, even though they were filed with the court. The second declaration, I submitted after the Trump administration pulled out of the INF Treaty. I itemized in that second declaration, which you could find on the Kings Bay Plowshares website, the dangers that are now facing us: that these INF weapons now will be deployed by the United States, first against Russia and also against China, and reduce warning time from the time they're set off to the time they land in either Moscow, St. Petersburg, or elsewhere to maybe 3 to 5 minutes. Clearly, they are there as part of an offense and first-strike nuclear weapons system against Russia and China. There's no question about it, if you study the literature.

Horton: Well, it's always been clear, from the project of putting the so-called "defensive" missiles in Eastern Europe, that that was part of an offensive package. It was meant to be able to shoot down their retaliatory strike to make it easier to launch a first strike in the first place.

Boyle: The Russians are now pointing out that in fact we could put Tomahawk cruise missiles in those launchers that are nuclear-capable, and that we therefore have an offensive capability and aren't just there for defense. The Russians say, "If you do that, we will have to respond in kind."

By the way, then you see this federal judge, this Bush, Jr. appointee, she knew full well that if I was able to present these arguments to the jury and Dan Ellsberg was able to send his arguments to the jury, there's a good chance we'd have gotten an outright acquittal for all of them, and that would have been a terrible blow against Trident II. So of course, she not only struck me, she also struck Dan, my declarations never got to the jury, Dan's affidavit never got to the jury, and the Kings Bay Plowshares were muzzled. You could read the transcript. They were threatened by the judge, that whenever they began talking about these things, she'd hold them in contempt.

Horton: Let me ask you this. Can you give us a bit of a summary of the religious freedom argument there? Because that was something that Liz McAlister had mentioned, but you were more specific. You said they were trying to invoke the Religious Freedom Restoration Act.

Boyle: The Religious Freedom Restoration Act was adopted by Congress to make sure that the United States government does not put undue burdens on people who are motivated by religious reasons. All they wanted to do here was to argue these points to the jury, and the judge stripped them of that argument too. So they were able to get up there and say they were motivated by religious reasons, but they could not tie in with this specific statute that gives special protections to people who are motivated by religious reasons. They were going to bring in a Catholic bishop, since they are all Catholic Workers, and a professor of theology from Fordham University, which is Catholic. I wasn't really involved in that; it was a novel defense. Bill Quigley has said publicly that he is going to try to appeal that precise issue, so we'll have to see what happens.

Horton: Well, how long until the sentencing is announced?

Boyle: I believe the sentencing will be in January. They each are facing 20 years. I suspect that the Department of Justice, pursuant to previous practice, will go for the maximum 20 years for all. That includes Liz McAlister, who is about to turn 80, so effectively, that will be a life sentence for her.

They tried to do this to her husband, Phil Berrigan, in his last Plowshares action (Plowshares vs. Depleted Uranium), and the prosecutor there publicly bragged that this is going to be Phil Berrigan's last hurrah, how he's going to put Phil away for the rest of his life. We started out facing 40 years, and at the end of the kangaroo court proceeding — his trial attorney was my friend Ramsey Clark — Phil and the others got two years. Then he got out and was then sent back on another Plowshares action for probation revocation, I think another six months. So he did get out. Soon thereafter, he was diagnosed with terminal cancer and died, but at least Phil Berrigan died at home surrounded by his wife Liz, their children, their grandchildren and their friends, not rotting away in some federal hellhole for crimes against the American empire.

That is what they're trying to do to Liz McAlister. Liz has been a pain in their neck for the last 50 years, since the Berrigans and McAlister launched their campaign against the Vietnam War. And they want to put Liz away for the rest of her life. I don't mean to diminish the significance of any of the others, but Liz has been at this for over 50 years.

Horton: Listen, I have to say — I told her this, too — that I've always known that there is this group of nuns and priests that do these actions and break into these nuclear weapons facilities in order to protest and bring attention to them. And that's not exactly the same thing as just a bunch of Haight-Ashbury hippies or something like that; it's a priest and his wife. I never knew a lot about it, but all my life growing up, I knew that these people existed, that they did this and that they went to prison to make this point. In fact, it was on the anniversary of Martin Luther King's assassination that they did this action, and it was part of that civil disobedience tradition.

But this stuff is really important. It was one of the things that got me interested in the subject as a young kid, and I know I'm not alone in that. So

I don't know how much that's worth, but it's not nothing, for all the sacrifice that they put into this.

Boyle: It's important to understand. Their slogan is: "They shall beat their swords into plowshares and their spears into pruning hooks." That comes from the Jewish prophets Micah and Isaiah, and these Plowshares — many of them are Catholic, but not all of them — are motivated by the example of the Jewish prophets. If you go back and read your Bible and the role they played in challenging the King and the Jewish emperors at that time, many of these prophets were killed or imprisoned. If you are interested in following that up, there are two books. The late Dan Berrigan, a Jesuit, wrote a series of books on the Jewish prophets, and the late Rabbi Abraham Heschel — he opposed the Vietnam War, supported civil rights for African Americans and marched with Dr. King at Selma — wrote a book on the Jewish prophets, too. This was Rabbi Heschel's doctoral dissertation for his PhD in theology. The Plowshares try to fit what they are doing into this biblical tradition of prophecy.

Horton: It's funny, because all these political figures pretend to be very religious, too, but when it comes down to it, the state ranks far higher than God or his son on their list. No question.

Boyle: Well, that's right. As Liz McAlister pointed out, this is a religion of nuclearism, and it's an idolatry. She's been very eloquent about it for all these years, and all the other Plowshares as well have tried to make this point the best they can in these trials. But again, the federal government and federal judges to a great extent have muzzled them and completely shut down their defenses, in violation of their rights of due process of law.

Horton: That's really one of the most outrageous parts of this, right? For a guy who was raised on *Matlock* like me, I know that the state's almost always going to win, but when the judge and the prosecutors conspire to strike down every witness that the defense wants to bring — when they're not even allowed to put on their case at all — that's the kind of thing that should never happen in America, especially in a case like this, where the prosecution could afford to be patronizing and insulting as hell and ask the judge to give them all six months, a $10,000 fine, make their point, shame them for jeopardizing our security, or some kind of thing, then let them go and treat this like a really bad misdemeanor. For them to go to these lengths to prevent even any semblance of a defense by these people who are clearly motivated by the highest purpose is sickening. It bothers me. It ain't right.

Boyle: Well, you're right, Scott, and your listeners have to understand that they were each facing 20 years. All they wanted were three witnesses. They wanted the bishop, the professor of theology for the religious defense and then me on the nuclear weapons aspect. Dan could not come because of a prior personal commitment, so sometimes in that type of situation, I do double duty — I've actually been qualified in court as an expert on U.S. nuclear targeting doctrines and strategy. But we were all stripped. That was it. They had no defense. The judge and the feds stripped them of all their defenses. That's why I said it was a kangaroo court with a rubber stamp and a railroad. Sure enough, that's what it turned out to be.

Horton: Well, listen, I'll let you go. Thank you so much for all your time here this afternoon on this very important story and all your great efforts. I wish you guys the best on all the appeals and all of this, and I wish I knew something better to say.

Boyle: Just thanks again for taking the time to go through all this so that people can understand what the Plowshares are all about and that they are acting as they see it as prophets in the tradition of the Jewish Bible.

Horton: Alright, well, thank you again.

In June 2020, Liz McAlister was sentenced to time served for their action at the Kings Bay naval base.

Joe Cirincione: U.S. Actions Don't Justify Putin's Attack, But Set Stage February 24, 2022

Scott Horton: Introducing Joe Cirincione from the Ploughshares Fund, and here he is, writing at the Quincy Institute. Welcome back to the show, Joe. How are you doing?

Joe Cirincione: Just great, Scott. Thanks for having me back on.

Horton: I don't think we've spoken since the Bush years, but I'm very happy to talk to you again.

Cirincione: Thank you. Great to be on. What do you want to talk about today?

Horton: Well, there's a few different important topics, but you wrote this really important piece for the Quincy Institute at ResponsibleStatecraft.org: "New Scientific Review Punctures the Myth of Missile Defense."

Are you trying to tell me that George W. Bush tore up the Anti-Ballistic Missile Treaty for no good reason, Joe?

Cirincione: That's exactly right. And particularly in light of the Ukraine crisis, we can look back at the last 20 years and see that tearing up these arms control agreements was not a very good idea. We really took down a lot of the guardrails that we wanted to have, and it began with the tearing up of the ABM Treaty. Remember, this is the treaty that Richard Nixon negotiated with Henry Kissinger as part of his effort to limit nuclear arms. The logic was that if you're going to limit the number of offensive nuclear weapons you could have, you have to limit defenses. And the reason is simple: Any defense can be overwhelmed by an offense. So if a country is going to build up a defensive system, the easiest thing for the adversary to do is increase the number of offensive weapons. That was the offensive/defensive cycle that was going on in the beginning of the 1970s when Nixon and Kissinger negotiated these treaties. We were going to reduce offensive, and we had to limit defensive.

In 2002, George W. Bush decided that no, we now have a technological solution to this. We have weapons that can intercept the other side's ballistic missiles. We're tearing up the treaty. We're going to deploy a defensive system. That was now 20 years ago, and we're no closer to having a system that works today than we were when George W. Bush made this terrible decision.

Horton: Can we rewind about a half a generation from that point and go back to the greatest tragedy that ever took place in the history of humanity, which was the Reykjavík Summit between Ronald Reagan and Mikhail Gorbachev, where they came — according to all accounts I've ever heard of or read — within a hair of abolishing all nuclear weapons from the face of the Earth? Then, isn't it the case that Reagan scotched the deal because his men told him this fantasy that we're going to be able to shoot down all the incoming Russian nukes, and we would rather build up defenses than get rid of all the missiles we need to defend from in the first place?

Cirincione: You're exactly right. I've spoken with the late Secretary of State George Shultz who was the only other American in the room with Reagan and Gorbachev and Gorbachev's translator at that time, and he recounts — and the transcripts of that meeting bear him out — that Gorbachev said, "Let's get rid of all strategic weapons, the long-range systems on missiles and bombers and subs."

Reagan said, "Yes, and let's go further. Let's get rid of the short-range ones too." And Gorbachev said, "Yes, let's get rid of them all." Reagan turned to George Shultz and said, "Can we do that?" George Shultz didn't hesitate. He said, "Yes, we can," and they walked out of that room with an agreement in principle to get rid of all nuclear weapons. But when Reagan ran this by his advisers, including Richard Perle — who many people call the Prince of Darkness — they were aghast at this. They didn't want to give up the nuclear weapons, and they walked Reagan back. Reagan then went back to Gorbachev and said, "Look, we'll do this, but we have to be able to deploy our missile defenses," what was called the Strategic Defense Initiative at that point — the Star Wars program — "I have to be able to do this."

And Gorbachev said, "No, no. I can't go back to my military and give up my weapons and have you deploying these other weapons. How about if we keep that in the laboratory for 10 years, just in the research stage?" Reagan wouldn't do it. He said, "No, I've got to have this insurance policy," and that's when the deal collapsed. That's when the Reykjavík Summit ended in failure producing those grim photos you see at the end of both men not smiling at all.

It turns out they both were wrong. Reagan was wrong. He could have given it up. There was no miracle technology, and nothing has advanced since then to allow us to intercept long-range ballistic missiles. And Gorbachev could have let Reagan pursue this fantasy because nothing would have come of it. So we missed this moment to solve one of the greatest threats facing humanity, the threat of nuclear annihilation.

Horton: The wisdom of the staircase, right? Gorbachev must have been absolutely kicking himself that he didn't just say, "But Mr. President, if we don't have any missiles to shoot down, you don't need anti-missile defenses, so let's just..."

Then here's the thing, you're really killing me, because I had never heard the very reasonable counter-offer: "Let's just study that together for 10 years." The Soviet Union didn't even exist for another two years after that, you know? And then, I've seen the pictures where Gorbachev is essentially following Reagan down the stairs from the building, where he's getting in his limo, and Gorbachev is saying, "Mr. President, stop, let's go back up there and talk about this one more time. Please don't go." But it's too late, that's it.

Cirincione: It's too late. We missed the opportunity. To their credit, neither one of them gave up. They didn't get the elimination they had hoped for, but a year later they did negotiate — which was one of Reagan's biggest accomplishments — the Intermediate Nuclear Forces Treaty. Which again has something to say about the current situation we're in. This treaty is where Russia and the United States agreed to destroy these brand-new nuclear weapons that they had been pouring into Europe that were so-called "intermediate-range" — not quite ocean-spanning, but bigger than short-range and medium-range missiles, such that Russia could hit European targets, while with our deployments in France and Germany and Italy could hit Russian targets. They said, "We're going to get rid of them all." And they eliminated almost 2,000 perfectly fine nuclear weapons attached to missiles and banned them globally.

Trump just tore that treaty up when he was in office because some of the people in our military wanted to deploy weapons like this again. He tore it up and said that we have nothing to worry about from Russia. Well, with Putin's invasion of Ukraine, once again we hear talk about Russia deploying these kinds of weapons back into Europe, and on the American side, us deploying these kinds of weapons against Russia and against China. We never should have torn that agreement up.

Horton: That's one of Putin's stated fears now: "What's keeping you from putting nuclear-tipped Tomahawk missiles in those dual-use Mk-41 missile launchers in Romania and Poland, now that you've torn up the INF Treaty?"

Cirincione: That's exactly right.

Horton: Let me go back to Reykjavík for one second here because this has always been a thing of mine. I think I know the answer to this, but I'd like to hear what you think about this. What about the idea that even if Reagan and Gorbachev agreed to get rid of all of their nukes, that still leaves Britain, France, China, Israel, India, Pakistan and, at that time, I guess, probably South Africa. So what about them? Would the idea be that America and Russia would just lean on their friends and allies and say, "If we're getting rid of our nukes, you're going to get rid of yours too"? The idea that the rest of the world would just go along with that if the U.S. and USSR were arm-in-arm on that?

Cirincione: Well, nobody talks about unilateral disarmament, just the United States giving them up. Or even bilateral disarmament, just the U.S. and Russia. But it's widely recognized that the United States and Russia have 90 percent of all the nuclear weapons in the world — we have about 5,500 weapons in our arsenal in various stages of readiness, and the Russians have about 6,200 in their arsenal. Everybody else has a couple hundred at most. France has about 300. China has about 300. So it's widely recognized that in order for us to have real progress towards nuclear disarmament for all nine nuclear nations, you've got to get the U.S. and Russia to come down lower, to about a thousand weapons each, even a thousand deployed weapons each. You do that, and now you're in the same ballpark as these other countries. Now you could have what some people talk about as a "freeze and reduce agreement."

So, you get China in particular, but also the others, to freeze their nuclear arsenals at their current levels and for the U.S. and Russia to continue talks about further reductions so that now you're having multilateral disarmament talks and everybody is in the mix, including India, including Pakistan, including Israel. They're all in the mix, and they're all talking about how to stabilize the situation so we don't get to the brink of nuclear war.

Horton: On the Intermediate Nuclear Forces Treaty, the narrative went that the Russians were breaking it first, we think, because they got these missiles that, by looking at them, seem like they fall within the prohibited range. The Trump administration's response to that was, "Oh yeah? We're just going to tear it up," rather than, "Let's sit down and start negotiating and make sure we can save this treaty in the name of Ronald Reagan," or something like that. But essentially what was going on here was that it looked like the Russians were violating the treaty, but they weren't deploying them in Europe, they were deploying them along their frontier with China. Or at least that's why they were doing this. I don't know if they were deploying them yet, but that was the purpose of it. And then that that's why America tore up the treaty, too. They weren't interested in putting Tomahawk H-bombs in Romania and Poland; they want these mid-range nuclear missiles for China, too.

So, America and Russia left this Intermediate Nuclear Forces Treaty that kept our medium-range nukes out of Europe since 1987, in order that they could both target China. Russia and China are getting along better than ever now — it seems like a real waste of effort on their part. I wonder what you think about that. Is that really right? We have, as you were just describing, a potential nuclear standoff in Europe where we just didn't need to have one at all.

Cirincione: Well, I would say two things. First, one of the most disturbing things about this Ukrainian crisis, in addition to the people being killed on the ground right now, is how Putin is talking about nuclear weapons. He is quite specific here. In his declaration of war, basically, he said that Russia has the largest nuclear arsenal in the world — that's true — and that anyone who dares to oppose them will be met with a level of destruction that they have never experienced. So he is directly making nuclear threats to the United States, to NATO, to anybody who would dare oppose his invasion. This is front and center in his mind. I know most Americans and most people in the world don't think about nuclear weapons, but Putin does. Putin does.

You realize that we have squandered so much time not pursuing more forcefully these efforts to reduce these weapons, to separate them from conventional conflicts, and while we haven't been doing that, the militaries in the United States, Russia and other countries have been moving to reintegrate nuclear weapons to their combat strategies. In other words, they are blurring and in some cases erasing the firebreak between nuclear weapons and conventional weapons, so that nuclear weapons just become another step up the ladder. The worry is that if Putin loses in Ukraine, or if he starts to lose, he may be tempted to use a nuclear weapon to prevent that loss. Which, for him, would not just be the loss of a battle or the loss of a war, but maybe the loss of his rule. Maybe the loss of his life, if the Russians rebel against this insane war that he started. So nuclear weapons are very much in the mix of this crisis that the invasion of Ukraine has started, and that brings us back to the INF Treaty.

I was on an advisory board of the State Department during that time, and I saw the information about Russia's cheating, and yes, they were, no question about it. They were testing a kind of prohibited intermediate missile, and testing was also prohibited by the treaty. The development of a system like that is prohibited anywhere in the world, not just in Europe. The failure of the Obama administration at that time was not to do something about it. I was appalled. We knew about this for several years, and the Obama administration could not come up with a way to enforce the requirements of this treaty and to prevent the deployment of that Russian system, wherever it was deployed.

Then Trump comes in, and he just rips it up. This is completely insane. Just because somebody is speeding, you don't repeal the speeding laws. Just because somebody kills somebody else, you don't repeal the laws against murder. But that's what we did with the INF Treaty. Because they were cheating — at the margins, but still cheating — we tore up the whole treaty. I believe that we're going to come to regret this if we see Russia once again deploying these systems that we could have stopped in Europe or on the border with China.

Horton: But so, is that really right? That in both cases, Russia and America really weren't even doing this for each other, but they were doing it for China?

Cirincione: It's certainly true of us. This is what we wanted. The day after Trump pulled out of the treaty, the military tested a missile that could go in these intermediate ranges, but they had been blocked. What they wanted to do was deploy it.

Horton: Speaking of violating the treaty against developing the things, right?

Cirincione: Right, you can't even develop it. You can't test a weapon at this range. It was a very good treaty.

Horton: Well, they had one ready. They had apparently developed one, if they had one ready to test the day after Trump tore the thing up, right?

Cirincione: It was an existing missile that had never been tested at that range, so now they could test it at that range. Now, I don't know what Russia's main purpose was. They were certainly looking at China, but I wouldn't say they were looking exclusively at China. For us, it was all about China; we weren't that concerned about Russia at that point. We are now.

Horton: I hate to keep going back to the 1980s on you, but I was just a kid then. I lived through this and was paying attention, but I was in elementary school, so I don't know, but the way I remember the story, and as I've kind of learned it since then, is that isn't it right that Reagan did this massive buildup of medium-range missiles in Europe, I believe in response to Russia deploying them first, and these were the Pershings, right? He spent a trillion dollars on this or whatever, and then he turned right around and negotiated them away again. As you said, they destroyed all these perfectly good missiles, and that essentially, he had played this extreme game of nuclear poker here, like, "Oh, you want to put medium-range missiles in Europe, huh? I'll show you medium-range missiles in Europe!"

This had people terrified in Reagan's first term, right? And then he did the right thing, got rid of the things and signed this magnificent treaty that only now we're talking about has been destroyed 30 years later.

Cirincione: You're killing me here, Scott, because I was working on the House Armed Services Committee in the 1980s. I was already dealing with these things while you were — were you born?

Horton: I was born in '76, so I was in 3rd and 4th grade during Hands Across America and all that stuff.

Cirincione: So I'm doing this, and I was deeply opposed to Ronald Reagan. I thought he was the Devil incarnate. I was working for the Democrats on the House Armed Services Committee. Remember that Reagan comes in on this myth of the "window of vulnerability." All the hawks had rallied around the growing Soviet threat, and the fear was that Russia was soon going to have a first-strike capability, meaning that they would have enough nuclear weapons to destroy all our weapons in a nuclear strike and to keep enough in reserves that they could deter us from launching whatever might remain. This was the "window of vulnerability," they said, and we had to build up our nuclear forces. Reagan comes in, and he does that. He starts building a new MX missile, the ICBM. A new bomber. A new sub. He just goes whole-hog. The defense budget skyrockets again.

Then in his second term, he makes a pivot, and it turns out that he really is a nuclear abolitionist. He really thinks that we could, as he said in his second inaugural address, eliminate nuclear weapons from the face of the Earth, and he tries to do it. He was influenced by things like *The Day After.* He was influenced by the intelligence he was getting, where he learned that the Soviets really thought we were going to attack first, and he never conceived that the Soviets could think that *we* would attack *them.* He's convinced that these weapons have to be eliminated, and he comes really close to doing it. He misses it — and I write about this in the Quincy issue brief on nuclear forces — but then he does the INF Treaty and he gets rid of those. Then he starts negotiations on strategic systems, and he starts what he calls the START talks — the first treaty that would actually eliminate weapons rather than just limit the arsenals. He negotiates a treaty that cuts U.S. and Russian — then Soviet — forces in half. His successor, George H. W. Bush, continues that work. He also cuts U.S. and Russian forces in half, and he goes further, unilaterally reducing, taking thousands of weapons off of our Navy ships, de-nuclearizing the Army, pulling the nuclear weapons out of Korea, etc. You really got this feeling like, "It's over. We've escaped. We're safe now. We didn't eliminate them all, but we're on the way."

Sure enough, here we are, 30 years later, and we are down by 80 percent from the heights of the Cold War. We had about 66,000 weapons during Reagan's term. We're down to 13,000 weapons globally now, but the reductions have stopped. The steam has gone out of nuclear disarmament. Obama failed in his vision of making the elimination of nuclear weapons the focus of his national security strategy. Trump reversed it and started beefing up again, pouring money into these systems. Now, Joe Biden is about to issue a Nuclear Posture Review that will basically stay the course. It will tweak Trump's policies around the edges, but it will not fundamentally change them.

The crisis in Ukraine is showing us how dangerous it is not to continue to reduce and work towards the elimination of these weapons. These are one of the three great threats that threaten destruction on a planetary scale. Climate change can do it over decades. Pandemics can kill millions in years. Nuclear weapons can destroy humanity in an afternoon. We ignore this danger at our peril.

Horton: Yeah, absolutely right there. You already absolutely condemned what Putin is doing. As we're recording this on Thursday morning, we're about 12 hours into a full-scale assault on Ukraine, and presumably they're going all the way to Romania. Maybe they're only going to take the eastern half of the country. We're at that point here that there's just no question whatsoever that this is — I won't say entirely unprovoked — absolutely unreasonable and unnecessary. So that's the question, to a guy like you who's been around this whole time, who knew better all along, and warned them all along. For example, "W. Bush, you shouldn't tear up this ABM Treaty." "Barack Obama, you should not put these anti-missile missiles into Romania and Poland because this guy Putin is dangerous." If you listen to what Putin says, he takes this threat, as you put it earlier, very seriously. We don't think about nukes. He thinks about nukes. This was in his rant the other day, "What are we going to do? We're going to let them put missiles in Ukraine pointed right at Moscow? No. Nyet. It's on now."

So, I'm not saying he's justified in any way. I'm 100 percent with you. I'm not saying it's reasonable, but I'm saying what he's doing is rational. And we know that it is because, as people like you have warned for 25 years, we should not be expanding NATO, tearing up our treaties, and getting in the Russians' faces in a way that puts them in the position to do something very horrible, which is I think what we're watching play out this morning. Right?

Cirincione: Yeah, it really does underscore that when you have a moment when you're able to move forward on reducing nuclear weapons, you really have to seize that moment because it opens and shuts very quickly. In this case, there's nothing that we did that justifies what Putin is doing, let's be clear about that, but a lot of what we did set the stage for what Putin is doing. The NATO expansion that George Kennan warned us at the time — in the 1990s when the Warsaw Pact collapsed and we were just starting to let those Warsaw Pact countries into NATO — he warned that this was going to stimulate the worst kind of militaristic and revisionist nationalist fears in Russia. Which it did. Obama was warned when he put in what he said were defensive missiles — which is true, we were putting interceptors in Poland and Romania to try to intercept an Iranian ICBM, should they develop a nuclear weapon or an ICBM. They have developed neither, but we still put those missile tubes in Poland and Romania.

The problem, as you point out, is that these are the same kinds of missile tubes we have on our Aegis cruisers and destroyers. In fact, it's called a land-based Aegis system. And while we have interceptors in them now, those same tubes could house offensive nuclear-tipped missiles like the Tomahawk cruise missile, and Russia wouldn't know what was in the tubes because there is no arrangement for inspections.

I was on the advisory board in the State Department during the Obama years. I heard the Russian complaints, and they were brushed off by State

Department officials as "ridiculous," the same way Reagan thought it was ridiculous that we would ever attack Russia. Obama officials said, "No, NATO is a defensive alliance." Well, it doesn't look defensive to Russia. And we don't need these weapons. They're serving no purpose.

So what is the point of keeping those interceptors in Poland and Romania? It looks like you're preparing for an attack on Russia. That's why one of Putin's demands, in the letters he sent back to the United States and Europe on this, was that those missiles had to be taken out; as well as, by the way, that the U.S. had to rejoin the INF Treaty. So these issues are front and center for him, even if they're not for us.

Horton: I have a friend who is very upset, who was saying to me, "Putin is insane. He's insane. He's insane." And I'm saying, "Listen, the guy is brutal — maybe that's the word that you're looking for — but this is strictly business." If you listen to his statements, he does talk about older humiliations and this kind of thing, but even in the middle of his rant, he's saying, "Look what happens. The communists gave away Ukraine and let them be independent. But the Americans want to have it. They won't let it be independent. Either it belongs to us or it belongs to them. Well, it belongs to us, not them." He even said, "Kiev is run out of D.C.," which is kind of true. So what he's doing is obviously a severe move, but it is rational if not reasonable.

Cirincione: I'm with you. I think it's very dangerous to brush this off as insanity. Is megalomania a type of insanity? Yes, it is. I got a degree in psychology from Boston College, so I can comment. [Laughing] Yes, it is. And is he a megalomaniac? Yes, he is. But is he crazy like they mean that he's irrational, cannot be reasoned with? No, because that excuses some of the mistakes we made that paved the road to this moment. Again, nothing we did justifies what Putin is doing, but it has certainly set the stage for what Putin is doing. You can look back and see these missed opportunities. These things that we thought were signs of strength on our part were very threatening not just to Putin, but to lots of Russians. The ignoring of Russian concerns, when we could have incorporated them fairly easily into our plans.

It really was the hubris of the "unipolar moment," when we thought we were the only remaining superpower in the world and we could do whatever the hell we wanted. If that meant intervening in Kosovo to end atrocities, we would do it. If that meant surrounding Iraq and then invading Iraq, we would do it. If that meant occupying Afghanistan for 20 years, we would do it. If that meant killing almost a million people over 20 years in this "War on Terror," we would do it.

Every step of the way, we thought this was justified and that it really was nobody else's business. But Russia didn't feel that way. You can imagine Putin looking at this and saying, "Look man, if you can do that, why exactly can't I do this? If you can invade another country, why exactly can't I? Oh, what? This is not a Middle Eastern country, so it doesn't count?"

Those are the kinds of norms we set in response to which others are now saying, "I'm just doing what you did," and we can see what it's like. We've got to rethink what it would have been like if we had spent the last 20 years enhancing global norms, strengthening international law, reducing nuclear arsenals, and reducing conflicts, instead of acting the way we did in our hubristic arrogance.

Horton: He was sort of parroting the Americans, but I'm surprised that he didn't outright say, "Listen, I'm concerned that they could develop nuclear weapons and give them to these Right Sector terrorists to use against us. And also, it's our Responsibility to Protect. We think they're going to wipe out the entire province of Luhansk. Just like you guys claimed about Gaddafi in Benghazi in 2011, Samantha Power said, 'I have to intervene to save the people.'" What did they do in Libya? In Kosovo, they went around the UN entirely. In Libya, they lied to the Russians and said, "We're just going to protect Benghazi, we swear," and made a chump out of Medvedev who went along with it. This apparently enraged Putin, who returned to the presidency after only one term instead of two as had been expected, not just because the Americans did that war in Libya, but that they screwed the Russians on it so that they could launch the war, too. Just one more little thing there.

Cirincione: That is exactly right, Scott, and people forget this. Again, it doesn't justify what Putin is doing, but you have to understand that we are not blameless here. The kind of policies we implemented set the stage for what's going on now. And you're right, he is echoing the same kind of charges that we made both about Iraq and about Iran. He said he will not allow Ukraine to have nuclear weapons. Well, that sounds a lot like George W. Bush's justification for the invasion of Iraq. It sounds a lot like what presidents from Bush to Obama to Trump to Biden are saying about Iran. Neither one of those countries has nuclear weapons, right? Iraq couldn't build a nuclear weapon. Iran can't build a nuclear weapon, or it would take them many years to do it. Ukraine does not have nuclear weapons, cannot build nuclear weapons, but you still make this ludicrous charge. Putin is saying, "Well, it worked for those guys. I'm going to pull the same play here and see if it works for me." And as you say, also adding in the Responsibility to Protect: "I'm protecting the Russians. That's why I'm going in." Again, standards that we invented. We set this up, the Responsibility to Protect. It's coming back to bite us.

Horton: You mentioned Iran, and I want to ask you a question about Iran, but first, I want to ask you about a debate that I know of. It never happened right in front of me, it's been a disjointed kind of a debate. My old friend Gordon Prather — who used to make nukes for Uncle Sam back in the day, was the chief scientist of the Army, a great anti-Iran-war-lies activist in the Bush years, and writer for us at Antiwar.com — he tells me, "Look, man. You want to take out incoming Russian nukes coming over the Poles in outer space? The only way to do that is with an enhanced radiation device" — in other words, a neutron bomb which doesn't explode with heat so much, as it has a thinner shell so that the fission and fusion take place mostly in a radioactive form rather than just heat — then that radiation is what you would use to take out incoming nukes. And it's a severe way to do it, but it's the only way to do it. Because anything else, you're trying to shoot a bullet with a bullet in outer space where everything is ice cold, so all your infrared and heat-seeking stuff is not going to work, and you're talking about such high speeds. Just forget about it.

But then I met a guy who said, "Actually, that was my job, building these anti-missile missiles, and I'm here to tell you that we can hit a bullet with a bullet. But the question is: How much are you willing to pay for these bullets we're firing at the bullets? Because it is a matter of cost. And if you're willing to pay enough of a price for an anti-missile missile, yes, I can shoot down an incoming attack." At least from North Korea, if they launched a handful. Russia shooting their whole arsenal on Doomsday? Forget it. So I wonder, what is your position on that?

Cirincione: No, we cannot shoot down a bullet with a bullet unless the other bullet is cooperating. That's the way to think about this. When we first started doing these kinds of systems in the 1960s and '70s, we couldn't come close to hitting a bullet with a bullet. We couldn't come close to having one of our interceptors actually physically hit the incoming warhead for all the reasons you said. So we armed them with nuclear warheads. Our first defensive systems that we deployed, the Sprint and Spartan systems in the late 1960s and early '70s, were nuclear-armed. These were the systems that Nixon was going to eliminate. The Russians were doing the same thing. So we limited those to two sites per country. As it turns out, the Russians deployed one around Moscow, we deployed one around an ICBM field for six months before the Army itself, who was in charge of this system, said, "This is stupid," and we took it down. So we have had no defensive systems.

In the 1980s and '90s there was the dream, which Reagan promoted, of new kinds of technologies — laser beams, particle beams, speed-of-light weapons — that could do the job. Reagan was convinced that it could work. In 1987, the American Physical Society, the nation's leading association of physicists, issued a study that said, "No. We're nowhere near being able to develop weapons like this, and we won't even know if we can do it for a good twenty years." So that took the steam out of that balloon, and we went back to what we call kinetic interceptors. Not laser beams, but regular interceptor missiles. And we got better at getting them to be able to hit, and guess what?

It turns out that under perfect conditions, you can do it. So we've had 19 tests of systems like this, and in half of them, we've been able to hit it.

But here's the thing. The system has to cooperate. In these tests — they're really demonstrations — we know exactly when and where and how fast the incoming target is coming at us. There is no attempt to deceive or suppress the defensive system. We have never tested them with the kind of simple countermeasures that our own intelligence services say any country can deploy, including simple balloons that look like a warhead — in the cold vacuum of outer space, a balloon travels just like a heavy metal warhead. We haven't tested them against chaffs, simple things that can confuse and blind the radar, or responders or jammers or any of those things.

Horton: Just to clarify again, you're saying that it's not that we haven't succeeded in a test, but that they haven't even bothered to try to test that.

Cirincione: There was one test where they put up a balloon that looked too close to what the warhead looked like, and the interceptor missed. So they stopped doing it. All this time we haven't been doing it. And this is what the new American Physical Society study says. They looked at the current interceptors — we have 44 interceptors based in Alaska and California designed to intercept a North Korean ICBM. Some in Congress and in the military say that we could deploy such a system to defeat a Chinese threat or a Russian threat. They looked at this, and they said, "No, it doesn't work."

The system itself is fatally flawed. The one that's in the ground doesn't work well enough. It doesn't have enough reliability for us to depend on it. Even if it worked perfectly, it still couldn't do it because of this countermeasure problem. By the way, it doesn't account for enemy suppression techniques. For example, knocking out the radars before they launch their missiles — the kind of thing that you would expect an enemy to do. Nor do any of the other proposed systems that are on the horizon, what they call "boost-phase" intercepts. These try to intercept the missile when it's just being launched, in that two minutes before it goes into outer space, when it's slow and fat and hot, when you can see it more clearly and might have a better chance. They say you have to be too close to be able to do that, and you'd only have at most a minute or two of warning. It's an impossible mission.

I hope this new study has the same kind of impact that the study did back in 1987, where it took the air out of the laser balloon. I hope it restores some kind of sense to the Biden administration and convinces them that they can't be spending $20 billion a year on these missile defense programs. You have to go bring them all back to the laboratory, all back to the basics, and don't start deploying systems or pushing systems into production until you know they can work. And that will be decades.

Horton: I mean, man, Joe, what you're telling me here! This entire era, the whole 21st century long so far, starting with W. Bush tearing up this treaty in December 2001, just a couple months after Putin was the first person to call on September 11th and say, "I am at your service."

Cirincione: Yeah, he announced it in 2001, and then he did it in 2002.

Horton: And this whole time, Obama was going along with this whole thing too, installing these radars and these anti-missile systems — again, from Mk-41 dual-use launchers that terrify the hell out of the Russians and provoke them this entire way. I was going to say, to Biden's credit, that he was climbing down a little in his counter-offer to Putin, saying, "Yeah, let's establish a verification regime for the missile stations in Poland and Romania so that you're not upset about that," but it's just too little, too late, man. What are they even doing there? They don't even work. They're just decorations.

So let me ask you this. Is the controversy purely just the dual-use launchers and how they could be a disguise for Tomahawks, or is the threat really that the defensive missiles might work and that what the U.S. is doing is wearing armor to a fistfight, where it's not defensive at all? It's putting America in the position of being able to attempt a first strike, where they can take out enough of Russia's nukes on the first hit that they feel like they can shoot down anything that survives for a retaliatory strike, because they have Russia ringed with these things. In Putin's interview with Oliver Stone, for example, I think Stone even says, "Come on, you know these things don't work, man. This is just a boondoggle for corporate America. You know how it goes."

And Putin says essentially, "Yeah, of course, Oliver Stone. However, I'm in charge of security around here, man. What am I supposed to do when you're ringing my country with anti-missile missiles? I've got to make better missiles, don't I?" And it was a couple of years after that that he debuted in his speech all of his new missiles.

Cirincione: That is exactly right, Scott. You've got it. So let's go back to the Poland and Romanian launch tubes that are there. I think there's three factors. As usual, it's never just one thing. So there's three factors, and one is the offensive threat. I think that's a genuine concern. The U.S. dismisses this as crazy, but if you're a Russian and you've been attacked a lot, this is something you've got to be aware of. So that's number one.

Number two is the point you're making, and the Russians and Chinese are both acting this way. The anti-missile systems are the worst of all worlds. They don't actually work, but the adversary can't count on that. They've got to calculate that eventually the United States — because we're so capable — might get them to work. So therefore, in order to preserve their deterrent, in order to keep their nuclear weapons capable of hitting the United States, they've got to make sure that they can penetrate any current or known defensive system. That's exactly what Russia is doing, and Putin has a point. In 2001, when Bush announced this and then implemented it in 2002, Putin said, "Don't do this. I'm going to have to respond to this."

It took two decades, but in 2018 he unveiled five new "super weapons," he called them — typical exaggeration, some of these things will never work — all designed to circumvent U.S. missile defensive systems and whatever we deploy in the future. Cruise missiles that could fly under them. Nuclear-powered torpedoes that we wouldn't even detect. Powerful ICBMs that could attack us from the south, that would fly not over the North Pole but over the South Pole because all our defensive radars are north-facing. Maneuverable glide vehicles traveling at hypersonic speeds. You don't need any of these unless you're going up against defenses.

And now China is doing the same thing. China has a very small nuclear force, about 300 weapons, but they are now afraid that the U.S. is moving to a first-strike posture — just what we feared the Soviets were doing in the 1980s — that the United States is aiming to be able to launch a strike against their weapons and take them all out and then have a defensive system that can mop up the 10 or 20 that might be left. So China is doing what? It's deploying new weapons. It might grow its force. It might double its force, maybe even triple it. It's going to deceptive basing modes, mobile launchers and multiple silos, and it's developing weapons that can evade our defenses — so not just balloons and decoys, but maneuverable warheads that you can't track and therefore can't hit — and hyper-velocity systems that could glide in through the atmosphere at supersonic speeds. All these things are happening.

Our response to this is to do what? Our response is to develop more nuclear weapons to counter China. And if this sounds like an arms race, you're getting the point. This is exactly the dynamics of an arms race. We think everything we're doing is defensive, everything our adversary is doing is aggressive and offensive, and we're in the middle of it. This Ukraine crisis is going to make it worse. There's not a chance in hell that Congress is going to reduce the military budget of the United States, even though we're spending more on the military than we have since World War II. There's not a chance that we're going to cut that in the face of a war in Europe. There's not a chance we're going to cut nuclear weapons. Why? Because the first thing people go to in a crisis like this is strength, they want to feel strong. And if you're a Democrat, you're scared to death of looking weak on defense. The Democrats' default position on defensive issues is to look strong. Whether they think it's smart or not, what they're concerned about is appearance, not capability.

Horton: Yeah, exactly. "No. What if they call us wimps? We can't have that. We've got to act tough" — to appease all the bullies who are bullying us. Meanwhile, out here in the real world, that's the only thing that's presumably a little bit better about the Democrats, that they're a little bit less hawkish than the Republicans. Which is not even really true, they're just absolutely horrible — but it's rumored to be the one good thing about them, you know?

Cirincione: I mean, obviously, there's Democrats who do want to cut the budget. In fact, the Chairman of the House Armed Services Committee, Adam Smith, is one. He thinks we have way too many nuclear weapons. But does he do anything about it? No, he does not. Because he knows that the committee won't go along with him. He couldn't get the Democrats in his own committee to vote for cuts to the budget or to cut nuclear weapons, so now he doesn't even try. And I've got to tell you, a lot of what is the problem here is not just this political perception, but the hammerlock that the military corporations have on the Congress and on the Pentagon. We think about nuclear weapons the way we've been talking about them, as strategic moves and countermoves and defensive systems, but they're also a product. They're a product that a corporation makes, and corporations, as we now know, will sell you products that will kill you. Whether it's tobacco or opioids or false cures for COVID, if they can make a buck off it, somebody's going to sell it to you. The same is true for nuclear weapons.

We are set to spend $634 billion this decade on nuclear weapons — $634 billion, that is a very large market. Northrop Grumman and Lockheed and Boeing and Raytheon make a lot of money on nuclear weapons, and they don't want anything to disrupt that market. They deploy an army of lobbyists in Washington. They contribute generously to members of Congress. They have a revolving door, where members of the military cycle in between programs to build these weapons and corporate offices that sell these weapons. They flood Washington think tanks with grants to mute criticism of their programs. They advertise heavily in Washington media markets to sell these as instruments of peace, not instruments of war, and they succeed. I mean, you don't have to think too far back to when you saw strategic bombers flying over a football game, it happens all the time. We've accepted these into our American vision, and it works, it's very profitable, and it's very hard to shake. So it's not just politics; it's profits that are keeping us on the nuclear knife's edge.

Horton: That's such an important point. I'll speak from my own ignorance, and I know it's therefore fair for me to project it onto the rest of the population too, but you and I could talk about any kind of crooked government boondoggle all day and all night, and still the part of my brain that says, "You know, nuclear weapons is the same racket." I put that off until last to understand. Even knowing everything I knew about the military-industrial complex and how they pushed their crappy fighter jets and every other thing, somewhere in the back of my mind was this fantasy leftover from 1980s movies or my 1980s government school education, that essentially, nuclear weapons are a completely demand-side economy. The military tells the Congress how many nukes they need, and the Congress buys them for them.

But I mean, what? Are you telling me that you have H-bomb salesmen like you have used car salesmen? Or you have, say, gun companies who want that military contract or that FBI contract or something like that? People out there beating the bushes to see if they can get rid of H-bombs like an inventory of furniture down at the outlet? Yeah, that's exactly what it is: "Man, I've got to get rid of these H-bombs, Senator. You've got to help me get rid of these H-bombs." And that's exactly what it is, no different than any other government racket in America. That's what you're telling me here?

Cirincione: It is. We think that some things are sort of immune to market forces, but no, I'm telling you…

Horton: Those aren't market forces; those are contract forces. That's different, right?

Cirincione: Yes, they've got a product to sell. I was on the staff of the House Armed Services Committee and the Government Operations Committee for almost 10 years, and I was regularly visited by these lobbyists, taken out to lunch, when we could still do that. I heard their briefings. I went out to their facilities. I saw where they built them. I went to the congressional districts which are heavily dependent on these contracts for jobs. I visited the missile fields. I'm telling you, it is a very large, very expensive, very powerful nuclear-industrial complex. It is very, very hard to go against it, which is why they win.

You have to have a president who's willing to counter this. You can't do it at the Assistant Secretary level or even, quite frankly, at the Secretary of Defense level. You've got to have a president who is determined to cut it. That's why the only time we've really made progress on this is when we've had Republican presidents cutting the nuclear weapons. Reagan did it. George H. W. Bush did it. W. Bush actually cut the nuclear force in half, and he was concerned about different things, not nukes. Clinton? Obama? Nope. They kept the nuclear arsenal pretty much flat during their eight years in office, and Joe Biden unfortunately looks like he's going to do the same with his Nuclear Posture Review coming out in a few weeks — changes at the margin. The president has not made this a priority, and if the president doesn't, the machine wins.

Horton: Here's the silver lining that probably isn't going to matter because we're all going to die, but Joe, what's going on with the JCPOA, are they actually going to save that thing?

Cirincione: We are, this is the good news. One of the issues that has embroiled American politics for almost two decades now. We might be back to restoring the deal that shrank Iran's program; that put it under a microscope; that froze it for a generation; that Donald Trump foolishly tore up, saying that he was going to give us a better deal and that he was going to hammer Iran with maximum pressure that was going to cause them to comply or collapse. None of that happened. The situation got worse. Iran is closer to being able to build the material for a bomb now than they were during any other previous period. It looks like we're going to get that deal back perhaps this week, but most likely next week.

And here, even though the Russians are at war with Ukraine, the Russians are still cooperating in these talks. They're part of the talks. China is cooperating with these talks as they did, by the way, in 2014, when there was also a conflict in Ukraine and also the talks going on. So I think we're going to get this deal back. It'll be a big win for Biden, and it actually might be good that it happens during this time of the Ukraine crisis because it may take some of the political heat against it. There are always hardliners ready to slam any president who negotiates an agreement as an "appeaser," as "weak." I think Biden is going to be able to withstand that quite easily because the forces have changed. The countries in the region want the deal back. The military and intelligence officials in Israel want the deal back. The resistance to the deal is still there, still well funded, but it is much smaller than it has been in the past. So I think that is the silver lining. At least we're going to solve that nuclear problem.

Horton: And boy, this thing going on with Russia and Eastern Europe sure makes the Ayatollah and little old Persia seem like a sideshow, anyway. Can somebody just get the Assistant Deputy Secretary of State to take care of that while we're dealing with important business over here? You know what I mean? Which hopefully means cooling things off and backing down and trying to negotiate, not escalating.

Can you tell us, does this mean that they're abandoning all their dumb tough talk about abolishing the sunset provisions and adding provisions about Hezbollah and missiles and everything under the Sun? And does it mean they're really willing to lift the sanctions in the deal? Which Obama sort of kind of did but never really finished following through on, I don't believe. Trump certainly never followed through on them. So, in other words, are the Biden guys climbing down a few rungs on this ladder, enough that the Ayatollah is going to welcome them back into the deal? Or is it that the Iranians are really seeking some compromise here on their side, too? I don't know what they've got to do. I know they're not promising to abandon their mid-range missile program.

Cirincione: Every successful international agreement requires two things: that the sides compromise and that the sides are all able to declare victory. So it's got to look like a win for both sides. This is not the Japanese on the deck of the USS *Missouri* unconditionally surrendering. This is an agreement that is reached, and there's something in it for both sides. We're going to go back to the agreement we had, which, I will remind you, was working. It was impossible for Iran to build a nuclear bomb under that agreement. It reduced Iran's nuclear capability significantly. Whatever else Iran was doing, at least they didn't have a nuclear bomb to back it up. It also provided sanctions relief for Iran so that they could start to do business with the rest of the world. That's what they want. We want the limitations, the restrictions and the reductions in the program, and the Iranians want to be able to sell their oil, and guess what? We actually want them to sell their oil. You see what's happening to the price of oil this week? That Iran deal is going to increase the supply of oil globally, it's going to lower oil prices, it's going to have a positive impact on inflation in the West, particularly in the United States. So this is actually going to be a good deal. This is going to be a win-win. We're going to win for the reasons I just cited, and Iran is going to win because they're going to start making money again.

And as we're making this deal, it isn't that the countries in the region are afraid that Iran is now going to go on a tear, flush with petrodollars, and is going to start funding Hezbollah. No, what you see is that diplomacy is increasing in the Middle East. All of Iran's rivals — Saudi Arabia, United Arab Emirates, Qatar, Oman — they're all in diplomatic talks with Iran right now. There's talks of the president of Iran visiting Saudi Arabia. Boy, would that be a region-changing event! So you see that there's more good news than just the JCPOA coming back. It looks like we might be able to start getting dialogues going on the regional level that can address the regional issues, like support for Hezbollah, if Israel is willing to talk; the Saudi-Iran rivalry; or limits on Iran's ballistic missile program. But all of these have to follow the restoration of the JCPOA. That's the foundation for any further diplomacy. You can't do it all in one deal. You've got to build a new security regime brick by brick. That may be what the restoration of the JCPOA allows us to do.

Horton: Great. Alright. Thank you so much, Joe. It's really great to have you back on the show. I really appreciate it a lot.

Cirincione: My pleasure, Scott.

Horton: Thank you. Alright, you guys, that's Joe Cirincione, formerly at Ploughshares and now at the Quincy Institute for Responsible Statecraft, ResponsibleStatecraft.org, and you've got to check out this really important piece: "New Scientific Review Punctures the Myth of Missile Defense."

Epilogue

"It would be a miracle if
the atmosphere were
ignited. I reckon the
chance of a miracle to
be about ten percent."
— Enrico Fermi

Ethan Siegel:
Hotter Than the Sun
May 16, 2022

Scott Horton: Alright, you guys. Introducing Ethan Siegel. He is the Editor of *Starts with a Bang!*, which is the science contributors group at *Forbes* magazine. Isn't that interesting? He wrote this piece a couple of years ago, called "Ask Ethan: How Can a Nuclear Bomb Be Hotter Than the Center of Our Sun?" Welcome to the show, Ethan. How are you doing?

Ethan Siegel: Hi there, Scott. I'm doing very well. Thanks for having me on, and thanks for being willing to take on such an interesting and unintuitive question.

Horton: Well, I'm very interested in it. In fact, I'll go ahead and tell you: I'm putting out a book called *Hotter Than the Sun: Time to Abolish Nuclear Weapons*, and it's a compilation of interviews that I've done over the last 15 years or so with different experts about nuclear weapons. It occurred to me that nowhere in the book does anybody talk about how nukes are hotter than the Sun, so I thought, "Well, you know what? I'll just interview the guy who wrote the entire article about it, and then we'll make this the afterword." So, welcome. You're being transcribed.

Siegel: Well, that's great. I hope your audio transcription software is flawless.

Horton: I've got human men to help. You know, I've talked with Daniel Ellsberg, and he's talked with the guys who were there for real. They told him that at least some of them gave it a 10 percent chance — 1 in 10 — that the Trinity test would ignite all of the nitrogen and hydrogen in the atmosphere and the oceans and burn it all off and kill every last living organism on Earth.

Then they did it anyway, and it turned out: "Don't worry about it, man. It's not going to ignite all the nitrogen and hydrogen in the atmosphere. It's fine." But they didn't know that, yet they were willing to risk that. Then they set off the thermonuclear bombs that, as you explain, evidently burn even hotter than that. Luckily, they have not burned off our entire atmosphere as of yet, but it's interesting to note that that's the kind of mad science that we're dealing with here.

To stop beating around the bush, is it really possible that a nuclear bomb could burn hotter than the Sun? The center of our solar system? Say it ain't so, Ethan. Come on.

Siegel: The answer to that is absolutely, yes. You have to remember that when we talk about temperature, we talk about it as occurring in a region of space. Now, the Sun is enormous, so the Sun is going to have more heat than even if we launched and detonated all of the nuclear bombs at once all over the Earth. If we had all the nuclear bombs we've ever made and we detonated them all at once, this would still be minuscule in terms of the total energy output compared to that of the Sun. But in terms of the total energy output in a specific region of space, yes, the Sun's core is really hot, but it's not like that energy is very concentrated. That energy is spread out over an enormous volume of space, whereas when you detonate a nuclear weapon, that energy all gets released in one tiny, tiny volume of space, and that's where you can exceed the temperature of not just the surface of the Sun, but even of the absolute center of the Sun.

Horton: Well, hang on. How do you know how hot it is in the absolute center of the Sun, Mr. Science Man?

Siegel: That's so wonderful that you introduced me as "Mr. Science Man," because we have the sciences of astronomy and physics that tell us how the Sun works. We know, "Okay, what's the Sun made out of?" Well, we have both theory and observation, and that tells us about 70 percent of the Sun is made out of hydrogen, about 28 percent is helium, and about two percent is everything else combined. So most of the Sun is hydrogen.

Then you can say, "Okay, well, what goes on with this hydrogen in the Sun to make it happen?" And I tell you, "Thankfully, we have the science of nuclear physics pretty well figured out such that we know the conditions under which different atomic nuclei will react, to either fizz apart or fuse together." So you can ask, "Well, what's going on inside the Sun?" And you say, "Well, we're going to have hydrogen fusing into, through a chain reaction, the element helium." So you'll get protons and protons fusing together, and they will make deuterium, the first heavy isotope of hydrogen. Then you say, "We're going to take deuterium and either a proton or another deuterium atom, and we're going to make either hydrogen-three or helium-three." Then you say, "Okay, we're going to build that up step by step." At the temperatures we release in the Sun, you're going to say, "Okay, look at what we get out." We should get hydrogen fusing into helium, and we should be able to calculate how often this fusion reaction occurs.

Then you'll say, "Well, but how sure are you that you're correct?" And I'll say, "Well, very." Because one of the things we can say is that when you have these nuclear reactions, one type of particle they also produce is called neutrinos. And we've been detecting neutrinos from the Sun since the 1960s. It turns out that when you understand neutrinos and you understand nuclear fusion, you can say, "Alright, hydrogen fusion into helium only occurs when you get past a threshold of about 4 million Kelvin. That takes place pretty much in the inner 50 percent of the Sun's radius." Then you can say, "If you go all the way down to the core" — based on things like neutrino energy, based on things like fusion rate — "how hot does the Sun get at its absolute hottest in the very center?" And the answer is an enormous number, but it's still pretty small. It's only about 15 million Kelvin.

Horton: So if I set off a thermonuclear bomb — say, a megaton — how hot do those burn?

Siegel: When you say thermonuclear, that tells me that we're not just doing nuclear fission here. What we're doing is having a little fission bomb that's going to trigger this thermonuclear reaction — that's going to trigger nuclear fusion. The way you can do this is by having a nuclear fusion bomb. It's going to blow up in an outward direction but still inside of the bomb, and it's going to have a little chamber that has something like a hydrogen pellet in there. When that hydrogen gets surrounded by this detonation fission reaction, it's going to compress it, and that's going to trigger the thermonuclear part. That's going to trigger the hydrogen bomb part, which is going to get the hottest of all. And as a bonus, sometimes that hydrogen bomb will emit neutrons, which will cause the fission reaction on the outside to proceed even faster.

So the physical explanation is that we have all this stuff going on in the Sun, but it's got an enormous volume to it. The majority of fusion is occurring in the innermost few percent of the Sun, but that innermost few percent of the Sun is still larger — hundreds of times larger — than Earth. So when you say, "Okay, that's how the energy is distributed in the Sun, but how is the energy distributed in a nuclear bomb?" It's that it only happens in this tiny, tiny volume of space. The number of particles that fuse together — the number of fusion reactions in a given volume of this nuclear explosion — is much greater than the number of nuclear reactions that occur in a given volume of the Sun. They also take place over a much shorter amount of time in a nuclear explosion. The thing about the Sun is that it's relentless. It's fusing all of these protons together all the time. It's a continuous thing, but in a nuclear reaction, it happens in this tiny volume of space all at once. That's how a nuclear explosion can out-heat even the center of the Sun.

Horton: So luckily, the cool Pacific air is empty of these hydrogen isotopes, and the chain reaction does not continue throughout the atmosphere, the ocean and the rest. How quickly does the temperature fall off to reasonable sub-Sun temperatures?

Siegel: That's kind of the beauty of it. I'll go back to physics again. Have you ever gone and decided to take your lips and make a very small opening like you were puckering them up and then tried to blow air out of them? What does that air feel like if you hold your hand in front of your lips, when you put them together with just a tiny opening and you blow onto your hand?

That air feels cool, doesn't it? That's a little weird, because when you open your mouth wide and you breathe out slowly, you have hot air. In fact, you have body temperature air at about 98 degrees Fahrenheit coming out of your mouth. So why does that air feel cool? Why is that air cool when you have your mouth just make a tiny, tiny opening?

The answer is something called adiabatic expansion. There are a lot of ways that gas can expand, but if you just allow it to expand freely in the environment of space, if you allow it to expand rapidly, adiabatic expansion is what it's going to do. It's going to cool it down. Same thing if you adiabatically compress something — it will heat up. That is how the pistons in your car engine work. When the piston presses down, it heats the gasoline up and the gasoline combines with oxygen in the air and ignites. That's why you get these little explosions when the piston presses down, and that's what powers your car. That's what makes the engine turn. That's what makes the power used to turn the drivetrain and the wheels.

What happens when you ignite this? When you ignite this hydrogen bomb, this thermonuclear device, immediately you will get this enormous temperature that doesn't just get hotter than the Sun; it gets maybe 20 times hotter than the Sun. Whereas the Sun gets up to about 15 million degrees Kelvin, the thermonuclear test detonations done by the USSR and the U.S.A. have been recorded at hundreds of millions of degrees. They've gotten up to 200 or even 300 million degrees. The thing is, they only get there for a small fraction of a second. As the milliseconds tick by, the rapid, rapid heating of the air and the material that's in the air expands so rapidly that it starts to cool off very, very quickly. So a second after this thing detonates, is it still hotter than the Sun? The answer is no. Only for a fraction of a second does it reach those temperatures hotter than the Sun. That's good for us because that prevents it from doing things like triggering spontaneous combustion of the atmosphere and the oceans. That would be bad. Thankfully, that doesn't happen.

So the reason why these hydrogen bomb blasts are so much hotter than the interior of the Sun is the short time scale on which the explosion happens, the concentrated volume within which this reaction happens and the fact that now, once it goes off, it starts expanding rapidly, which means it cools rapidly. Therefore, it only stays at that ultra-high temperature for a very, very short period of time.

Horton: So it's still enough to kill all of Houston in one shot, but luckily not turn it into its own separate star.

Siegel: Right, and if you talk about the Trinity blast site, you can discover that there was a new type of mineral that was made during the atomic bomb explosion. The material is called Trinitite. The explosion was so hot that it baked the sand and baked the impurities within the sand, creating this radioactive material that we call Trinitite, which is this green, radioactive glass. There are only a few days a year that they allow you onto this site, but if you rent a Geiger counter and you go onto that site, you can still find these pieces of glass lying around.

Not just at millions of degrees, but even at thousands of degrees, the whole surrounding landscape can get baked. So you can say, "Oh, yeah. I'm not worried about being hotter than the Sun," but being way too hot for humans, for life, for the buildings we build, that's a very real concern. That's why these atomic bombs are so destructive over such a large region on Earth. Like you said, if you put a well-placed bomb over a city like Houston, *devastation* is the only word to describe it. It's going to be absolutely terrible.

Horton: When they're testing these things under the ocean, were they taking any greater risk of a chain reaction of the hydrogen or that kind of thing, since it's so much denser than the air and unable to cool off in the same kind of way?

Siegel: You say, "Okay, if I'm injecting all of this energy into the ocean, what's going to happen? Can I trigger the chemical reaction that turns water into hydrogen gas and oxygen gas?" Yes, you can. Then would that hydrogen gas explode under the heat by reacting with the oxygen gas? And you'll say, "Yeah, it will do that too," but you don't gain extra energy out of that, because the energy of the atomic bomb blast made the hydrogen and oxygen fuel out of water. Then when the hydrogen and oxygen react again, it just goes back into becoming water. So you don't get any extra net energy out.

The big problem you get from doing underwater tests is — remember, I talked about it earlier — that when you have these fusion devices, they produce free neutrons. Fission also relies on neutrons. If you have neutrons entering your water, that is going to produce deuterium when one neutron hits a hydrogen nucleus. Deuterium is no big deal — it's stable, and it's not poison. You don't want to drink D_2O instead of H_2O for a long time, but getting a little bit of deuterium in your water isn't bad. The problem comes when you add a second neutron to hydrogen and you make something called Tritium. Tritium is a radioactive waste product with a half-life of about 12 years. So you do run the risk of injecting an enormous amount of radioactivity into your water when you perform an underwater test. This is actually a very big problem for nuclear power plants. Nuclear power plants use water to cool their reactors, and one of the byproducts is tritiated water, which is this radioactive water.

If you're a fan of *The Simpsons*, you might remember Monty Burns's three-eyed fish, Blinky. The show was lampooning it, but the radioactivity of your water is also a realistic problem that you do have to worry about. So even though we decided that we don't want open-air tests because of radioactivity, it turns out that underwater nuclear tests are also pretty bad for the environment. If you have to conduct a nuclear test — and I hope you don't, but if you have to — your best bet is to conduct it underground so that any radioactivity remains safely buried as well as possible.

Most of the byproducts you produce do decay pretty fast. The ones that decay fast and wind up in the air tend to be what we call alpha emitters. They tend to emit helium nuclei, and you can stop those with pretty much the outer layer of your skin cells. The danger is when you inhale them, because if you inhale them and they sit in your lungs and are radioactively decaying in there, that's a wonderful recipe for cancer. I would recommend against doing something like that. So if a nuclear bomb goes off, don't go outside and take a bunch of deep breaths to try and get as much of that debris into your lungs as possible. That would be bad. But hopefully, we won't have World War III where it's Vladimir Putin versus the entire world and we won't suffer from mass nuclear fallout.

Horton: Well, against the West anyway. But that's a good question: How far out of town should I be if they nuke Austin?

Siegel: Assuming that they can aim properly, there's the initial blast radius, where if you're caught in that, you're going to die immediately. Outside of that, there's a radiation zone where you will die that horrible, painful death over a long period of time like many of the residents of Hiroshima did over the decade following the nuclear attack on them in 1945. So you want to get beyond the blast zone and the radiation zone, and although that typically depends on the type of bomb used, I would say that if you are more than about 20 or 30 miles away from the detonation site, you should consider yourself safely out of the radiation zone.

Horton: Then as far as the fallout and all that, you think a few weeks would be good enough before you can go outside?

Siegel: A lot of that will depend on weather conditions, and that's not something that's really predictable at this point more than a week out, under the best of circumstances. So I would say that just like we have monitoring where we know toxic things are going to occur — like at Hawaii, where they have the Hawaiian Volcano Observatory that tells you when it's safe and when it's not safe to breathe in that combination of volcanic gases and smog — I think we would need to wait for the announcements for something like that. I have a lot of faith that that is something that the United States government and the scientists working for the government would take an interest in broadcasting in order to keep people safe in such an event.

Horton: Yeah, if any of them are still alive after getting us into such a mess.

Siegel: It's not a pleasant situation to think about.

Horton: That would be funny — power devolves to the head of the National Weather Service.

Siegel: It's important, though, because you get these nuclear fallouts and you get these radioactive particles going in various places. Just like different types of pollution can land in different places, if something went off in Austin and then came around the world and started landing in Anchorage, you would want to make sure that the residents of Anchorage are not all dying of radiation poisoning because you failed to track where the nuclear debris went.

Horton: Alright. Well, I wonder how prepared they are for the National Weather Service after the nuclear apocalypse to keep track of the fallout clouds. They'd better be planning ahead over there at the RAND Corporation.

Siegel: [Laughing] You're getting all Howard Hughes on me.

Horton: They're the guys who had the plan for, "Here's how we can provoke Russia without probably leading to too bad of a backlash." So I hope they're the ones who are getting the National Weather Service prepared for Dr. Strangelove days, here.

Siegel: Well, may the future be kind to us, and may we never need to find out.

Horton: Seriously. Alright, you heard him. It's Ethan Siegel. The bad news is the H-bombs. Yeah, they burn at a few hundred million Kelvin, but only for a moment, and then it's cool after that, so we can rest assured. Thank you so much, Ethan.

Siegel: Oh, thank you. It was my pleasure to be here.

Appendix A
The Interviews

Part 1: The Threat of Nuclear War

Hans Kristensen: the Bleak Outlook for Nuclear Arms Control
June 26, 2020

https://scotthorton.org/interviews/6-26-20-hans-kristensen-on-the-bleak-outlook-for-nuclear-arms-control/

Chas W. Freeman: the Threat of Nuclear War with China
February 22, 2019

https://scotthorton.org/interviews/2-22-19-chas-w-freeman-on-the-threat-of-nuclear-war-with-china/

Gilbert Doctorow: Avoiding Nuclear War with Russia
February 22, 2019

https://scotthorton.org/interviews/2-22-19-gilbert-doctorow-on-avoiding-nuclear-war-with-russia/

Lawrence Wittner: Trying a Little Nuclear Sanity
March 22, 2012

https://scotthorton.org/interviews/3-22-12-lawrence-wittner-on-trying-a-little-nuclear-sanity/

Tom Collina: America's Dangerous "Nuclear Sponge"
December 6, 2019

https://scotthorton.org/interviews/12-6-19-tom-collina-on-americas-dangerous-nuclear-sponge/

Colleen Moore: America's Nuclear Arms Nightmare
April 10, 2020

https://scotthorton.org/interviews/4-10-20-colleen-moore-on-americas-nuclear-arms-nightmare/

Conn Hallinan: the New Nuclear Arms Race
May 8, 2017

https://scotthorton.org/interviews/5-8-17-conn-hallinan-on-the-new-nuclear-arms-race/

Daniel Ellsberg: *The Doomsday Machine* and Nuclear Winter
August 31, 2018
https://scotthorton.org/interviews/8-31-18-daniel-ellsberg-on-the-doomsday-machine-and-nuclear-winter/

Andrew Cockburn: How Easy It Is to Start a Nuclear War
July 20, 2018
https://scotthorton.org/interviews/7-20-18-andrew-cockburn-on-how-easy-it-is-to-start-a-nuclear-war/

Michael Klare: the Threat of War with North Korea
November 22, 2017
https://scotthorton.org/interviews/112217-michael-klare-on-the-threat-of-war-with-north-korea/

Conn Hallinan: the Risk of War Between India and Pakistan
December 13, 2016
https://scotthorton.org/interviews/121616-conn-hallinan-on-the-risk-of-war-between-india-and-pakistan/

Ray McGovern: Russia's Latest Nuclear Weapons
March 6, 2018
https://scotthorton.org/interviews/3-6-18-ray-mcgovern-on-russias-latest-nuclear-weapons/

Part 2: The Nuclear-Industrial Complex

Darwin BondGraham: the New START Treaty
November 2, 2010
https://scotthorton.org/interviews/11-2-10-darwin-bond-graham-on-the-new-start-treaty/

Kelley B. Vlahos on the Insidious Nuclear Weapons Industry
July 17, 2017
https://scotthorton.org/interviews/71717-kelley-b-vlahos-on-the-insidious-nuclear-weapons-industry/

Len Ackland: Obama Breaking His Pledge Not to Build Nukes
August 3, 2015
https://scotthorton.org/interviews/8-3-15-len-ackland-on-obama-breaking-his-pledge-not-to-build-nukes/

Michael Klare: the Catastrophic Consequences of Nuclear AI
December 19, 2018

https://scotthorton.org/interviews/12-19-18-michael-klare-on-the-catastrophic-consequences-of-nuclear-ai/

William Hartung: the Existential Threat of the Nuclear Weapons Lobby
October 22, 2021

https://scotthorton.org/interviews/10-22-21-william-hartung-on-the-existential-threat-of-the-nuclear-weapons-lobby/

Part 3: About Those Nuclear "Rogue" States

Grant F. Smith: the Niger Uranium Forgeries
May 24, 2019

https://scotthorton.org/interviews/5-24-19-grant-smith-on-the-niger-uranium-forgeries/

Joe Cirincione: Syria, North Korea, Pakistan, Iran and the Bomb
October 8, 2008

https://scotthorton.org/interviews/10-8-08-joe-cirincione-on-syria-north-korea-pakistan-iran-and-the-bomb/

Grant F. Smith: Israeli Theft of U.S. Nuclear Material
April 14, 2010

https://scotthorton.org/interviews/4-14-10-grant-f-smith-on-israeli-theft-of-us-nuclear-material/

Grant F. Smith: U.S. Violating Its Own Laws for Israeli Nukes
July 2, 2018

https://scotthorton.org/interviews/7-2-18-grant-f-smith-on-us-violating-its-own-laws-for-israeli-nukes/

Gordon Prather: Iran and North Korea's Nuclear Programs
August 3, 2019

https://scotthorton.org/interviews/8-3-09-gordon-prather-on-iran-and-north-koreas-nuclear-programs/

Lt. General Robert G. Gard: Loose Nukes and Iran's Program
April 15, 2010

https://scotthorton.org/interviews/4-15-10-lt-general-robert-g-gard-jr-on-loose-nukes-and-irans-program/

Seymour Hersh: Iran and the IAEA
November 22, 2011
https://scotthorton.org/interviews/11-22-11-seymour-hersh-on-iran-and-the-iaea/

Gareth Porter: the Ayatollahs' Fatwas Against WMD
October 17, 2014
https://scotthorton.org/interviews/10-17-14-gareth-porter-on-the-ayatollahs-fatwas-against-wmd/

Gareth Porter: the Iran Nuclear Agreement
August 18, 2015
https://scotthorton.org/interviews/8-18-15-gareth-porter-on-the-iran-nuclear-agreement/

Jim Lobe: Pro-Israel Supporters Working Against the Iran Nuclear Agreement
September 4, 2015
https://scotthorton.org/interviews/9-4-15-jim-lobe-on-pro-israel-supporters-working-against-the-iran-nuclear-agreement/

Andrew Cockburn: U.S. Support for Pakistan's Nuclear Weapons Program
December 15, 2009
https://scotthorton.org/interviews/12-15-09-andrew-cockburn-on-us-support-for-pakistans-nuclear-weapons-program/

Doug Bandow: North Korea's Nuclear Weapons
December 13, 2017
https://scotthorton.org/interviews/121317-doug-bandow-on-north-koreas-nuclear-weapons/

Tim Shorrock: the Prospects for Peace with North Korea
December 18, 2020
https://scotthorton.org/interviews/12-18-20-tim-shorrock-on-the-prospects-for-peace-with-north-korea/

Gareth Porter: Israeli Fabrication Almost Led to War with Iran
May 1, 2020
https://scotthorton.org/interviews/5-1-20-gareth-porter-on-the-israeli-fabrication-that-almost-led-to-war-with-iran/

Part 4: Hiroshima and Nagasaki

Daniel Ellsberg: Hiroshima and the Danger of 100 Holocausts
August 5, 2011

https://scotthorton.org/interviews/8-05-11-daniel-ellsberg-on-hiroshima-and-the-danger-of-100-holocausts/

Peter Van Buren: the Hiroshima Myth
August 23, 2019

https://scotthorton.org/interviews/8-23-19-peter-van-buren-on-the-hiroshima-myth/

Gar Alperovitz: the Decision to Nuke Japan
August 27, 2019

https://scotthorton.org/interviews/8-27-19-gar-alperovitz-on-the-decision-to-nuke-japan/

Anthony Weller: My Father's Lost Dispatches from Nagasaki
August 10, 2007

https://scotthorton.org/interviews/8-10-07-anthony-weller-on-his-fathers-lost-dispatches-from-nagasaki/

Greg Mitchell: the Real History of Hiroshima and Nagasaki
July 31, 2020

https://scotthorton.org/interviews/7-31-20-greg-mitchell-on-the-real-history-of-hiroshima-and-nagasaki/

Josiah Lippincott: the Wholesale Slaughter of Japanese Civilians in WWII
September 11, 2020

https://scotthorton.org/interviews/9-11-20-josiah-lippincott-on-the-wholesale-slaughter-of-japanese-civilians-in-wwii/

Part 5: Protesting Omnicide

Paul Kawika Martin: the New Treaty to Ban Nuclear Weapons
March 29, 2017

https://scotthorton.org/interviews/03-29-17-paul-kawika-martin-on-the-new-treaty-to-ban-nuclear-weapons/

Frida Berrigan: a Childhood Ruined by Nuclear Weapons
April 26, 2019

https://scotthorton.org/interviews/4-26-19-frida-berrigan-on-a-childhood-ruined-by-nuclear-weapons/

Elizabeth McAlister: Nuclear Winter and the Kings Bay Plowshares 7
October 18, 2019

https://scotthorton.org/interviews/10-18-19-elizabeth-mcalister-on-nuclear-winter-and-the-kings-bay-plowshares-7/

Francis Boyle: In Defense of the Kings Bay Plowshares 7
November 7, 2019

https://scotthorton.org/interviews/11-7-19-francis-boyle-in-defense-of-the-kings-bay-plowshares-7/

Joe Cirincione: U.S. Actions Don't Justify Putin's Attack, But They Set the Stage for It
February 24, 2022

https://scotthorton.org/interviews/2-24-22-joe-cirincione-us-actions-dont-justify-putins-attack-but-they-set-the-stage-for-it/

Epilogue

Ethan Siegel: Hotter Than the Sun
May 16, 2022

https://scotthorton.org/interviews/5-16-22-ethan-siegel-hotter-than-the-sun/

And finally, Scott's ten-part, six-hour interview of Gareth Porter on his definitive book *Manufactured Crisis: The Untold Story of the Iran Nuclear Scare*, from March 2014. (Not featured in this book, unfortunately, but for obvious reasons.)

https://scotthorton.org/interviews/gareth-porter-interview-series-on-his-new-book-manufactured-crisis-the-untold-story-of-the-iran-nuclear-scare/

For the full playlist of these interviews and more information go to HotterThanTheSun.com.

Appendix B
Who Opposed Nuking Japan?

"The Japanese were ready to surrender and it wasn't necessary to hit them with that awful thing."

— General Dwight D. Eisenhower

"In 1945 Secretary of War Stimson, visiting my headquarters in Germany, informed me that our government was preparing to drop an atomic bomb on Japan. I was one of those who felt that there were a number of cogent reasons to question the wisdom of such an act. … The Secretary, upon giving me the news of the successful bomb test in New Mexico, and of the plan for using it, asked for my reaction, apparently expecting a vigorous assent. During his recitation of the relevant facts, I had been conscious of a feeling of depression and so I voiced to him my grave misgivings, first on the basis of my belief that Japan was already defeated and that dropping the bomb was completely unnecessary, and secondly because I thought that our country should avoid shocking world opinion by the use of a weapon whose employment was, I thought, no longer mandatory as a measure to save American lives. It was my belief that Japan was, at that very moment, seeking some way to surrender with a minimum loss of 'face.' The Secretary was deeply perturbed by my attitude."

—Dwight D. Eisenhower

"The use of the atomic bomb, with its indiscriminate killing of women and children, revolts my soul."

—President Herbert Hoover

"[T]he Japanese were prepared to negotiate all the way from February 1945… up to and before the time the atomic bombs were dropped. … [I]f such leads had been followed up, there would have been no occasion to drop the bombs."

— Herbert Hoover

"I told [General Douglas] MacArthur of my memorandum of mid-May 1945 to Truman, that peace could be had with Japan by which our major objectives would be accomplished. MacArthur said that was correct and that we would have avoided all of the losses, the Atomic bomb, and the entry of Russia into Manchuria."

— Herbert Hoover

"MacArthur's views about the decision to drop the atomic bomb on Hiroshima and Nagasaki were starkly different from what the general public supposed. When I asked General MacArthur about the decision to drop the bomb, I was surprised to learn he had not even been consulted. What, I asked, would his advice have been? He replied that he saw no military justification for the dropping of the bomb. The war might have ended weeks earlier, he said, if the United States had agreed, as it later did anyway, to the retention of the institution of the emperor."

— Norman Cousins

"General MacArthur definitely is appalled and depressed by this Frankenstein monster. I had a long talk with him today, necessitated by the impending trip to Okinawa. He wants time to think the thing out, so he has postponed the trip to some future date to be decided later."

— Weldon E. Rhoades
Gen. Douglas MacArthur's pilot

"[Gen. Douglas] MacArthur once spoke to me very eloquently about it, pacing the floor of his apartment in the Waldorf. He thought it a tragedy that the bomb was ever exploded. MacArthur believed that the same restrictions ought to apply to atomic weapons as to conventional weapons, that the military objective should always be limited damage to noncombatants. … MacArthur, you see, was a soldier. He believed in using force only against military targets, and that is why the nuclear thing turned him off."

— President Richard Nixon

"The Japanese were ready for peace, and they already had approached the Russians and the Swiss. And that suggestion of giving a warning of the atomic bomb was a face-saving proposition for them, and one that they could have readily accepted. In my opinion, the Japanese war was really won before we ever used the atom bomb."

— Ralph Bird, Under Secretary of the Navy

"I concluded that even without the atomic bomb, Japan was likely to surrender in a matter of months. My own view was that Japan would capitulate by November 1945. Even without the attacks on Hiroshima and Nagasaki, it seemed highly unlikely, given what we found to have been the mood of the Japanese government, that a U.S. invasion of the islands scheduled for 1 November 1945 would have been necessary."

— Paul Nitze, director, later Vice Chairman of the
Strategic Bombing Survey

"[E]ven without the atomic bombing attacks, air supremacy over Japan could have exerted sufficient pressure to bring about unconditional surrender and obviate the need for invasion. Based on a detailed investigation of all the facts, and supported by the testimony of the surviving Japanese leaders involved, it is the Survey's opinion that certainly prior to 31 December 1945, and in all probability prior to 1 November 1945, Japan would have surrendered even if the atomic bombs had not been dropped, even if Russia had not entered the war, and even if no invasion had been planned or contemplated."

— U.S. Strategic Bombing Survey, 1946

"The use of this barbarous weapon at Hiroshima and Nagasaki was of no material assistance in our war against Japan. The Japanese were already defeated and ready to surrender because of the effective sea blockade and the successful bombing with conventional weapons. ... The lethal possibilities of atomic warfare in the future are frightening. My own feeling was that in being the first to use it, we had adopted an ethical standard common to the barbarians of the Dark Ages. I was not taught to make war in that fashion, and wars cannot be won by destroying women and children."

— Fleet Admiral William D. Leahy
Chief of Staff to President Truman

"Truman told me it was agreed they would use it, after military men's statements that it would save many, many American lives, by shortening the war, only to hit military objectives. Of course, then they went ahead and killed as many women and children as they could, which was just what they wanted all the time."

—William D. Leahy

"The Japanese had, in fact, already sued for peace before the atomic age was announced to the world with the destruction of Hiroshima and before the Russian entry into the war. The atomic bomb played no decisive part, from a purely military standpoint, in the defeat of Japan."

— Fleet Admiral Chester W. Nimitz
Commander in Chief of the U.S. Pacific Fleet

"The Japanese position was hopeless even before the first atomic bomb fell, because the Japanese had lost control of their own air."

— General "Hap" Arnold, commanding general of
U.S. Army Air Forces in World War II

"It always appeared to us that, atomic bomb or no atomic bomb, the Japanese were already on the verge of collapse."

— "Hap" Arnold

"[Gen.] Arnold's view was that the dropping of the atomic bomb was totally unnecessary. He said he knew the Japanese wanted peace. There were political implications in the decision, and Arnold did not feel it was the military's job to question them. … [Arnold's view was]: when the question comes up of whether we use the atomic bomb or not, my view is that the Air Force will not oppose the use of the bomb, and they will deliver it effectively if the Commander-in-Chief decides to use it. But it is not necessary to use it in order to conquer the Japanese without the necessity of a land invasion."

— General Ira Eaker, Deputy Commander of U.S. Army Air Forces

"The war would have been over in two weeks without the Russians entering and without the atomic bomb. … The atomic bomb had nothing to do with the end of the war at all. The war would've ended anyway."

— Major General Curtis LeMay XXI Bomber Command

"[LeMay said] if we'd lost the war, we'd all have been prosecuted as war criminals. And I think he's right. He, and I'd say I, were behaving as war criminals. LeMay recognized that what he was doing would be thought immoral if his side had lost. But what makes it immoral if you lose and not immoral if you win?" [A reference to firebombing Tokyo, but still. –Ed.]

— Lieutenant Colonel, later Secretary of Defense Robert McNamara

"Just when the Japanese were ready to capitulate, we went ahead and introduced to the world the most devastating weapon it had ever seen and, in effect, gave the go-ahead to Russia to swarm over Eastern Asia. Washington decided it was time to use the A-bomb. I submit that it was the wrong decision. It was wrong on strategic grounds. And it was wrong on humanitarian grounds."

— Ellis Zacharias, Deputy Director of the Office of Naval Intelligence

"The first atomic bomb was an unnecessary experiment. … It was a mistake to ever drop it. … [The scientists] had this toy and they wanted to try it out, so they dropped it."

— Fleet Admiral William Halsey Jr.

"When we didn't need to do it, and we knew we didn't need to do it, and they knew that we knew we didn't need to do it, we used them as an experiment for two atomic bombs. Many other high-level military officers concurred."

— Brigadier General Carter Clarke, Military Intelligence officer in charge of preparing summaries of intercepted Japanese cables for President Truman and his advisers

"The commander in chief of the U.S. Fleet and Chief of Naval Operations, Ernest J. King, stated that the naval blockade and prior bombing of Japan in March 1945, had rendered the Japanese helpless and that the use of the atomic bomb was both unnecessary and immoral."

— Carter Clarke

"In the light of available evidence I myself and others felt that if such a categorical statement about the retention of the dynasty had been issued in May 1945, the surrender-minded elements in the Japanese government might well have been afforded by such a statement a valid reason and the necessary strength to come to an early clear-cut decision. If surrender could have been brought about in May 1945, or even in June, or July, before the entrance of Soviet Russia into the Pacific war and the use of the atomic bomb, the world would have been the gainer."

— Under Secretary of State Joseph Grew

"I proposed to Secretary Forrestal that the weapon should be demonstrated before it was used. ... The war was very nearly over. The Japanese were nearly ready to capitulate. ... My proposal... was that the weapon should be demonstrated over... a large forest of cryptomeria trees not far from Tokyo. ... [This] would lay the trees out in windrows from the center of the explosion in all directions as though they were matchsticks, and, of course, set them afire in the center. It seemed to me that a demonstration of this sort would prove to the Japanese that we could destroy any of their cities at will. ... Secretary Forrestal agreed wholeheartedly with the recommendation. ... It seemed to me that such a weapon was not necessary to bring the war to a successful conclusion, that once used it would find its way into the armaments of the world."

— Lewis Strauss, Special Assistant to the Secretary of the Navy

Acknowledgements

Thank you very much especially to Jared Wall for transcribing virtually all of these interviews. Thanks also to my copy editor Ben Parker, and, as always, Mike Dworski and Grant F. Smith for their assistance in the preparation of the final drafts for publication.

Many thanks to all my assistants and producers over the years who have helped me arrange and edit all these interviews, including Angela Keaton, Michael Couvillion, Gage Counts, Nico Palimino, Aaron Vaughan, Ed Huff, August Wagele, Adam McDonald, Mark Jokel, Damon Hatheway, Phil Brown, Sam Hage and Connor O'Keefe.

And I am eternally grateful also of course to all my colleagues at the Libertarian Institute and Antiwar.com, and all my great guests and listeners over the years for your great insights and support.

About the Author

Scott Horton is director of the Libertarian Institute, editorial director of Antiwar.com, host of *Antiwar Radio* on Pacifica, 90.7 FM KPFK in Los Angeles, California and podcasts the *Scott Horton Show* from ScottHorton.org. He's the author of the 2021 book *Enough Already: Time to End the War on Terrorism*, the 2017 book, *Fool's Errand: Time to End the War in Afghanistan* and editor of the 2019 book *The Great Ron Paul: The Scott Horton Show Interviews 2004–2019*. He's conducted more than 5,700 interviews since 2003.

In 2007, Horton won the *Austin Chronicle*'s "Best of Austin" award for his Iraq war coverage on *Antiwar Radio*.

Scott's articles have appeared at Antiwar.com, *The American Conservative* magazine, the History News Network, *The Future of Freedom*, *The National Interest* and the *Christian Science Monitor*. He was featured in the 2019 documentary *An Endless War: Getting Out of Afghanistan* and contributed a chapter to the 2019 book, *The Impact of War*.

He is a regular guest on *Kennedy Nation* on Fox Business Channel.

Scott lives in Austin, Texas with his wife Larisa Alexandrovna Horton.

The Scott Horton Show and The Libertarian Institute

Listen to *Antiwar Radio* every Sunday morning on 90.7 FM KPFK in Los Angeles and sign up for Scott's daily email, the *Scott Horton Show* podcast feed and check out the full interview archive, at scotthorton.org. To interview Scott, email his producer Ed Huff, ed@scotthorton.org.

Regular donors of $5 or more per month by way of Patreon or Paypal get access to the /r/scotthortonshow group on Reddit.com. Info at scotthorton.org/donate.

Also, check out The Libertarian Institute at libertarianinstitute.org. It's Scott Horton, Sheldon Richman, Kyle Anzalone, Keith Knight and the best libertarian writers and podcast hosts on the internet. We are a 501(c)(3) tax-exempt charitable organization. EIN 83-2869616.

Help support our efforts — including our project to purchase wholesale copies of this book to send to important congressmen and women, antiwar groups and influential people in the media. We don't have a big marketing department to push this effort. We need your help to do it. And thank you.

libertarianinstitute.org/donate or

The Libertarian Institute
612 W. 34th St.
Austin, TX 78705

Check out all of our other great Libertarian Institute books at libertarianinstitute.org/books:

Enough Already: Time to End the War on Terrorism by Scott Horton
Fool's Errand: Time to End the War in Afghanistan by Scott Horton
The Great Ron Paul*: The Scott Horton Show Interviews 2004–2019*
No Quarter: The Ravings of William Norman Grigg edited by Tom Eddlem
Coming to Palestine by Sheldon Richman
What Social Animals Owe to Each Other by Sheldon Richman

Keep a look out for more great titles to be published in 2022.

Selected praise for Scott Horton's previous book *Fool's Errand*

"Scott Horton's *Fool's Errand* is a deeply insightful and well-informed book on America's longest war, explaining why it remains as unwinnable as it ever was." — Patrick Cockburn, Middle East correspondent for the *Independent*, author of *The Age of Jihad: Islamic State and the Great War for the Middle East*

"An incisive, informative analysis of the Afghan fiasco and how we got there, scrubbed clean of propaganda and disinformation. Horton captures the situation very well indeed. I much enjoyed reading it." — Eric S. Margolis, author of *War at the Top of the World: The Struggle for Afghanistan, Kashmir and Tibet* and *American Raj: Liberation or Domination? Resolving the Conflict Between the West and the Muslim World*

"Scott Horton's *Fool's Errand: Time to End the War in Afghanistan* is a definitive, authoritative and exceptionally well-resourced accounting of America's disastrous war in Afghanistan since 2001. Scott's book deserves not just to be read, but to be kept on your shelf, because as with David Halberstam's *The Best and Brightest* or Neil Sheehan's *A Bright Shining Lie*, I expect Horton's book to not just explain and interpret a current American war, but to explain and interpret the all too predictable future American wars, and the unavoidable waste and suffering that will accompany them." — Capt. Matthew Hoh, USMC (ret.), former senior State Department official, Zabul Province, Afghanistan

"*Fool's Errand* is a hidden history of America's forgotten war, laid bare in damning detail. Scott Horton masterfully retells the story of America's failed intervention, exposes how Obama's troop surge did not bring Afghanistan any closer to peace, and warns that the conflict could go on in perpetuity — unless America ends the war. Horton shows why the answer to a brutal civil war is not more war. *Fool's Errand* is a scintillating and sorely needed chronicle of the longest war in American history." — Anand Gopal, journalist and author of *No Good Men Among the Living: America, the Taliban, and the War Through Afghan Eyes*

"Scott Horton's *Fool's Errand: Time to End the War in Afghanistan* is a brilliant achievement and a great read. I recommended it to the faculty at the Army Command and General Staff College to be part of the course work. It's that important." — Col. Douglas Macgregor, U.S. Army (ret.), author of *Warrior's Rage: The Great Tank Battle of 73 Easting*

"Scott Horton's *Fool's Errand* is really a great book, the best one I know on this convoluted (and irritating) topic." — Noam Chomsky, author of *Hegemony or Survival: America's Quest for Global Dominance*

"A lot of people think of the war in Iraq as the bad war, but Afghanistan as the good and justifiable war. That convenient view does not survive Scott Horton's careful and incisive demolition." — Thomas E. Woods Jr., author of *Nullification: How to Resist Federal Tyranny in the 21st Century* and *Rollback: Repealing Big Government Before the Coming Fiscal Collapse*

"America's longest war — in Afghanistan — has until now been among America's least documented. Horton brings together far more than 16 years of conflict, drawing in sources from well before most Americans even heard of Osama bin Laden to show how the Afghan quagmire's roots are deep. The title tells it all, however: this war cannot be won, and 'victory' will be in the form of escape. Meticulously researched and footnoted, *Fool's Errand* is required reading." — Peter Van Buren, retired foreign service officer and author of *We Meant Well: How I Helped Lose the War for the Hearts and Minds of the Iraqi People* and *Hooper's War: A Novel of World War II Japan*

"Why is the United States still fighting in Afghanistan? In this timely new book, Scott Horton explains why America's longest war is strategically misguided and why getting out would make the United States safer and advance America's broader national interests. Even readers who do not share Horton's libertarian world-view are likely to find themselves nodding in agreement: the war in Afghanistan has indeed become a 'fool's errand.'" — Stephen M. Walt, professor of international affairs, Harvard University, co-author of *The Israel Lobby and U.S. Foreign Policy*

"Scott Horton's *Fool's Errand* makes a well-researched and compelling case that American policies in Afghanistan were ill-conceived from the outset and doomed to fail. Chronicling one unsuccessful American initiative after another in a seemingly endless war, Horton argues that our continued military efforts now yield little more than revenge-minded blowback. There is no light at the end of the tunnel, and no one can imagine realistically that one will appear. It will no doubt be a bitter pill to acknowledge that the Beltway fantasy of a united, democratic, pluralist and feminist Afghanistan will never be achieved, but it is one we will have to swallow eventually. Since the cost of pursuing unrealistic goals is exorbitant in blood and treasure, the sooner our failure is faced up to, the better." — Scott McConnell, founding editor of *The American Conservative* magazine

Selected praise for Scott Horton's previous book *Enough Already*

"If you only read one book this year on America's unending 'War on Terror,' it should be this persuasive and devastatingly damning account of how the United States created the original al Qaeda terrorism threat by its own actions and then increased that threat by orders of magnitude by its wanton killings in one country after another in the name of 'counterterrorism.' Once I started reading it, I couldn't stop!" — Daniel Ellsberg, *Pentagon Papers* whistleblower and author of *The Doomsday Machine: Confessions of a Nuclear War Planner*

"Nothing has fueled the abuse of government power in the last 20 years like the 'War on Terrorism.' Scott Horton's essential new book, *Enough Already*, is the key to understanding why it's not too late to end the wars and save our country. Three administrations in a row have promised us a more restrained foreign policy. It is time we insisted on it." — Ron Paul, M.D., former congressman and author of *Swords into Plowshares: A Life in Wartime and a Future of Peace and Prosperity*

"With outstanding scholarship, research and analysis, Scott Horton's new book, *Enough Already*, lays bare the logical absurdity and self-defeating nature of America's permanent-war establishment. It might have taken its title from a line in the book's introduction: America's war policy since at least the Carter Administration has been 'a policy in search of a reason.' As Horton painstakingly explains, in virtually none of the military conflicts the United States has chosen to fight in since the 1970s was our security ever genuinely threatened. His ultimate solution is the only one that has any chance of preserving American security and giving us a chance to be ready in case we do face a genuine threat in the future: end the pointless and self-defeating forever wars. All of them." — Lt. Col. Daniel L. Davis, USA (Ret.) four-time combat deployer, two-time winner of the Bronze Star and author of *Eleventh Hour in 2020 America: How American Foreign Policy Got Jacked Up and How the Next Administration Can Fix It*

"Scott Horton is a walking encyclopedia of U.S. foreign policy, and he has managed to chronicle the entire history of the government's meddling in the Middle East in this well-written and important volume. *Enough Already* explains why still today the U.S. government continues to bomb countries at its caprice, long after the perpetrators of the September 11, 2001, attacks have disappeared." — Laurie Calhoun, author of *We Kill Because We Can: From Soldiering to Assassination in the Drone Age*

"I finished reading *Enough Already* the same week I had to attend the funeral of a sailor from one of my Iraq deployments. He killed himself leaving behind a wife and three boys. Nothing in this book is simply historical or abstract to tens of millions of families. Scott Horton has written an incredible accounting of the wars of the last twenty years. This book should be used to hold accountable those who purposefully committed these crimes and to remember the generations of Iraqis, Afghans, Somalis, Yemenis, Pakistanis, Palestinians, Libyans, Iranians, Syrians, sub-Saharan Africans and Americans whose lives have been forever damaged and destroyed." — Capt. Matthew Hoh, USMC (ret.), senior fellow at the Center for International Policy

"Scott Horton is one of the best informed and hardest hitting critics of the War on Terror. His new book is a gold mine for anyone seeking to learn about the frauds and failures of U.S. foreign policy." — Jim Bovard, columnist at *USA Today* and author of *Public Policy Hooligan*

"There is no better title for Scott Horton's ambitious new book than *Enough Already* because honestly, those are the two words that floated through my head the entire time I was reading it. U.S. foreign policy for the last 20 years has been an endless parade of regime changes, useless attempts to 'fix' our blunders, extra-legal killing and detention, and military interventions that have made troubled states failed states. It never ends. Whether it be our destruction of Iraq, the illegal drone wars, JSOC manhunting or Hillary Clinton's last stand in Libya, Scott has expertly harnessed mountains of detail here in a compelling narrative that underscores what a rotten mess the War on Terrorism has become. More importantly it shows that no matter how righteous the mission to go after the 9/11 perpetrators seemed at the time, the American government has managed to twist those goals into something perverse and ultimately more dangerous to the world than the 19 hijackers who changed everything that fateful day. The book is grim but persuasive, and a must-read for anyone who is interested in learning how we got to this place two decades later." — Kelley Beaucar Vlahos, senior adviser at the Quincy Institute for Responsible Statecraft and contributing editor at *The American Conservative*

"Scott Horton has put together a devastating, deeply-researched account of how the U.S. war system betrayed the American people's trust in carrying out the so-called 'War on Terror.' He shows convincingly that it actually served other objectives and represents an unforgivable treachery that has inflicted incalculable harm on the United States. Readers across the entire deeply divided U.S. political spectrum will find truth in it that they can trust." — Gareth Porter, Martha Gellhorn Award-winning journalist and author of *Manufactured Crisis: The Untold Story of the Iran Nuclear Scare*

"*Enough Already* is essential reading for anyone who wants to know how we got here." — Andrew Cockburn, investigative reporter and author of *The Spoils of War: Power, Profit and the American War Machine*

"Scott Horton's book courageously investigates the deception that is the 'War on Terror.' It provides an impressive and wide-ranging examination of the misguided and costly U.S. foreign policy decisions which led to morally indefensible, strategically useless and militarily catastrophic interventions and wars in Somalia, Afghanistan, Iraq and other parts of the world. *Enough Already* must be read by every American who cares about the future of his country because the cost of these imprudent wars has proven detrimental to the nation's moral compass, global reputation, economic well-being and, indeed, national security. *Enough Already* is an eloquently written book. Using accessible language, exhaustive research and indisputable arguments, Horton's latest volume is a damning and impassioned case against war." — Ramzy Baroud, editor of Palestine Chronicle and author of *These Chains Will Be Broken: Palestinian Stories of Struggle and Defiance in Israeli Prisons*

"*Fool's Errand: Time to End the War in Afghanistan* was by far the best single account of the Afghan mire. Yet Afghanistan is only one of the conflicts unleashed by the U.S. and Western interventions in the 'War on Terror.' Scott Horton has now given us *Enough Already: Time to End the War on Terrorism*, a masterly history of these chaotic, tragic and above all futile conflicts, ranging with his usual excoriating accuracy from Mali to Pakistan, from Iraq to Yemen by way of Libya and Syria. Millions are dead, disabled or languish desperately far from their homes as the direct result of our blunders, bewilderment and outright malicious stupidity. Thousands of our own soldiers have died or are disabled. Hundreds more of our citizens have died in the U.S. and Europe in what Horton calls the 'backdraft' of our disastrous actions. Ignore the self-serving memoirs or grandiose academic tomes; if you read only one book on the so-called 'War on Terror,' this must be that book." — Frank Ledwidge, former Royal Navy Reserve intelligence officer, "Justice Advisor" to the UK Mission in Afghanistan's Helmand Province and author of *Investment in Blood*

"*Enough Already* is amazing. I have traveled to more than 25 countries around the world and spent time in the Middle East. Horton's book is the truest depiction of what is going on there that I have ever read. No one has dared to tell the truth that he has told in this book. He has clearly shown that we are the bullies of the world." — Shuja Paul, actor and documentary filmmaker

"*Enough Already* makes the compelling case against our endless and phony 'War on Terror' the way *Fool's Errand* made the case against our war on Afghanistan. War comes home — and nobody explains this simple truth better than Scott Horton." — Jeff Deist, president of the Ludwig von Mises Institute

"The United States must end its foolish empire building. In *Enough Already*, Scott Horton clearly demonstrates the dangers of following those that profit from the deaths of American service men and women fighting in undeclared and endless wars. We cannot wage conventional war on an unconventional enemy. Enough Already!" — Sgt. Dan McKnight, Idaho Army National Guard (ret.), Afghanistan war veteran, founder and chairman of BringOurTroopsHome.us

"I really recommend Scott Horton's indispensable *Enough Already* to anybody who wants to understand what the U.S. does around the world and the mountains of propaganda that are used to sustain it." — Aaron Maté, reporter for the Grayzone and contributing editor at *The Nation*

"Scott Horton wrote the definitive book on Afghanistan, and why Americans should be out of that endless civil war. The misnamed war on terrorism has been an even greater disaster, costlier to Americans and foreign peoples as well. Buy it, read it, and act on it!" — Doug Bandow, senior fellow at the Cato Institute, author of *Foreign Follies: America's New Global Empire*

"Scott Horton's book is an indispensable contribution to the record of imperial madness. Its careful documentation and moral outrage are as much historical accountability as we can expect, for now." — Philip Weiss, journalist, editor of Mondoweiss.net

Learn more about *Fool's Errand: Time to End the War in Afghanistan* and *Enough Already: Time to End the War on Terrorism* at LibertarianInstitute.org/books.

Made in the USA
Middletown, DE
29 October 2023